GROUNDWATER AGE

GROUNDWATER AGE

GHOLAM A. KAZEMI
Faculty of Earth Sciences
Shahrood University of Technology, Iran

JAY H. LEHR, Ph.D.
Senior Scientist, Technical Advisory Board, EarthWater Global, LLC
and Senior Scientist, The Heartland Institute

PIERRE PERROCHET
Centre of Hydrogeology
University of Neuchâtel, Switzerland

A JOHN WILEY & SONS, INC., PUBLICATION

Copyright © 2006 by John Wiley & Sons, Inc. All rights reserved

Published by John Wiley & Sons, Inc., Hoboken, New Jersey
Published simultaneously in Canada

No part of this publication may be reproduced, stored in a retrieval system, or transmitted in any form or by any means, electronic, mechanical, photocopying, recording, scanning, or otherwise, except as permitted under Section 107 or 108 of the 1976 United States Copyright Act, without either the prior written permission of the Publisher, or authorization through payment of the appropriate per-copy fee to the Copyright Clearance Center, Inc., 222 Rosewood Drive, Danvers, MA 01923, (978) 750-8400, fax (978) 750-4470, or on the web at www.copyright.com. Requests to the Publisher for permission should be addressed to the Permissions Department, John Wiley & Sons, Inc., 111 River Street, Hoboken, NJ 07030, (201) 748-6011, fax (201) 748-6008, or online at http://www.wiley.com/go/permission.

Limit of Liability/Disclaimer of Warranty: While the publisher and author have used their best efforts in preparing this book, they make no representations or warranties with respect to the accuracy or completeness of the contents of this book and specifically disclaim any implied warranties of merchantability or fitness for a particular purpose. No warranty may be created or extended by sales representatives or written sales materials. The advice and strategies contained herein may not be suitable for your situation. You should consult with a professional where appropriate. Neither the publisher nor author shall be liable for any loss of profit or any other commercial damages, including but not limited to special, incidental, consequential, or other damages.

For general information on our other products and services or for technical support, please contact our Customer Care Department within the United States at (800) 762-2974, outside the United States at (317) 572-3993 or fax (317) 572-4002.

Wiley also publishes its books in a variety of electronic formats. Some content that appears in print may not be available in electronic formats. For more information about Wiley products, visit our web site at www.wiley.com.

Library of Congress Cataloging-in-Publication Data:

Kazemi, Gholam A.
    Groundwater age / Gholam A. Kazemi, Jay H. Lehr, Pierre Perrochet.
      p. cm.
    Includes index.
    ISBN-13: 978-0-471-71819-2 (cloth)
    ISBN-10: 0-471-71819-X (cloth)
    1. Groundwater—Analysis. 2. Radioactive tracers in hydrogeology. I. Lehr, Jay H. II. Perrochet, Pierre. III. Title.
    GB1001.72.R34K39 2006
    551.49—dc22

                                                              2005030770

Printed in the United States of America

10 9 8 7 6 5 4 3 2 1

Wholeheartedly, I dedicate this book to **Professor Hossein Mehdizadeh Shahri**, *the founder of Geology at Shahrood University, a human giant and an endless source of capital support for all around him.*
G. A. Kazemi

My part of this book is dedicated posthumously to **Dr. Paul Damon**, *University of Arizona, who was a pioneer in the geologic use of radio isotopes in the 1960s. He passed away in April 2005.*
J. H. Lehr

My contribution to this book is dedicated to these special friends and **groundwater modelers of all ages**—*André, Attila, Christine, David, Dominique, Ellen, Fabien, Ghislain, Hans, Jahan, Jaouhar, Jean-Christophe, László, Laurent, Lloyd, Nari, Olivier, Peter, Philippe, Rachid, Ray, Rob, Vincent and Yvan—who devote so much of their time and energy to the mathematical analysis of hydrogeological processes.*
P. Perrochet

CONTENTS

Preface xiii

Acknowledgments xvii

1 Introduction 1

    1.1 Age and Lifetime, 1
    1.2 Age Determination in Geology (Geochronology) and in Other Disciplines, 3
        1.2.1 Absolute Age and Relative Age, 3
        1.2.2 Determination of Absolute Age of Rocks, 4
        1.2.3 Geological Time Table, 4
    1.3 Groundwater Age and Groundwater Residence Time, 6
        1.3.1 Young, Old, and Very Old Groundwaters, 9
        1.3.2 Dead Water and Active Water, 10
        1.3.3 Age Gradient, 11
        1.3.4 Age Mass, 11
        1.3.5 Mixing, Dispersion, and Transport of Groundwater Age, Mean Age, and Distribution of Ages, 12
        1.3.6 Average Residence Time of Water in Various Compartments of the Hydrologic Cycle, 14
        1.3.7 Hydrogeochronolgy, Interdisciplinary Groundwater Age Science, and Hydrologic Time Concept, 15
        1.3.8 Event Markers, 15
    1.4 Life Expectancy, 17
    1.5 Isochrone and Life Expectancy Maps, 18
    1.6 Some Groundwater Age-Related Terms, 19

1.6.1 Isotopic Age, Radiometric Age, and Decay Age, 19
1.6.2 Hydraulic Age, 21
1.6.3 Piston-Flow Age, Streamtube Age, and Advective Age, 22
1.6.4 Model Age and Apparent Age, 23
1.6.5 Storage Time, Mean Transit Time, Turnover Time, Flushing Time, and Travel Time, 23
1.6.6 Reservoir Theory and Its Relation with Groundwater Residence Time, 24

2 History of Groundwater Age-Dating Research 25

2.1 Pioneer of Groundwater Age Discipline—Sequence of the Earliest Publications, 25
2.2 Laboratories Worldwide for Dating Groundwater Samples, 31
2.3 Major Contributors to Groundwater Age-Dating Discipline, 31
2.4 Names Familiar in the Groundwater Dating Business, 36
2.5 Important Publications, 37
   2.5.1 Book Chapters, 38
   2.5.2 Ph.D. and M.Sc. Theses, 40
   2.5.3 Journals, 43
   2.5.4 Reports (mainly by the USGS), 45
2.6 Aquifers Subjected to Extensive Dating Studies, 50

3 The Applications of Groundwater Age Data 51

3.1 Renewability of the Groundwater Reservoirs, 51
3.2 An Effective Communication Tool for Scientists and Managers—and Curiosity to Laymen as Well, 53
3.3 Age Monitoring for the Prevention of Overexploitation and Contamination of Aquifers, 54
3.4 Estimation of the Recharge Rate, 55
3.5 Calculation of the Groundwater Flow Velocity, 58
3.6 Identification of the Groundwater Flow Paths, 59
3.7 Assessing the Rates of Groundwater and Contaminants Transport Through Aquitards, 61
3.8 Constraining the Parameters of Groundwater Flow and Transport Models (Estimation of Large-Scale Flow and Transport Properties), 64
3.9 Identification of the Mixing Between Different End Members, 65
3.10 Study of the Pre-Holocene (Late Pleistocene) Climate, 67
3.11 Evaluation of the Groundwater Pollution, 68
3.12 Calculation of the Travel Time of the Groundwater Plume to the Points of Interest, 71
3.13 Mapping Vulnerability of the Shallow Aquifers, 73
3.14 Performance Assessments for Radioactive Waste Disposal Facilities, 75

CONTENTS ix

 3.15 Site-Specific Applications, 76
  3.15.1 Identification of the Seawater-Level Fluctuations, 77
  3.15.2 Calculating the Timescale of Seawater Intrusion, 78
  3.15.3 Disposal of Wastes into the Deep Old Saline Groundwater Systems, 78
  3.15.4 Management of the Dryland Salinity in Australia, 79
  3.15.5 Hydrograph Separation, 80

4 Age-Dating Young Groundwaters   81

 4.1 Important Points, 82
 4.2 Tritium, 84
  4.2.1 Production of Tritium, 84
  4.2.2 Sampling, Analyzing, and Reporting the Results, 86
  4.2.3 Age-Dating Groundwater by Tritium, 87
  4.2.4 Advantages and Disadvantages, 90
  4.2.5 Case Studies, 90
 4.3 $^3H/^3He$, 92
  4.3.1 Sources of ^3He, 93
  4.3.2 Sampling, Analysis, and Reporting the Results, 94
  4.3.3 Dating Groundwater by $^3H/^3He$, 94
  4.3.4 Advantages and Disadvantages, 96
  4.3.5 Case Studies, 97
 4.4 Helium-4, 99
 4.5 Krypton-85, 99
  4.5.1 Production of ^{85}Kr, 100
  4.5.2 Sampling and Analyzing Groundwater for ^{85}Kr, 101
  4.5.3 Age-Dating Groundwater with ^{85}Kr, 103
  4.5.4 Advantages and Disadvantages, 105
  4.5.5 Case Studies, 105
 4.6 CFCs, 106
  4.6.1 Sampling and Analyzing Groundwater for CFCs, 108
  4.6.2 Dating Groundwater by CFCs, 108
  4.6.3 Limitations and Possible Sources of Error in CFCs Dating Technique, 111
  4.6.4 Advantages and Disadvantages, 113
  4.6.5 Case Studies, 113
 4.7 SF_6, 118
  4.7.1 Sampling and Analyzing Groundwater for SF_6, 119
  4.7.2 Age-Dating Groundwater with SF_6, 119
  4.7.3 Advantages and Disadvantages, 121
  4.7.4 Case Studies, 122
 4.8 $^{36}Cl/Cl$, 124
  4.8.1 Dating Groundwater by $^{36}Cl/Cl$ Ratio and Case Studies, 125

4.9 Indirect Methods, 129
    4.9.1 Stable Isotopes of Water, 129
    4.9.2 Case Study, 131

5 Age-Dating Old Groundwaters — 134

5.1 Silicon-32, 135
    5.1.1 Production of ^{32}Si, 135
    5.1.2 Sampling and Analyzing Groundwater for ^{32}Si, 136
    5.1.3 Dating Groundwater with ^{32}Si, 137
    5.1.4 Advantages and Disadvantages, 138
    5.1.5 Case Studies, 138

5.2 Argon-39, 139
    5.2.1 Production and Sources of ^{39}Ar, 140
    5.2.2 Sampling and Analyzing Groundwaters for ^{39}Ar, 141
    5.2.3 Age-Dating Groundwater by ^{39}Ar, 142
    5.2.4 Advantages and Disadvantages, 144
    5.2.5 Case Studies, 145

5.3 Carbon-14, 146
    5.3.1 Production of ^{14}C, 146
    5.3.2 Sampling, Analysis, and Reporting the Results, 148
    5.3.3 Groundwater Dating by ^{14}C, 149
    5.3.4 Advantages and Disadvantages, 154
    5.3.5 Case Study, 155

5.4 Indirect Methods, 157
    5.4.1 Deuterium and Oxygen-18, 157
    5.4.2 Conservative and Reactive Ions, 160

6 Age-Dating Very Old Groundwaters — 165

6.1 Krypton-81, 166
    6.1.1 Production of ^{81}Kr, 166
    6.1.2 Sampling, Analysis, and Reporting the Results, 166
    6.1.3 Age-Dating Groundwater by ^{81}Kr, 167
    6.1.4 Advantages and Disadvantages, 168
    6.1.5 Case Studies, 169

6.2 Chloride-36, 170
    6.2.1 Production of ^{36}Cl, 171
    6.2.2 Sampling, Analysis, and Reporting the Results, 173
    6.2.3 Groundwater Dating by ^{36}Cl, 174
    6.2.4 Advantages and Disadvantages, 177
    6.2.5 Case Studies, 177

6.3 Helium-4, 181
    6.3.1 Production and Sources of ^{4}He, 181
    6.3.2 Sampling, Analysis, and Reporting the Results, 183
    6.3.3 Age-Dating Groundwater by ^{4}He, 184

        6.3.4 Advantages and Disadvantages, 185
        6.3.5 Case Studies, 185
    6.4 Argon-40, 188
        6.4.1 Sampling, Analysis, and Reporting the Results, 189
        6.4.2 Age-Dating Groundwater by ^{40}Ar and Obstacles, 189
        6.4.3 Case Studies, 190
    6.5 Iodine-129, 191
        6.5.1 Production of ^{129}I, 192
        6.5.2 Sampling, Analysis, and Reporting the Results, 194
        6.5.3 Age-Dating Groundwater by ^{129}I, 195
        6.5.4 Advantages and Disadvantages, 196
        6.5.5 Case Studies, 197
    6.6 Uranium Disequilibrium Series, 198
        6.6.1 Sampling, Analysis, and Reporting the Results, 200
        6.6.2 Dating Groundwater by UDS, 200
        6.6.3 Case Studies, 203

7 Modeling of Groundwater Age and Residence-Time Distributions   204

    7.1 Overview and State-of-the-Art, 206
    7.2 Basics in Groundwater Age Transport, 211
        7.2.1 The Reservoir Theory, 211
        7.2.2 Determination of Age and Residence-Time Distributions, 227
    7.3 Selected Typical Examples, 244
        7.3.1 Aquifer with Uniform and Localized Recharge, 245
        7.3.2 Hydro-Dispersive Multilayer Aquifer, 246
        7.3.3 The Seeland Phreatic Aquifer, 248

8 Issues and Thoughts in Groundwater Dating   254

    8.1 The Need for More Dating Methods and the Currently Proposed Potential Methods, 254
    8.2 Translating Simulation of Groundwater Ages Techniques into Practice—More Applications for Age Data, 259
    8.3 Worldwide Practices of Groundwater Age-Dating, 260
    8.4 Proposal for a Groundwater Age Map—Worldwide Groundwater Age Maps, 260
    8.5 Works That Can and Need to Be Done to Enhance Groundwater Age Science, 263
    8.6 Major Problems Facing Groundwater Dating Discipline, 264
    8.7 Some Thoughtful Questions—Concluding Remarks and Future of Groundwater Dating, 264

References   266

Appendix 1.	Decay Curves of Groundwater Dating Isotopes. That of Tritium Is Shown in Chapter 4	299
Appendix 2.	Some Useful Information for Groundwater Dating Studies and Table of Conversion of Units	300
Appendix 3.	Concentration of Noble Gases (Used in Groundwater Dating) and Some Important Constituents of the Atmosphere	302
Index		303

PREFACE

In the important fields of hydrogeology and groundwater resources, new scientific, technical, and legal questions are constantly being posed, and the resolution of a large number of these problems requires some understanding of the residence time/age of groundwater. It is now half a century that groundwater age dating is carried out regularly and forms an important part of some of the hydrogeological/groundwater investigations. During these years "groundwater age" has steadily been coalescing into a potent and much respected field. The application of groundwater age data, being traditionally concerned with the flow rate calculation, recharge estimation, and renewability of the groundwater systems, has now expanded to include items such as the calibration of groundwater flow and transport models, the management of dry land salinity and the study of groundwater pollution. There are now over 20 direct and indirect methods based upon a large number of isotopic and conservative tracers to age date young, old, and very old groundwaters with an age range of a few days to tens of millions of years. In addition, simulation of groundwater age distributions has recently received considerable attention for monitoring the quality of groundwater and for predicting the fate of solutes and contaminants in the aquifer systems.

This important field of knowledge, however, has not received adequate attention in terms of books published to describe its many aspects. Although there are a number of scattered book chapters that include limited aspects of this topic, no single book has so far been produced to include its many fascinating angles in an integrated manner. Our intent is to fill this gap, appropriately presenting a half-century of progress and thus to play an important role in the further development of this still relatively new discipline. It should act as a major reference for any work that includes groundwater age or related subjects. The book includes a thorough discussion of groundwater age, which will help the audience fully digest the topic and plan research upon it, and it will form reference for anyone who has an interest in advanced subjects such as groundwater tracers. It is the first book devoted solely to this topic.

Groundwater Age is the first book to incorporate and synthesize the entire state of the art of knowledge about the business of groundwater dating including historical development, principles, applications, various methods, and likely future progress in the concept. It is intended for professionals, scientists, graduate students, consultants, and water sector managers who deal with groundwater and seek a comprehensive treatment of the subject of groundwater age. No similar book currently exists. Primary readers include graduate students, researchers, and university faculties working in water resources, earth sciences, and environmental studies. It will be useful for advanced undergraduate students, consultants, and high school teachers. We have attempted to present modern knowledge and cutting-edge research in a simple and clear language, such that it will be of interest to anyone with a passion for new technologies and modern sciences.

Chapter 1 discusses a variety of concepts and aims to provide essential definitions of the basics needed in understanding groundwater dating science. Topics covered in this chapter are wide-ranging and hence they may seem less coherent and somehow scattered. This is the nature of introductory chapters of books of any kind. The history of groundwater dating science and the pioneers, major contributors, well-known people, worldwide related laboratories, important publications and theses, and extensively dated aquifers are dealt with in Chapter 2. Chapter 3 covers a large number of applications and uses that are made from the groundwater age-data. This chapter is specifically useful for consulting companies and resource managers. Chapters 1 to 3 are original and their subjects have not been previously covered in any book. How to do dating is the subject covered by Chapters 4 to 6. These chapters have been written with this in mind that for the majority of the dating methods, the available literature is vast. Even general groundwater textbooks such as Freeze and Cherry (1979), Domenico and Schwartz (1990), and Hiscock (2005) as well as a number of public domain websites contain good sections on this subject. Therefore, age-dating methods are comparatively brief. Each method could have been expanded into a chapter leading to a larger book, yet not offering information presently available in the literature. Also, we have in dating chapters minimized our attention to the hostile parameters such as mixing, dispersion, and corrections, which are covered in Chapter 7. These three chapters are considered partially original. Chapter 7 is a mathematical description of the dating subject, which covers some very recently acquired knowledge related to age and life expectancy distributions. It discusses the deficiencies and incompleteness that are associated with reporting a single value as the age of a groundwater sample. New mathematical approaches are introduced in this chapter to model age and residence time distributions in samples comprising billions of water molecules originating from various groundwatersheds. Chapter 8 aims to cover stimulating new topics and those topics that have not been dealt with in other parts of the book, due to the specialized nature of Chapters 1 to 7. It tries to take the reader beyond the present-day knowledge and discusses a wide range of issues we hope will inspire creative thinking about the subject. The last two chapters are also original, not available in any book before the present one. Finally, decay curves of various dating isotopes and some useful information about noble gases and various measurement units are placed in the appendices. There is much here to satisfy and stimulate readers!

ACKNOWLEDGMENTS

We are indebted to the following people or organizations, across the world, who in one way or another helped us to compile this book.

J. Rose from John Wiley and Sons for his editorial handling and for his very quick reactions to the many questions asked by the authors. R. Purtschert from the University of Bern, for his time to show the underground laboratory at the University of Bern and for the discussions. D. J. Goode from the USGS for giving permission to use materials from his Ph.D. thesis. K. McGuire from the Georgia Technical University and R. E. Criss from California Institute of Technology for providing some papers and information about the use of stable isotopes of water in dating young groundwaters. L. Sebol from Golder Associates, Canada, F. Cornaton from the University of Waterloo, D. Pinti from the University of Québec, D. Hunkeler from the University of Neuchatel, and W. Deutsch from the Pacific Northwest National Laboratory for their review of a chapter of the book. T. Vitvar from IAEA Vienna for information about his Ph.D. thesis. L. D. Cecil from the USGS for reviewing part of the manuscript and for providing an electronic copy of his Ph.D. thesis. B. G. Katz from USGS for his help in obtaining CFCs and SF_6 data. E. Busenberg for providing CFCs and SF_6 data. D. Nelms from the USGS for his information about the susceptibility/vulnerability mapping using age data. C. Bethke from the University of Illinois for his initial positive response to an email by G. A. Kazemi confirming the need for such a book and also for his assessment of the proposal later. P. G. Cook from the CSIRO, Australia, for providing some information regarding age-dating young groundwater by $^{36}Cl/Cl$. B. T. Verhagen from the University of Witwatersrand for his information about dating laboratory facilities in South Africa and elsewhere.

Elsevier, Springer, Nature publishing group, Geological Society of America, Geological Society, London, American Institute of Hydrology (AIH), TPS, and many authors for permission to reproduce copyrighted materials. Janet Lehr for her help to gather information and references. The friendly and scientific environment at CHYN, University of Neuchatel, has helped greatly in delivering the manuscript on time. We especially have to thank Ms. Elisabeth Kuster, Librarian, for timely provision of books and papers and for her help in translating from German texts.

GROUNDWATER AGE

1

INTRODUCTION

The aim of this chapter is to provide the reader with the basic definition of some of the terms and concepts essential to understanding groundwater age science. We start with a brief introduction about age and lifetime (in general), geochronology, and geological time table. We then attempt to clearly define and illustrate a number of core concepts such as groundwater age, groundwater residence time, groundwater life expectancy, isochrone map, turnover time, etc. A table is presented that includes mean residence time of water in various compartments of the hydrologic cycle, such as in oceans, in rivers, in the atmosphere, and in groundwater systems. Definition and preparation of groundwater's isochrone and life expectancy maps are a short topic followed by the definition of a large number of terms synonymous to groundwater age and residence time. It is recommended that the reader starts with the preface of this book, where a general discussion about groundwater dating can be found.

1.1 AGE AND LIFE TIME

Every living and nonliving entity on the planet Earth, and indeed within the entire universe, whether natural or artificial, has a particular age. For example, an animal has an age, a car has an age, a bolt has an age, a factory has an age, a tree has an age, a soil particle has an age, a mountain has an age, a road has an age, and so on. Age for some entities is the length of time that has elapsed

Groundwater Age, by Gholam A. Kazemi, Jay H. Lehr, and Pierre Perrochet
Copyright © 2006 John Wiley & Sons, Inc.

since the moment they were created until the present time if they still exist, that is, if they have not died or have not been transformed into another form with another name. For instance, the age of a second-hand car is from the year it was built by the factory until the day one wants to know its fabrication date to purchase it. For some other entities, the reference date of the age "birthday" is not their creation time, but their appearance on the surface of the Earth. The age of a child, for example, is the time passed since he or she was born. In this case, the age of the child in question does not include the 9 months' duration he or she spends as an embryo in the womb.

Similarly, all living and nonliving entities are assigned lifetimes (there is no single thing that lasts forever). The lifetime is the length of the time between their birth/creation and their death/demolition. It is, therefore, impossible to know the lifetime of a person before his or her death. The concepts of age and lifetime are illustrated in Figure 1.1. For human-made equipment, the lifetime can be defined as the time interval between its fabrication and the time it is no longer capable of performing the function it was built to fulfill. The lifetime of "a bolt" equals the time it was built by the factory until it is no longer good enough to be used as a bolt, perhaps it is broken or its threads are not in good shape. One should also be aware of the term "Date" and avoid confusing it with age or lifetime. Date is usually employed to specify a particular day; for example, 23 August 1996 is a date. It does not show any particular age or lifetime. Some other particulars are associated with the age and lifetime of humans. These are place of birth, place of death, and place of burial. Not all these are applicable to, or important for, entities other than human beings.

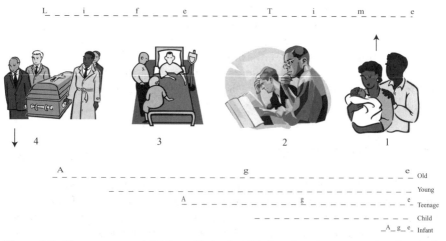

Figure 1.1. Human age and lifetime. Note that lifetime is only one number for each person, but age varies depending on the date we want to know it. In other words, each person has many ages but one lifetime. Also note that terms like "infant," "child," and "young" are semiquantitative ages.

1.2 AGE DETERMINATION IN GEOLOGY (GEOCHRONOLOGY) AND IN OTHER DISCIPLINES

Hydrogeology and groundwater are primarily a branch of geology and Earth sciences; hence a brief overview of the field of age-dating in geology needs to be initially presented. Attempts to determine the origin of Earth, a topic of interest for many scientists, for common people, and for philosophers, started with the beginning of civilization (Ringwood, 1979). However, it was not until the late 17th and early 18th centuries that observation-based knowledge led Abraham Werner, James Hutton, and especially William Smith to lay the foundation of the historical geology (Kummel, 1970), a branch of geology that discusses the age and history of the Earth. Despite the long history of intensive research, the precise age of the Earth is still a topic of intense discussion (Dalrymple, 1991). It is now estimated to be over 4.5 billion years (Faure, 2001). Further, considerable uncertainty surrounds the age of the universe, which recent estimates put at between 11.2 to 20 billion years (Krauss and Chaboyer, 2003). Geochronology is the title of the discipline that deals with age-dating of the Earth, rocks, minerals, and perhaps rocks of other planets.

Age-dating also has a long history in archeology and in many other fields such as estimating the age of various materials like woods, trees, glasses, wines, etc. See, for example, Wagner (1998) and Arnold and Libby (1951). A large number of methods based upon physical, biological, and chemical principles have been developed for this purpose. In fact, estimating the age of people is also a recognized expertise and is required in some criminal investigations, in estimating the age of a vehicle's occupant mainly for complying with children's safety requirements (e.g., Moeller, et al., 2002), in determining athletic eligibility (whose official reported age is in doubt) to play in some categories of sports (e.g., for playing for the FIFA World Youth Championship), in identifying young people who are in the age range of compulsory military service but have not done it, an exercise called *Nazar Begiry* (summoning by observation) in Iran, and in doing much more.

1.2.1 Absolute Age and Relative Age

In historical geology, two equally important concepts are the key to evaluating the issue of timescale in geological processes and stratigraphy. These are absolute age and relative age. **Absolute age** is the age of a rock sample, a stratum, or a geological formation (in numbers) that has been determined, in most cases, using radiometric decay principles. Absolute age is usually reported as a number, e.g., 20.6 million years, 128 million years, etc. On the other hand, the **relative age** of a geological stratum is inferred by its position in the stratigraphical sequence or its relation with a particular geological structure. The principles of stratigraphy (such as superposition), tectonics, and structural geology are employed to obtain the relative ages of various geological entities. Relative age is reported with terms like younger or older than a particular age, that is, relative age studies do not yield a specific age number. They indicate whether a particular geological formation or a geological event is older or younger than another one. For instance, Permian is above, and therefore younger than Carboniferous, but this tells us nothing about the age or the duration of either of these two periods.

1.2.2 Determination of Absolute Age of Rocks

A large number of well-established methods are now in place to age-date rocks as the main material constituting the Earth. These techniques, extracted from Attendorn and Bowen (1997), are listed in Table 1.1. They were initiated earlier than groundwater dating methods, at the beginning of the 20th century by Arthur Holmes (Holmes, 1913), because of the much greater interest in, and more importance of, knowing the age of the Earth as a whole. All methods, with the exceptions of fission track and thermoluminescence techniques, are based on the radioactive decay principle. The radioactive methods either measure the remaining of a specific disintegrating radioactive element (carbon-14 method, for instance) in the rock sample or count the daughter product of a radioactive reaction and compare it with the concentration of the parent element (potassium-argon dating, for example). Fission track dating is based on the amount of the densities of tracks created by spontaneous fission of uranium-238 in a rock sample. The thermoluminescence method is based on the emission of light from the samples as a result of heating.

1.2.3 Geological Time Table

Relative ages and absolute ages obtained by geologists during the last two centuries for worldwide rock formations have helped to construct what is called

TABLE 1.1. Various Techniques for Dating Rock Samples

Ionium (^{230}Th) dating of deep-sea sediments	^{187}Os/^{188}Os dating
^{234}U-^{238}U secular equilibrium dating of carbonates	Tritium dating
^{230}Th-^{238}U and ^{230}Th-^{234}U dating	I-Xe dating
Uranium-Xenon dating	La-Ce dating
Uranium-Krypton dating	Lu-Hf dating
Rubidium-Strontium dating	^{53}Mn dating
Potassium-Argon dating	^{59}Ni dating
Argon-Argon dating	Po-Pb dating
^{230}Th-^{231}Pa dating	K-Ca dating
Carbon-14 dating	Re-Os dating
Sm-Nd dating	^{137}Cs/^{135}Cs-A chronometer tracer
^{187}Os/^{186}Os dating	^{81}Kr dating
Fission track dating	Thermoluminescence technique

Source: Attendorn and Bowen (1997).

the **geological time table**. The geological time table, also referred to as the geologic column or geologic time scale, is regarded as a founding block of geology and shows the sequence of geological formations in order of decreasing age. As illustrated in Table 1.2, younger formations are on top and older ones are at the bottom of the table. Timescale in geology is very large; hence geologists

TABLE 1.2. Geological Time Table (Extracted with Some Modifications From Websites—See Footnote)

Era	Period, Mya	Distinctive characteristics
Cenozoic–Quaternary	Holocene, 0.01–present	The last major ice age ends and the sea level rises by 80–110 m.
	Pleistocene, 1.6–0.01	Ten major ice ages of 100 ky each. Last ice age from 100 kya to 10 kya.
	Pliocene, 5–2	Continued uplift and mountain building, extensive glaciation in N. Hemisphere.
Cenozoic–Tertiary, 66.4–1.6 Mya	Miocene, 25–5	Moderate climate; extensive glaciations begins again in S. Hemisphere.
	Oligocene, 38–25	Rise of Alps and Himalayas.
	Eocene, 55–38	Australia separates from Antarctica; India collides with Asia.
	Paleocene, 65–55	Mild to cool climate. Wide, shallow continental seas largely disappear.
	66.4–65	Extinction of dinosaurs
Mesozoic, 245–66.4 Mya	Cretaceous, 144–66.4	
	Jurassic, 208–144	
	208	Mass extinction
	Triassic, 245–208	
	248	Mass extinction
Paleozoic, 540–245 Mya	Permian, 286–245	
	Carboniferous, 360–286	
	367	Mass extinction
	Devonian, 408–360	
	Silurian, 438–408	
	438	Mass extinction
	Ordovician, 505–438	
	Cambrian, 540–505	
Precambrian, 3960–590 Mya	Proterozoic, 2500–540	
	Archean, 3960–2500	
Azoic 4500–3960 Mya		Planet forms

Modified after http://www.seafriends.org.nz/books/geotime.htm.
Mya = million years ago. Kya = thousand years ago.

use a finer division of eras, periods, epochs, and ages to differentiate among various stages of the geological time table. Note that the starting and ending ages of various temporal divisions (eras, periods, etc.) in Table 1.2 have some uncertainty, increasing toward earlier times. This is why age values shown here are slightly different from the values of a recently prepared geological time scale by Gradstein et al. (2004), which may be considered as the latest version. Although the geological time table was proposed long ago by Holmes in 1913, books and papers are still being written about it, revising and fine-tuning the boundary and age range of different times (e.g., Gradstein et al., 2004; Geyer and Shergold, 2000). It should be noted that in the hydrogeological time scale, we deal primarily with the uppermost section of the geological time table, namely Quaternary.

1.3 GROUNDWATER AGE AND GROUNDWATER RESIDENCE TIME

A large number of terms and phrases in the groundwater literature all represent the age and lifetime of a groundwater molecule. Throughout this book, we try to use only two terms that are most common and well established: groundwater age and groundwater residence time. Some other terms, briefly touched upon at the end of this chapter, more or less reflect the same meaning and have been employed or invented by the researchers to describe the subject in a more relevant manner, and in a more familiar language, to their specific research in hand.

Groundwater age is defined as the amount of time that has elapsed since a particular water molecule of interest was recharged into the subsurface environment system until this molecule reaches a specific location in the system where it is either sampled physically or studied theoretically for age-dating. On the other hand, **groundwater residence time** is the time it takes for particles of water to travel from the recharge area to the discharge area of the aquifer (Modica et al., 1998). It is the time interval between infiltrating into, and exfiltrating out of, the subsurface media. By this definition, what most researchers measure is the age of groundwater, not its residence time. One could also suggest that groundwater age is the groundwater residence time, but only at the discharge area (i.e., at the discharge area, groundwater age equals groundwater residence time). Groundwater age and groundwater residence time are comparable to the age and lifetime of a person, respectively. These are illustrated in Figure 1.2. It should be noted that some researchers confuse these two concepts by not clearly differentiating between them. It is also extremely important and necessary to point out here that the groundwater ages we measure by various methods as described in the later parts of this book are not necessarily identical to what we have defined here. The reasons for such a general statement include, but are not limited to, travel time of water in the unsaturated zone, uncertainties in the available tech-

GROUNDWATER AGE AND GROUNDWATER RESIDENCE TIME

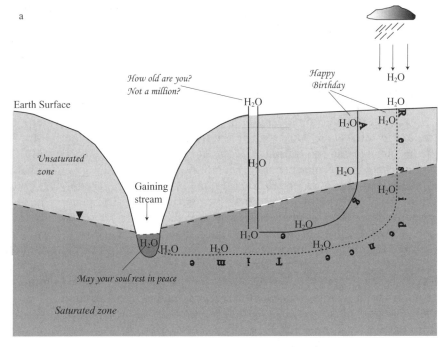

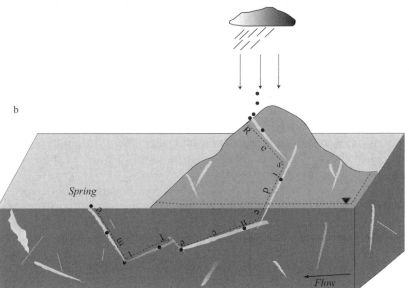

Figure 1.2. (a) Groundwater age and groundwater residence time. (b) Groundwater residence time.

niques, flow and transport complexities and mixing phenomena, difficulty in obtaining representative data, the common problem of hydrogeology "heterogeneity" and the fact that most isotopes used for age-dating are not part of the water molecule; rather, they are solutes. These are better explained as we proceed in the book. This is why some references, in order to reflect on the uncertainties involved, refrain from using terms like "groundwater age" and prefer "mean residence time of groundwater" instead (Clark and Fritz, 1997).

Based on the preceding definition, the date of birth of a groundwater molecule (age zero) is the date in which the water molecule enters the subsurface media through the infiltration process. In other words, by entering into the ground, water molecules become groundwater; before that, it is water, that is, it belongs to surface waters. Hence, day one of the life of a groundwater molecule is its very first day in the subsurface environment. The place of birth of groundwater is usually the recharge area of the aquifer, or simply any location on the Earth's surface where water molecules are provided with the required conditions to infiltrate into the ground. Groundwater molecules die when they leave subsurface environments through various means such as direct discharge, evaporative discharge, or abstraction in wells. Commonly, groundwater discharge areas are the place of death of groundwater molecules, that is, groundwater is no longer groundwater when it reaches a discharge area; it becomes a *water* molecule again. Terms like *rebirth* or *regeneration* may be more appropriate (compared to death) for describing such situations. One may also refer to groundwater abstraction by a pumping well as a premature death for the groundwater molecule.

A Few Points About Groundwater Age and Groundwater Residence Time
Some researchers (e.g., Metcalfe et al., 1998) differentiate among solute, or tracer, residence time and groundwater residence time. Solute residence time is the amount of time the solute in question resided in the groundwater system. In contrast, groundwater residence time is the amount of time the flowing groundwater molecules resided in the groundwater reservoir (aquifer). These two residence times may be fairly different in cases of tracers submitted to degradation or absorption processes.

Relatively recent researchers (Goode, 1996; Etcheverry and Perrochet, 2000; Bethke and Johnson, 2002) have added a new dimension to the concept of groundwater age. They explain that groundwater age is an intrinsic property of the groundwater molecule, like other parameters such as electrical conductivity, EC, and temperature. This means that age and groundwater are not two separable components. The reason is that water starts aging from the first day it enters the subsurface. This means it is impossible to find a single molecule of groundwater that has no age. As soon as a water molecule enters the subsurface, it becomes groundwater and it has an age. It is easy to understand this concept if we compare the groundwater age with the groundwater EC. Pure rainwater infiltrating into the subsurface ground has a very low, or in some circumstances zero, EC. As soon as it enters the ground, its EC start to increase. Age is similar. Before entering the subsurface, a groundwater molecule has no age. After entrance, it starts aging.

1.3.1 Young, Old, and Very Old Groundwaters

As far as age is concerned, in this book we divide groundwaters into three different age groups: young, old, and very old. This type of grouping has been previously used in the literature, but in a noncohesive and scattered form. Some textbooks divide groundwaters into modern and old (Clark and Fritz, 1997); others place all dating methods into one group with the exception of those methods suitable for very large timescales such as helium-4 (Mazor, 2004); a large number of papers have introduced and dealt with young groundwaters (Cook and Solomon, 1997; Schlosser et al., 1998; Busenberg and Plummer, 2000), and a few others have dealt with very old groundwaters (Bentley et al., 1986b; Fröhlich et al., 1991; Bethke et al., 2000). Therefore, all three different groundwater groups we are proposing in this book have appeared in previous publications in one form or another. The term *young groundwater* is now a widely established one, which most if not all specialists agree on. Dating young groundwater has also received wider attention and application, especially during the past 15 years. Quite recently, very old groundwaters have gained increasing attention, and new methods to date them are increasingly being developed (e.g., Sturchio et al., 2004; Lehmann et al., 2003). As an additional note, Back (1994) divides the aquifers, based on the reaction time to various stresses, into three categories: aquifers with reaction time of 100s to 1000s of years; those with 10s to 100s of years; and those with 1 to 10s of years. These three correspond to regional, local, and site in terms of space scale.

The range of ages covered by available dating methods is the key criterion in differentiating among and classifying various age groundwaters. Young groundwaters can be dated using techniques whose dating range extends from less than a year to about 50 or 60 years—post-thermonuclear bombs or CFC-free groundwaters. Old groundwaters can be dated with methods whose range is between 60 to 50,000 years. Finally, very old groundwaters can be dated using techniques whose coverage ranges from 50,000–100,000 years to more than a few tens of millions of years. These categories of groundwater dating methods are discussed in Chapters 4, 5, and 6, respectively. [The hydraulic age method (Section 1.6.2), calculating the age of groundwater based on Darcy's law using measured hydraulic gradients, effective porosity, and hydraulic conductivity, is a method by which we can date groundwater of all ages. It is hence treated separately. However, such a technique is not generally considered as a dating method. Similarly, the use of computer codes to simulate the distribution of groundwater ages of all ranges in various parts of an aquifer is dealt with in Chapter 7.]. It is generally accepted in the scientific community that there is no age technique to satisfactorily date groundwaters whose age is in the range of 60–200 years. This is what some references refer to as the *age gap*. However, as will be shown in later chapters, the techniques for this age range exist, but due to technical inadequacies have not been applied extensively.

In our system of classification, there is a resounding similarity between the upper limit of young groundwaters (about 50–60 years) and the current average life expectancy of humans, which is circa 63.8 years (*World Fact Book*, 2001). The lower and upper age boundaries of the three classified groups of groundwaters are not sharp. The margin of blurriness increases toward older groups. No surprise; this is very similar to age categories for humans, where we have infancy, childhood, and adolescences. The boundary between any two of these is not quite sharp. Childhood may start at 2 or 3 years of age; the starting age of adolescence may vary from 8.5 to 17 years (http://www.ship.edu/~cgboeree/genpsychildhood.html from C. G. Boeree, e-book, general psychology). The age range in each division varies depending on gender, race, geographical position, and living conditions. In spite of these large variations and discrepancies, these classifications are made because some usage is made of this practice. It is easier to study the behavior of a few children and then generalize these behaviors to other children within the same category. A similar rationale applies to groundwater. One can study the characteristics of a specific age groundwater, say a few 20-year-old groundwater samples, and make some useful generalizations that are equally applicable to other ages in the same range, e.g., for 10-year-old and 40-year-old groundwaters. For example, a 20-year-old groundwater is similar to a 40-year-old groundwater as far as recharge and susceptibility to contamination are concerned.

1.3.2 Dead Water and Active Water

The total volume of world groundwater resources is about $23 \times 10^6 \text{km}^3$ (Gleick, 1993). About $4–8 \times 10^6 \text{km}^3$ of this is in the form of active groundwater that circulates continuously within the hydrologic cycle (Freeze and Cherry, 1979; Gleick, 1993). The rest is sometimes referred to as **dead water**. Dead groundwaters include connate water (water entrapped in the sediments during deposition), magmatic water (water contained within magmas deep in the Earth), metamorphic water (water produced as a result of recrystallization during metamorphism of minerals), and marine water (seawater intruded into coastal aquifers). Dead water also refers to stagnant waters in isolated envelopes in deep regional aquifers that are not in full hydraulic connection with the surrounding groundwater (Mazor and Native, 1992). They may inflict substantial errors when calculating the hydraulic groundwater ages of these aquifers. With regard to groundwater dating, **active waters** are of greater interest because they are the main source of water used for various purposes. Further, it is as yet not possible to date some of the dead waters although they have an age. A recent reference (Mayo et al., 2003, page 1456) characterizes active and inactive groundwater zones (which corresponds to active and dead water) as

Active zone groundwater flow paths are continuous, responsive to annual recharge and climatic variability, and have groundwater resident times "ages" that become progressively older from recharge to discharge area. In contrast, inactive zone groundwater has extremely limited or no communication with annual recharge and has groundwater mean residence times that do not progressively lengthen along the flow path. Groundwater in the inactive zone may be partitioned, occur as discrete bodies, and may occur in hydraulically isolated regions that do not have hydraulic communication with each other.

In spite of this description, one could argue that, comparatively, in a very long time scale such as what we deal with in a geological one, there is no inactive groundwater system.

1.3.3 Age Gradient

Age gradient shows the rate of increase in the age of groundwater with depth (or distance), e.g., 3 year/m or 100 year/m. It is similar to any other gradient such as the geothermal gradient, which depicts the rate of increase in the Earth's temperature with depth. Age gradient is the inverse of the flow velocity and can be used to calculate the recharge rate of an aquifer if the specific porosity is known. The recharge rate is equal to specific porosity divided by the age gradient (recharge rate = specific porosity/age gradient). In such a case, steeper age gradient indicates a lower groundwater recharge. A horizontal age gradient may also be introduced to portray the rate of increase in the age of groundwater along its flow path. In such a case, a steeper age gradient reflects slower groundwater velocity.

1.3.4 Age Mass

The age mass concept was introduced by Goode (1996). The product of the mean age of a groundwater sample with its mass is called the **age mass** (Equation 1.1):

$$\text{Age mass} = \text{age} \times \text{density} \times \text{volume}$$
$$A_m = A \times \rho \times V \tag{1.1}$$

The unit of age mass would, therefore, be year $\times m^3 \times kg/m^3$ = year kg. The idea of age mass is to show that for groundwater samples with many fractions of different ages, the average of ages may be misleading, and what is crucial and should be counted is the average of age masses. The age mass of a two-component groundwater mixture is $A_1 \times V_1 \times \rho_1 + A_2 \times V_2 \times \rho_2$. If the density of water is taken as constant (hence removed from the equation) the mean age of such a mixture is calculated by the volume weighted average, e.g., Equation 1.2:

$$A = \frac{A_1 V_1 + A_2 V_2}{V_1 + V_2} \qquad (1.2)$$

For example, if 1 kg of a 20-year-old groundwater mixes with 2 kg of a 40-year-old groundwater, the resulting mean age would be 3 kg of 33.3-year-old groundwater. In terms of age mass, the mixture would be a 100 kg-yr age mass. If 1 liter of a 200,000-year-old water mixes with 2,000 liters of 10-year-old water, the mean age and the age mass of the mixture would be

$$\text{Mean age} = \frac{0.001 \text{m}^3 \times 200,000 \text{ years} + 2\text{m}^3 \times 10 \text{ years}}{0.001\text{m}^3 + 2\text{m}^3} = 109.95 \text{ years old}$$

Age mass = 220,000 kg-year age mass

1.3.5 Mixing, Dispersion, and Transport of Groundwater Age, Mean Age, and Distribution of Ages

All of the above terms and phrases are used to express a more or less similar issue.

Mixing As shown in Figure 1.3(a), a groundwater sample may contain waters that have originated from various recharge areas and input points. The result of such a situation is a sample that consists of many fractions with different ages. This phenomenon is referred to as **mixing** and represents a—if not *the*—major challenge to age-dating practice.

Dispersion and Transport of Groundwater Age The dispersion and transport of groundwater age have a role similar to mixing. Age is a groundwater property associated to the moving fluid particles. As suggested by the concepts of age mass (Goode, 1996), age is therefore submitted to the various processes governing mass transport in aquifers, such as advection, dispersion, and mixing. When these processes are active, the age of a groundwater sample cannot be a single number, because the sampled mixture may consist of numerous fractions with different ages. In fact, dispersion and transport of groundwater age fulfill the same function as mixing, but at a microscale level. As discussed by Weissmann et al. (2002), based on field experiments, the variance, or the dispersion, of groundwater ages (or residence times) is primarily due to heterogeneous groundwater velocities and hydrodynamic dispersion.

Mean Age and Distribution of Ages In a groundwater sample, there are billions of groundwater molecules. In a well-mixed groundwater system, any one of these molecules may have its own distinct particular age. The ages present in the sample are therefore described by a statistical distribution, yielding the frequency of occurrence of one given particular age in the sample. Mean age, or the age measured by isotopic and chemical methods, is practically the

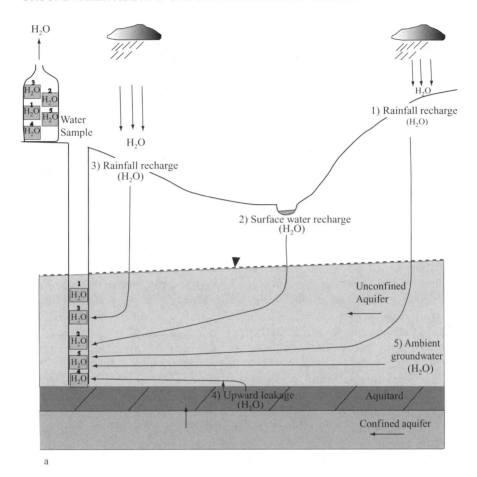

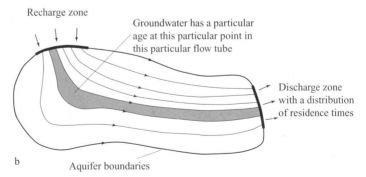

Figure 1.3. (a) Mixing; a groundwater sample consists of waters that have originated from different places. (b) The concept of mean age or mean residence time. At the discharge zone, there are a number of residence times.

average of ages of all molecules in the sample. Statistically speaking, mean age is the first moment (i.e., the average) of the age distribution. Many different age distributions may indeed result in the same mean age. Understanding the type of statistical age distribution of various ages in a groundwater sample is a must if one has to describe the age of a given sample. This can only be done by simulation through mathematical models. In situations where groundwater flow can be modeled accurately, the first moments (e.g., mean, variance, and skewness) of the age or residence time distributions can be simulated with temporal moments equations of the advective-dispersive type (e.g., Harvey and Gorelick, 1995; Varni and Carrera, 1998). Other recent mathematical approaches combining advective-dispersive equations and the reservoir theory (Etcheverry and Perrochet, 2001; Cornaton and Perrochet, 2005) yield the full age distributions at any point of an aquifer and the residence time distributions at its outlets in a deterministic manner. Depending on the flow configurations, groundwater age distributions can either resemble a standard functional (such as singular, uniform, exponential, normal, etc.) or be of a more complex shape, as is generally the case when groundwater moves through large and deep heterogeneous formations. For example, at a spring producing 1000 L/s, of which 999 L/s have an age of 1 y and 1 L/s has an age of 100,000 y, isotopic measurements would indicate an average age of 101 y. Without the knowledge of the actual age distribution, which in this case is bimodal-singular, the above average age would be particularly misleading since 99.9% of the spring discharge is 1 year old. The notions of residence time distribution and mean age at an aquifer outlet are illustrated in Figure 1.3(b) representing an aquifer as a collection of flow tubes associated to the flow paths from the recharge zone to the discharge zone. As mentioned, the distribution of groundwater ages at an aquifer outlet (i.e., the distribution of residence times) is a function of the organization of the flow paths inside the aquifer, each flow path yielding one specific age, as well as one specific discharge at the outlet. The mean residence time at the outlet is therefore an average of each specific age weighed by the corresponding specific discharge.

1.3.6 Average Residence Time of Water in Various Compartments of the Hydrologic Cycle

Mean residence time of water molecules in different compartments of the water cycle are presented in Table 1.3. The largest variability is associated with groundwater residence time. This is further highlighted when we add to Table 1.3 the 20,000-year average residence time of groundwater estimated by Fitts (2004). This shows that calculating the residence time of groundwater is a complicated process and requires intensive research and a variety of techniques. The residence time of surface waters in a watershed is partly similar to the residence time of groundwater and has received considerable attention recently (Rodgers et al., 2004; McGuire et al., 2005; Aryal et al., 2005). A watershed residence time is already short (in the order of weeks), so there is no

TABLE 1.3. Mean Residence Time (R_t) or Residence Time Range of Water in Various Compartments of the Hydrologic Cycle

Reservoir	R_t (Langmuir, 1997)	R_t (Freeze and Cherry, 1979)
Oceans	3,550 years	~4,000 years
Ice caps and glaciers	10s–1,000s years	10–1,000 years
Lakes and reservoirs	10 years	~10 years
Groundwater	1,700 years	2 weeks–10,000 years
Rivers	14 days	~14 days
Atmosphere	11 days	~10 days
Water in the biosphere		~7 days
Swamps		1–10 years
Soil		14 days–1 year

need to propose watershed age, which could potentially be a fraction of the watershed residence time.

1.3.7 Event Markers

Tracers used in dating groundwaters are sometime classified into three groups:

1. Clocks or radioactive tracers: These decay radioactively in the subsurface, e.g., ^{81}Kr. Half-life, $t_{0.5}$, the time it takes for half of the atoms to decay, mean life ($t_{0.5}/\ln2$), and decay constant, λ, ($\ln2/t_{0.5}$) are principal specifications of these tracers used in age-dating calculations.
2. Accumulating tracers: These increase in concentration in the subsurface environment as time goes by, e.g., ^{4}He.
3. Signals or event markers: These are used for dating young groundwaters only and are further described ahead.

Event markers are defined as the distinct signatures of some human activities on the environment that can represent a specific time. These are mainly the results of either nuclear or industrial activities. Event markers may also be referred to as historical tracers or slug tracers (Cook and Solomon, 1997). Some processes in the saturated and unsaturated zones such as diffusion, dispersion, mixing, and radioactive decay (in some cases) blur the results obtained from the study of the event markers in the groundwater system. Important event markers and some of their characteristics are shown in Table 1.4.

1.3.8 Hydrogeochronology, Interdisciplinary Groundwater Age Science, and Hydrologic Time Concept

It has been suggested to refer to the groundwater age-dating practice as "hydrochronology" (Cook and Böhlke, 2000) as an analogy to "geochronol-

TABLE 1.4. Event Markers and the Related Information (Compiled Based on Various Sources)

Event marker	Production		Present status	Remarks
	Start	Peak		
CFCs[1]	1930	1992[2]	Decreasing	Worldwide impact
SF_6	1953	1996	Increasing	Worldwide impact
Bomb tritium	1951	1963	Almost diminished	Some lag time for Southern Hemisphere—worldwide impact
Bomb Chlorine-36	1952	1957	Almost diminished	Mostly local impact
Bomb Carbon-14	1955	1964	Almost diminished	Mostly local impact
HCFCs	1980	1995–1996	Increasing	
Chernobyl accident[3]	1986		Diminishing	Local impact
Krypton-85	1952	1985, 1989[4]	Increasing	North America, Europe, Russia

[1] Concentration of some species still continues to rise.
[2] Some references put it at late 1980s (e.g., Montzka et al., 1999).
[3] Released up to 3 Exa Bq of activity into the atmosphere, mainly ^{137}Cs and ^{90}Sr (see UNESCO, 1992).
[4] Based on measurements in Belgium (Clerk et al., 2002)—worldwide concentration is still rising.

ogy," which deals with age-dating of rocks. This terminology may not be quite correct because in groundwater age-dating, only the dating of groundwater samples is of concern. Hydrochronology has a wider arena and can cover dating of any type of water such as river water, rainwater, lake water, some of which are already under intense investigation in fields other than groundwater science. For instance, determination of residence time in Lake Bled, Slovenia, was a major project of the International Association of Tracer Hydrology in early 1990s (Leibundgut and Zupan, 1992). Further, residence time of surface water in lakes and in watersheds is equally attracting attention from hydrologists (e.g., Kumar et al., 2000; Sakaguchi et al., 2005). Therefore, terms such as "groundwater chronology," "hydrogeochronology," or "groundwater age science" may be more focused and more appropriate. Future researchers may choose any of these terminologies or create one of their own. Dating groundwater is primary related to the field of Earth sciences. Fields such as geochemistry, hydrogeology, and groundwater hydrology have the most immediate connection with groundwater dating. Some other disciplines, however, also have close ties with groundwater dating. For examples, physical principles are the fundamental of groundwater dating because isotopic dating is based upon radioactive decay. For this reason, physics institutes and physicists have been involved in groundwater dating from the beginning. A number of groundwater modelers have also researched in this field, which has resulted in the knowledge presented in Chapter 7. Therefore, one could argue that groundwater age science or hydrogeochronology is an interdisciplinary subject.

Through a philosophical discussion, William Back (1994), a pioneer in the fields of groundwater and hydrogeology, argues that measurement of short intervals of time and water, are coeval events and commenced about 6,000 BC in Mesopotamia. He believes that early short-interval measurements of time were perhaps undertaken for hydrological purposes, most likely to measure the volume of water for scheduling farmer irrigation water. Similarly, Marks (2005) believes that for the greater part of the human history, time was calculated through the use of "water clocks." Therefore, the hydrologic time concept can be regarded as a historical-philosophical issue that extends far beyond the present-day hydrogeochronology or hydrochronology terms we propose here.

1.4 LIFE EXPECTANCY

The term **life expectancy** was first introduced in the groundwater literature by Etcheverry (2001). It shows the amount of time left before a "presently" groundwater molecule leaves the aquifer, reaches the discharge point, and ends its journey in the subsurface environment. In other words, how long will it take for a current groundwater molecule to become a plain water molecule again—to die, to be reborn, that is, for how many more years will a groundwater molecule remain a groundwater molecule? More clearly, life expectancy is the time taken by a groundwater molecule somewhere in the aquifer to leave the system and become a surface water molecule again. Figure 1.4 shows the life expectancy concept. By this definition groundwater residence time = groundwater age + groundwater life expectancy, and on recharge areas life

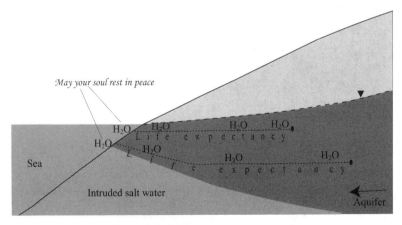

Figure 1.4. Life expectancy concept.

expectancy equals the residence time. In the definition of life expectancy the assumption is that the flow of groundwater is under natural condition and there is no artificial withdrawal. Production from the aquifers significantly reduces the life expectancy of the groundwater molecules. No chemical or physical method exists to measure the life expectancy of a groundwater sample. Such information can only be obtained mathematically based on groundwater flow models with reverse flow fields.

1.5 ISOCHRONE AND LIFE EXPECTANCY MAPS

First of all, it should be pointed out here that in the scientific literature, "isochrone" is also written as "isochron," without the "e" at the end. For example, see page 332 of Appelo and Postma (1999) and page 69 of Dickin (1995). Throughout this text, the *isochrone* version is adapted. Also, strictly speaking, isochrone means iso-time, not iso-age. However, here it is treated (as in the literature) as iso-age.

Isochrone maps depict the age of groundwater in various parts of the aquifer through age contour lines (Figure 1.5) similar to a topography map

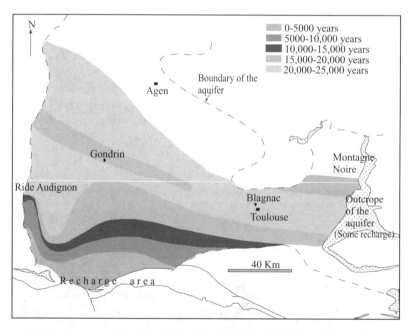

Figure 1.5. Isochrone map of the Infra-Molassic Sands aquifer, Southern Aquitaine Basin, southwest France (Modified after Andre, 2002). This map was prepared using ^{14}C and tritium ages, as well as hydrogeological interpretation. The groundwater flow is both northward and westward, and geological structures, not shown here, play a significant role in the chemical and isotopic composition of the groundwater.

where elevations of the Earth's surface are shown by means of elevation contours. They can be constructed if all other classical maps of the aquifers such as transmissivity map, water level map, and base of aquifer map are available. Life expectancy maps illustrate the same concept but in the reverse direction, where, zero time is at the aquifer discharge area/outlet. One such map for the unconfined Superficial Aquifer of Perth is shown in Figure 1.6. As stated, to prepare the isochrone or life expectancy maps, a variety of other maps and information are required, including

1. Saturated thickness of the aquifer, which is prepared by overlaying water table contours over the base of the aquifer contours and subtracting base of aquifer elevation from water table elevation;
2. Hydraulic conductivity of the aquifer, which is prepared by overlaying transmissivity contours over saturated aquifer thickness, i.e., item 1 above, and dividing transmissivity by the thickness;
3. Hydraulic gradients in the aquifer, which is calculated by dividing water table elevation contours to the contour spacing interval; and
4. Effective porosity (specific yield) of the aquifer, which is estimated by analysis or by searching the literature for values for similar types of material.

Using the above maps and the effective porosity values, the pore velocity of groundwater in different parts of the aquifer is deduced from Darcy's law by $v = KI/\theta$, where v is the pore velocity, K is the hydraulic conductivity, I is the hydraulic gradient, and θ is the specific yield of the aquifer. The life expectancy of groundwater at the point of interest is calculated by $T = L/v$, where T is the life expectancy and L is the distance from the discharge area of the aquifer. The age of groundwater is calculated in a similar manner with the exception that L is the distance of the point of interest from the "recharge area" of the aquifer.

1.6 SOME GROUNDWATER AGE-RELATED TERMS

As we stated, a large number of terms have been invented and employed to address the issue of timescale in groundwater science, or more simply the age of groundwater. Some of these terms were in place before the "age" concept, but some were introduced afterward. Most of these are used in other sciences and disciplines, but a few are specific to groundwater.

1.6.1 Isotopic Age, Radiometric Age, and Decay Age

Isotopic age is the age measured by various isotopic methods; but one may also use it to highlight the difference between the true age of the groundwater

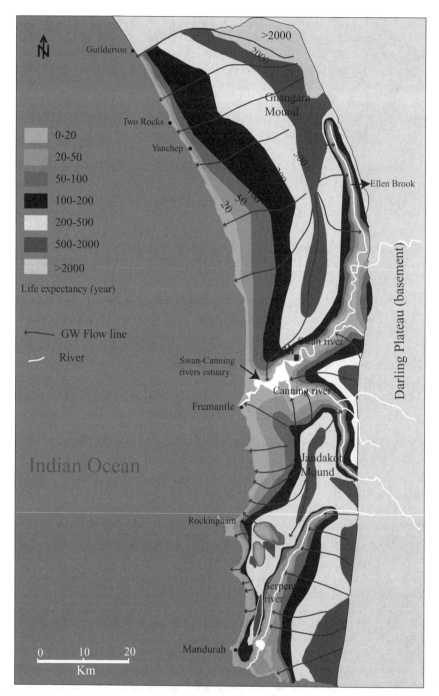

Figure 1.6. Life expectancy map of the Superficial Aquifer, Perth, Western Australia (modified after Kazemi et al., 1998a). Note that in the original paper, this map was called an isochrone map, which, although incorrect, is not vastly different.

molecule and the age derived on the basis of isotopic decay principles. These two, isotopic age and true age, may be different because the isotopes used to measure the age of groundwater are not part of the water molecule (excluding tritium). **Radiometric age** is similar to isotopic age but involves exclusively radioactive isotopes, that is, it does not include ages determined using ^{4}He, CFCs, stable isotopes of water, and other stable environmental isotopes. Similar to radiometric age, **decay age** is derived on the basis of radioactive decay. Isotopic age assumes that tracers and solutes dissolved in the groundwater move with the same velocity of the water molecule, which is not always true. For instance, Rovery and Niemann (2005) have shown that solutes may travel up to 10 times slower than pore water molecules if the aquifer is nonhomogeneous and the flow is transient. Such a large difference still has to be proved, but it is thought to be possible because the steady and the transient flow behave differently with respect to heterogeneity in the 3D flow systems.

1.6.2 Hydraulic Age

Based on Darcy's law and using measured hydraulic gradients, effective porosity, and hydraulic conductivity, we can calculate the age of all groundwaters, from young to very old. Here we present a brief overview of some of the formulas to calculate hydraulic age. In an aquifer with horizontal flow and no recharge along the flow line, such as a confined aquifer, the age of groundwater at any point along the flow line equals distance from the recharge area divided by the velocity of the groundwater. In an unconfined aquifer with uniform recharge and constant thickness, hydraulic age, T, at any depth below the water table, Z, can be calculated by Equation 1.3. The mathematical procedure to arrive to such an equation is explained in Chapter 7, but the base is the Darcy law.

$$T = \frac{H\theta}{R} \ln\left(\frac{H}{H-Z}\right) \quad \text{(Vogel, 1967)} \quad (1.3)$$

where H and θ are the aquifer thickness and specific porosity, respectively, and R is the recharge rate. For example, if the recharge rate to a 40-m-thick aquifer with a specific porosity of 0.1 is 50 mm/year, then the groundwater age at the depth of 5 m below the water table would be

$$T = \frac{40 \times 0.1}{0.05} \ln\left(\frac{40}{40-5}\right) = 80 \times \ln 1.14 = 10.8 \text{ years}$$

To solve Equation 1.3, the challenging task is the estimation of the recharge rate. For other, more complicated (natural) aquifer configuration, a number of equations have been developed in Chapter 7; also Cook and Böhlke (2000) is a good reference in this regard. The simplest situation—for an unconfined aquifer whose thickness increases downgradient—Equation 1.3 changes to Equation 1.4:

$$T = \frac{Z\theta}{R} \qquad (1.4)$$

In practice, we use Equation 1.4 to calculate recharge rate using groundwater age data, not vice versa.

1.6.3 Piston-Flow Age, Streamtube Age, and Advective Age

These terms are all synonymous and assume that groundwater molecules move by advection only, that is, their displacement is only governed by the overall velocity field; other transport processes such as hydrodynamic dispersion, molecular diffusion, and mixing are insignificant or absent. As shown in Figure 1.7, following this approach, the groundwater molecule travels steadily from the recharge area of the aquifer to the discharge area as an isolated parcel, with minimal turbulence and without any interference and exchanges with the surrounding molecules across the flow path. In such a situation, the age of any groundwater molecule only depends on the evolution of the pore velocity along a given flow path, or within a given flow tube, and ignores groundwater age evolution in its vicinity. It equals the distance from the recharge point (the length of the flow path) divided by the groundwater flow velocity.

In nature, there are not many instances where piston, streamtube, or advective flow theory conditions are fully met. However, several studies have shown that approaches of this kind often yield a good approximation of groundwater flow through deep, unsaturated zones (Zimmerman et al., 1966) and in thick,

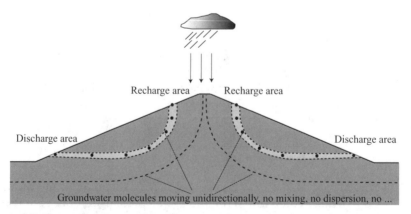

Figure 1.7 Piston-flow theory. Groundwater molecules migrate as isolated packets. Age of groundwater molecule in such a system is called the piston-flow age.

low-salinity aquifers with insignificant cross-formational flow and confined by aquitards with a small diffusion coefficient (Park et al., 2002). Piston-flow, streamtube, or advective age essentially reflects the streamlike nature of steady-state flow regimes. In a first approximation, this concept is very useful to gain insights into theoretical age distributions for simple, typical flow configurations. Moreover, filtering out dispersion and mixing processes, the advective age approach provides a valuable tool to assess the relations between residence time distributions and the internal organization of flow patterns, particularly in multiple flow systems and heterogeneous configurations (Etcheverry and Perrochet, 2000).

1.6.4 Model Age and Apparent Age

Groundwater ages can be determined by a variety of methods, as we will discuss in the subsequent chapters of this book. These can be classified into two broad categories: chemistry-isotope-based techniques and hydraulic-based techniques. **Model ages** are referred to ages derived by hydraulic-based techniques that employ the principles of groundwater hydrodynamics and groundwater flow and transport modeling. However, various adjusting models and methods have also been developed for correcting the ages obtained by isotope-based techniques in order to reduce the uncertainties involved. Hence, model age may also refer to the age obtained after such revisions have been applied. **Apparent age** is a term and a way to imply that ages derived by various methods, whether isotopic or hydraulic, are not the true ages of groundwater molecules as has been defined in this book or other publications. Adaptation of the term "apparent age" is a widespread practice in the literature and is an advance warning to highlight that ages reported, though corrected, are still an approximate average of the many real ages that exist in a single sample.

1.6.5 Storage Time, Mean Transit Time, Turnover Time, Flushing Time, and Travel Time

Initial publications used **storage time** to describe what we here refer to as groundwater residence time (e.g., Eriksson, 1958). At that time, the reservoir theory was under development and groundwater age had yet to be introduced into the literature. In fact, storage time is the old version of residence time. **Mean transit time** is a synonym for mean residence time. Other early concepts and definitions are summarized in simple and rigorous form in Bolin and Rodhe (1973). **Turnover time** shows how quickly a specific reservoir is depleted. It is the same as residence time also defined by Maloszewski and Zuber (1982) as $T = V/Q$, where V is the volume of mobile water in the system and Q is the average steady-state volumetric flow rate. It might be more appropriate to use "turnover time" for fractured rock aquifers where flow rate is high and storativity is low; thus, the term reflects a short time as compared to residence time where a long time may be imagined. **Flushing time** has not been adapted widely in groundwater studies, but it is a term to represent the same concept as storage time or residence time. It is more suited to surface watershed studies, where the residence time is very short.

Instead of using groundwater age, some researchers prefer **travel time**. However, depending on the spatial reference, the two terms may be the same or different. In groundwater age, the spatial reference for time zero is the groundwater recharge area. In travel time, the spatial reference could be any conspicuous location within the aquifer; it could also be the recharge area. Travel time shows the amount of time it takes for a groundwater molecule to move from one point of the aquifer to the other. It is often used to show the migration time of contaminants to a specific location such as a production bore (e.g., Clark et al., 2005; Ceric and Haitjema, 2005).

1.6.6 Reservoir Theory and Its Relation with Groundwater Residence Time

The reservoir theory was first introduced in chemical engineering to evaluate the time substances remain in contact within a chemical reactor (Danckwerts, 1953; Zwietering, 1959; see Cornaton, 2004, for more information). A **reservoir** is defined as any part of nature with clearly defined boundaries. To calculate the average residence time in any reservoir, τ_r, the volume of fluid in the reservoir is divided by the net flux through it:

$$\tau_r = \frac{V}{Q} \qquad (1.5)$$

where V is the volume of the fluid in the system in m^3 and Q is the net flux in m^3/year (identical to turnover time described in Section 1.6.5). Residence time can be calculated for reservoirs of any size (or subreservoirs) such as for oceans, lakes, or any small compartment with well-defined boundaries. It corresponds to the amount of time that any particular portion of the substance in question remains in any one arbitrary space within the reservoir. In Equation 1.5, it is also possible to calculate the residence time by having the weight of the substance (fluid, for example) in the reservoir and the flux rate in g/time, that is, units can be weight units rather than volume units. With reference to the groundwater residence time studies, the reservoir in question is the entire aquifer system. As a consequence, Equation 1.5 can be used to estimate the volumetric recharge rate to the aquifer (Q) if the residence time (τ_r), and the volume (V) of the mobile water in the aquifer are known. For this, we need to multiply specific porosity and volume of the saturated zone, and divide by residence time: volumetric recharge = porosity × saturated volume / residence time.

2

HISTORY OF GROUNDWATER AGE DATING RESEARCH

This chapter includes a chronology of groundwater dating studies using isotopic and geochemical tracers from the earliest methods/applications like tritium to the latest ones such as CFCs and SF_6. It names various methods and explains how they have evolved. All book chapters, M.Sc. and Ph.D. theses, and professional reports that have been produced in relation to groundwater age are cited and a few back-bone journal papers are introduced as comprehensive references. Worldwide institutions where groundwater samples can be dated with relevant information are introduced, and well-known people in this field and their accomplishments in terms of the evolution of various methods of age-dating are described. This chapter has been designed to equip the reader with a full historical perspective and a complete picture of the topic. It is necessary because this is the first ever book devoted solely to the groundwater age concept.

2.1 PIONEER OF GROUNDWATER AGE DISCIPLINE—SEQUENCE OF THE EARLIEST PUBLICATIONS

In 1952, Willard Frank Libby (December 17, 1908–September 8, 1980), a Chemistry Noble laureate, published his famous book *Radiocarbon Dating*. In that book, which was re-edited soon after in 1955 and has been translated into German, Russian, and Spanish, Libby neither discusses nor mentions the term

Groundwater Age, by Gholam A. Kazemi, Jay H. Lehr, and Pierre Perrochet
Copyright © 2006 John Wiley & Sons, Inc.

"groundwater age." However, in a two-and-a-half-page article he wrote a year later in P.N.A.S. (Libby, 1953), Libby introduced the term "storage time" of water. Though he did not specifically mention storage time of "groundwater," he actually meant it because he went on to elaborate further by adding "the determination of 'tritium in wells' should reveal whether their ultimate supplies are closely connected to the rainfall." It must be added that in the earlier section of the same article, Libby introduces an Age of Rain concept and defines it as the "time elapsed since precipitation." However, no words like groundwater age are used in Libby's article. For clarity and due to the importance of this article, its first page has been scanned and reproduced here as shown in Figure 2.1.

In our estimation, the first ever scientific piece of writing to include/cite the term "age" for groundwater is most probably that of Begemann and Libby (1957). Again, they do not specifically mention "groundwater age," but they write "average age of the main body of water," by which they really meant the average age of the groundwaters emanating from the thermal springs, i.e., the main body of water was thermal groundwaters. The last 6 lines of the abstract of their article (page 277) read (reproduced with permission from Elsevier):

Studies of the circulatory pattern of hot springs have shown that the waters of several hot springs studied are rainwater that has been stored for relatively brief periods. Study of groundwater has shown that in large areas the water issuing from wells dug for normal use is older than 50 years. It appears that the technique of studying the tritium content of well-water is quite likely to prove to be of real value in studying underground water supplies and in the prediction of their susceptibility to drought as well as depletion by pumping and the possibility of replenishment from rain or snow.

In page 292 of their article, Begemann and Libby define average age: "Average age is understood to be the age corresponding to the average tritium content and represent the time elapsed since the waters last fell as rain or snow."

Based on the above explanations, it is, to a high degree of certainty, possible for us to call both Friedrich Begemann and Willard Frank Libby cofounders of the groundwater age science. These two were from The Enrico Fermi Institute for Nuclear Studies, University of Chicago, USA. However, it should be reiterated that W. F. Libby published a large number of articles as well as a book on the subject of age-dating by radiocarbon. As pointed out earlier, Libby won the Noble Prize in chemistry in 1960 for his method to use carbon-14 for age determination in archaeology, geology, geophysics, and other branches of science. It should immediately be added that the paper by K. O. Münnich (1957) in German is the oldest research that we can find to include measurement of ^{14}C in groundwater samples. Münnich's work, however, does not include words like groundwater "age" or groundwater "dating." Therefore, K. O. Münnich, although an authority in the field of groundwater dating through his substantial later publications, cannot be specifically called the

THE POTENTIAL USEFULNESS OF NATURAL TRITIUM*

By W. F. Libby

INSTITUTE FOR NUCLEAR STUDIES AND DEPARTMENT OF CHEMISTRY, UNIVERSITY OF CHICAGO

Communicated February 25, 1953

The production of tritium by the cosmic radiation in the earth's atmosphere[1-5] probably can now be taken as being established though many quantitative aspects remain unknown. Rainwater definitely contains tritium in amounts corresponding to between $5 \cdot 10^{-18}$ and $40 \cdot 10^{-18}$ tritium atoms per hydrogen atom. This is a small but definitely measurable quantity, requiring between one quart and five gallons of water as sample. The possible potential uses appear to be worth consideration.

A. Age of Rain and Agricultural Products.—In the first place it is clear that the tritium content of rain water will decrease with time elapsed after precipitation, since the cosmic ray source can introduce fresh tritium atoms only in the atmosphere where the new tritium atoms react with the atmospheric oxygen to form radioactive water. The time scale is set by the decay constant for tritium—12.5 years half life and 18.0 years average life. We expect therefore that whereas freshly fallen rain or snow will show a definite radioactivity due to tritium, old water will show none, and that the age of the water in the sense of the time elasped since precipitation may be determinable from the specific radioactivity. Further, agricultural products should be datable by the tritium content of their constituent hydrogen, the principle being that a considerable fraction of the hydrogen utilized in plant growth originates as rain water and is incorporated in the first few months after precipitation. The time of harvest therefore is close to the time of precipitation. Consequently, it seems possible that the age of agricultural products in the practical sense of the time elapsed since harvest probably can be measured by determination of the tritium content of the product itself. A series of tests is in progress in our laboratory using water of known age and wines of known vintage to test these postulates.

B. Hydrology and the Identification of Water Sources.—A second type of application occurs in the identification of water sources. The mean storage time of rain or snow water before it appears in a given water supply should be given directly by the tritium content of the given supply relative to that of average rain in the region involved. The determination of mean storage times of between about three and twenty-five years should prove possible, these times corresponding to between 15 and 75% loss of tritium due to radioactive decay. The determination of tritium in wells should reveal whether their ultimate supplies are closely connected to the rainfall and therefore are subject to seasonal fluctuations in rainfall.

Figure 2.1. Scanned version of page 245 of Libby's (1953) paper.

Father of the groundwater age-dating discipline. Some references (Freeze and Cherry, 1979; Bentley et al., 1987) mistakenly refer to K. O. Münnich (citing his above paper) as the first scientist who proposed age-dating groundwater by radio carbon. Perhaps the second oldest publications with regard to groundwater age are those by Brinkmann et al. (1959), where they applied the ^{14}C method to date groundwater of the till plains in Germany; and von Buttlar (1959) by which the tritium ages were calculated in New Mexico, USA. For New Mexico, there is another similar work with an almost identical title by von Buttlar and Wendt (1958), which the authors had no access to, but it is safe to assume both references as one. Also, the reference by Eriksson (1958) deals briefly with the travel time of water parcels in the groundwater "region," but cannot be referred to as a publication that contains groundwater age elements. In 1961, two short publications dealt with groundwater age. *Selected Bibliography of Groundwater* by van der Leeden (1971) cites only these two as the examples of publications dealing with groundwater age. These include

> Clebsch, A. Jr. (1961) Tritium age of groundwater at the Nevada test site, Nye county, Nevada. U.S. Geological Survey Prof. Paper 424-C, pp. 122–125.
>
> Thatcher, L., Rubin, M., Brown, G. F. (1961) Dating desert groundwater. *Science*, Vol. 134, pp. 105–106. (In this paper, age of groundwater was defined as the time since exposure of water to atmospheric supply of tritium.)

It will be explained later in the book that the use of tritium for age-dating groundwater could not have taken place prior to the nuclear tests, the first of which occurred in 1952. Further, it was only in 1951, long after the discovery of tritium in 1932, that van Gross and his colleagues found out that tritium occurs naturally in the atmospheric precipitation. It is safe to say that if age-dating of the other materials such as rocks or archeological objects (and age determination of the Earth as a whole in the 18th century) was not an established practice prior to 1950, it would have not been possible to initiate groundwater dating as early as in 1950s, immediately after the nuclear testing. In other words, dating of other materials has served as a background for groundwater dating, and it has especially paved the way for the carbon-14 measurement of groundwater age.

The first modern groundwater textbook, i.e., Tolman (1937), does not include a section on the groundwater quality or groundwater chemistry. So, one would not expect to see a section or a page about groundwater age in that book. In the first edition of the earliest groundwater textbook with some chapters devoted to the chemistry of groundwater (which contain some groundwater chemistry chapters), i.e., *Groundwater Hydrology* of D. K. Todd published in 1959, the words "dating groundwater" appeared on page 61, but the word "age" was not used.

The hydrogeology textbook of Davies and De Weist (1966) presents a very interesting (considering the old date of the publication) discussion of radionuclides in groundwater in Chapter 5 and includes approximately two pages of writings about dating groundwater by ^{14}C and ^{3}H methods and potential ^{36}Cl dating. Words like groundwater age or residence time, however, have not been used in this book. Freeze and Cherry's 1979 outstanding groundwater textbook devotes about 3 pages to groundwater age dating out of a total 8 pages allocated to environmental isotopes, covering only two techniques of tritium and carbon-14. This book is most probably the first major textbook to include the term "groundwater age." There is no single textbook, prior to the present one, devoted completely to the groundwater age discipline. One might be interested to compare this with the number of books allocated to the age of Earth. Few examples with a rhythmic 10-year interval are York and Farquhar (1972), Young (1982), and Dalrymple (1991).

Table 2.1 provides a list of all groundwater age-dating methods and includes the date, the pioneering scientist/s, and the country where these methods were first introduced or applied. In providing Table 2.1, we took a number of points into account:

a. The researchers who were selected as the pioneer of the methods are those who either initiated application of, or clearly proposed, the method.
b. It has been tried to consider journal papers or IAEA proceedings as the basis of the selection, i.e., presentation at workshops or internal reports were not given serious consideration.
c. There are some uncertainties about exactly who for the first time proposed a particular method. Because the proposal is usually in the form of a few words or a few lines in a paper or in a book. Initially, it was intended to have two separate columns in Table 2.1, one column for the names who proposed the methods and one for those who applied the methods.
d. Only publications in English were, inevitably, studied with a few well known in German. It is well possible that scientists from the former USSR were exercising groundwater dating behind the Iron Curtain at dates earlier than those suggested here. It is difficult to access such publications, and we tend to believe that they are limited, and not at milestone level. If these were plenty, we should have come across them cited in the mainstream literature.
e. It is believed that this table still contains some inaccuracies, and the authors wish to apologize in advance, if any name that should have, has not appeared in this table.
f. The country specified is the country where at the time of the report of the method, the authors were working, not where they work at present.
g. Hydraulic age calculation, or age-dating hydraulically, has not been mentioned in Table 2.1 because it is not generally considered as a dating method (it is a method but rather uses completely different principles

TABLE 2.1. Various Age-Dating Methods, the Commencement Dates, the Pioneering Scientists, and the Country of Origin

Method	Commencement date	Source	Country
Age-dating young groundwater			
^{3}H	1957	Beggeman and Libby	USA
^{3}H/^{3}He	1969	Tolstikhin and Kamensky	Former USSR
^{4}He	1996	Solomon et al.	USA
^{85}Kr	1978	Rozanski and Florkowski	Poland
CFCs	1974	Thompson et al.	USA
SF$_6$	2000	Busenberg and Plummer[1]	USA
^{36}Cl/Cl	1982	Bentley et al.	USA
^{18}O-^{2}H	1983	Maloszewski et al.	Germany
Age-dating old groundwater			
Silicon-32	1966	Nijamparkar et al.	India
Argon-39	1974	Oeschger et al.	Switzerland
Carbon-14	1959	Brinkmann et al.	Germany
Oxygen-18 and deuterium	1981	Fontes	France
Inert and active elements	2000	Edmunds and Smedley	UK
Age-dating very old groundwater			
Krypton-81	1969	Loosli and Oescher	Switzerland
Chlorine-36	1966	Davis and DeWiest[2]	USA
Helium-4	1979	Marine; Andrew and Lee[3]	USA-UK
Argon-40	1987	Zaikowski et al.	USA
Iodine-129	1985	Fabryka-Martin et al.	USA
Uranium Disequilibrium series	1974	Kronfeld and Adams[4]; Osmond and Cowart	USA

[1] Note that this method was first briefly introduced by Busenberg and Plummer (1997) and Fulda and Kinzelbach (1997).
[2] Proposed by Davis and De Weist and applied for the first time by Bentley et al. (1986).
[3] Based on Torgersen and Clarke (1985), a USSR scientist named Savchemko proposed this method in 1937.
[4] Initially proposed by USSR scientists, P. I. Chalow, T. V. Tuzova, Y. A. Musin, in 1964, but for dating closed drainage basins (source: Osmond et al., 1968).

as compared to tracer methods). Similarly, the use of computer codes to simulate the distribution of groundwater ages, discussed in Chapter 7, is not regarded as a dating method and hence there is no mention of it in Table 2.1.

Table 2.1 shows that a large number of age-dating methods have been developed or proposed by the scientists from the USA, Germany, and Switzerland.

2.2 LABORATORIES WORLDWIDE FOR DATING GROUNDWATER SAMPLES

Table 2.2 provides a comprehensive list of worldwide analytical facilities for analyzing groundwater samples for environmental isotopes used in the age-dating practice. This information was gathered mainly from personal contact and websites. Therefore, laboratories without a website may have not appeared here. In addition to Table 2.2, the following web addresses contain very useful information.

1. AMS laboratories worldwide at: www.univie.ac.at/Kernphysik/VERA/ams_list.htm
2. A full list of all isotope laboratories world wide can be found at: www.gwadi.org/GWADI_Isotope_Labs.pdf. It was not possible to find the website address of all laboratories cited in this website.
3. Website of the Chlorofluorocarbon Laboratory of the USGS, Reston, Virgina, at http://water.usgs.gov/lab/cfc/ contains a large amount of useful information about a variety of analytical methods.
4. List of all known radiocarbon laboratories at http://www.radiocarbon.org/Info/lablist.html

2.3 MAJOR CONTRIBUTORS TO THE GROUNDWATER AGE DISCIPLINE AND THEIR ACHIEVEMENTS

Over the past 50 years, a large number of very talented scientists have contributed to, and developed many methods for, groundwater age-dating. Due to the complexity and scientifically advanced nature of the topic, most groundwater dating researches have been carried out jointly by a number of scientists. One is fascinated with the number of researchers working cooperatively on a single project that includes only one or two dating methods (see, for example, Sturchio et al., 2004, a paper with 18 authors!; or Lehmann et al., 2003, a paper with 14 authors!). Here, emphasis is on those who have had a long-lasting career in the age-dating business. Some researchers, though distinguished and highly influential, worked in this field for a short period of time and then left for another field. These are not included in this section just because of their relatively shorter experience as compared to those who are cited here. For their achievements, the scientists in this field have so far been rewarded with three Meinzer Awards, one Darcy Distinguished Lecture award, and two Geological Society of America, Birdsall-Dreiss Distinguished Lecture, i.e., Fogg, 2002 and Phillips (1994). [The O. E. Meinzer Award, which

TABLE 2.2. Worldwide Institutions with Laboratory Facilities for Groundwater Age Measurements (Due to space limitations, http//www have been omitted from the addresses. Also, word "laboratory/ies" has been omitted from the names in the first column)

Laboratory	Institution	Country	Website	Analysis
Adelaide	CSIRO Land and Water	Australia	clw.csiro.au/services/isotope/index.html	^{14}C, CFCs
Department of Nuclear Physics	Australian National University	Australia	wwwrsphysse.anu.edu.au/nuclear/ams/ams.html#hydro	^{36}Cl, ^{129}I
ANTARES AMS	Australian Nuclear Science and Technology	Australia	ansto.gov.au/ansto/environment1/ams/ams_Radiocarbondating.htm	^{14}C, ^{3}H
Isotope hydrology	IAEA, Vienna	Austria	iaea.org/programmes/rial/pci/isotopehydrology/	^{3}H
VERA	University of Vienna	Austria	univie.ac.at/Kernphysik/VERA/vera1.htm	^{14}C, ^{36}Cl, ^{129}I
Testmark Laboratories	Garson, Ontario	Canada	testmark.ca/research.htm#agedating	CFCs, HCFCs
Environnemental isotope	University of Waterloo	Canada	science.uwaterloo.ca/research/eilab/	^{37}Cl, ^{3}H
IsoTrace	University of Toronto	Canada	physics.utoronto.ca/~isotrace/	^{14}C, ^{129}I
Radiocarbon dating	GSC	Canada	http://gsc.nrcan.gc.ca/lab_e.php	^{14}C
AMS ^{14}C Dating Centre	University of Aarhus	Denmark	phys.au.dk/ams/	^{14}C
ARTEMIS	CNRS	France	www2.cnrs.fr/en/291.htm	^{14}C
CRPG	CNRS	France	crpg.cnrs-nancy.fr/	^{4}He, ^{3}He
Leibniz-Labor	Christian Albrechts Universität	Germany	uni-kiel.de/leibniz/indexe.htm	^{14}C
Institute of Environmental Physics	Bremen University	Germany	noblegas.unibremen.de/index_eng.html	^{3}H, He, CFCs
Hydroisotope GMBH	University of Heidelberg	Germany	iup.uni-heidelberg.de/	
Department of Isotope hydrology	UFZ Center for Environmnetal research	Germany	hydroisotop.de/isometheng.html	^{14}C, ^{3}H, ^{85}Kr
NIH	University of Roorkee	Germany	halle.ufz.de/isohyd/index.php?en=1430	^{14}C, ^{3}H
Isotope hydrology division	Bhabha atomic research center	India	angelfire.com/bc/nihhrrc/Roorkee/	^{14}C, ^{3}H
		India	http://www.nih.ernet.in/nccih/BARC-prj.htm	^{14}C, ^{3}H
Radiocarbon	University College Dublin	Ireland	ucd.ie/radphys/page2.htm	^{14}C
Earthquake chemistry	University of Tokyo	Japan	eqchem.s.utokyo.ac.jp/index_e.html	^{4}He
Environmental Chemistry Division	National Institute for Environmental Studies	Japan	nies.go.jp/chem/index-e.html	^{14}C

Lab	Institution	Country	URL	Isotopes
Center for Isotope Research	University of Groningen	Netherlands	cio.phys.rug.nl/HTML-docs/cio-us/frb06.htm	^{14}C, ^{3}H
	Institute of Geological and Nuclear Sciences	New Zeland	gns.cri.nz/services/groundwater/	CFCs, SF_6, ^{3}H, ^{13}C, ^{32}Si
PINSTECH	Inst. of nuclear Sci. & Tech.	Pakistan	paec.gov.pk/pinstech/index.htm	^{14}C, ^{3}H
Tritium dating	Instituto Tecnológicoe Nuclear	Portugal	itn.pt/Facilities/tritium_dating_laboratory.htm	^{3}H
Environmental isotope	University of Witwatersrand	South Africa	www.tlabs.ac.zacontact: butler@src.wits.ac.za	^{3}H, ^{14}C, ^{222}Rn
Laboratory of isotope hydrology	CEDEX	Spain	cedex.es/ingles/presentacion/datos/instalaciones/ins10_ing.html	^{14}C, ^{3}H
Environmental Radioactivity Laboratory	Autonomous University of Barcelona	Spain	http://cc.uab.es/lra/lab.htm	^{3}H
Institute of Physics	University of Bern	Switzerland	climate.unibe.ch/~jcor/Group/Group.html	^{39}Ar, ^{37}Ar, ^{85}Kr, ^{3}H, ^{14}C
	Swiss Federal Institute of Science and Technology (EAWAG, Paul Scherrer and ETH)	Switzerland	eawag.ch/research_e/w+t/UI/methods_e.html	^{4}He, ^{40}Ar, $^{3}He/^{4}He$, ^{3}H, $^{40}Ar/^{37}Ar$, SF_6, CFCs, ^{36}Cl
Environmental tracers	Queens University, Belfast	UK	prbnet.qub.ac.uk/eerg/Facilits/traclab.htm	Ar, Kr, CFCs, SF_6
Cosmogenic isotope laboratory	NERC-SUERK	UK	gla.ac.uk/suerc/research/nercicosmo.html	^{40}Ar, ^{14}C
Groundwater tracing unit	University College London Champaign, Illinois	UK	es.ucl.ac.uk/research/hydro/staff.htm	CFCs, SF_6
Isotech Labs		USA	isotechlabs.com/	^{14}C, ^{3}H
PRIME	University of Purdue	USA	physics.purdue.edu/primelab/	^{36}Cl, ^{129}I, ^{14}C, ^{35}Cl
Earth and Atmospheric Sciences	University of Purdue	USA	eas.purdue.edu	^{3}H
National Superconducting Cyclotron	Michigan State University	USA	nscl.msu.edu/	^{81}Kr

TABLE 2.2. Continued

Laboratory	Institution	Country	Website	Analysis
Lawrence Livermore National	University of California	USA	Llnl.gov str/Davisson.html	^{129}I, ^{36}Cl, ^{14}C, ^{3}H, ^{3}H/^{4}He, ^{4}He
IRIM	University of Tennessee	USA	http://web.utk.edu/~irim/	^{85}Kr and ^{81}Kr
CFCs	USGS Reston, Virginia	USA	water.usgs.gov/lab/shared/guidance.htm	CFCs, SF$_6$, Diss.gases, ^{3}H/^{3}He
SAHRA	University of Arizona	USA	sahra.arizona.edu/programs/isotopes.html	Helium isotopes
RSMAS	University of Miami	USA	rsmas.miami.edu	^{3}He, ^{3}H, CFCs
LDEO	Columbia University	USA	ldeo.columbia.edu/res/fac/etg	He, Ne, Ar, Kr, Xe, SF$_6$
Water Sciences (Water centre)	University of Nebraska	USA	waterscience.unl.edu	CFCs
Argonne National	University of Chicago	USA	mep.phy.anl.gov/atta/index.html	^{81}Kr, ^{85}Kr
Arizona AMS and Geosciences	University of Arizona	USA	physics.arizona.edu/ams/service/sub_info.htm	^{14}C, ^{3}H
BETA	Beta Analytical Florida	USA	radiocarbon.com/groundwater.htm	^{14}C
National Ocean Sciences AMS Facility	Woods Hole Oceanographic Institution	USA	http://nosams.whoi.edu	^{14}C
Isotope (Tritium)	USGS–Menlo Park	USA	http://menlocampus.wr.usgs.gov	^{3}H
Isotope geochemistry	USGS–Illinois	USA	isgs.uiuc.edu/isotope/radio.html	^{14}C
Centre for applied isotope	University of Georgia	USA	http://www.uga.edu/cais/	^{3}H, ^{14}C
Noble gas	University of Utah	USA	mines.utah.edu/geo/	Dis. gases, ^{3}H, CFCs
Cosmogenic isotope	University of Rochester	USA	earth.rochester.edu/fehnlab/	^{14}C, ^{3}He, ^{36}Cl, ^{129}I
Geochron Laboratories	Krueger Enterprises	USA	geochronlabs.com/	^{14}C, ^{3}H

was established by the GSA in 1963, is named for Oscar Edward Meinzer (1876–1948), whom hydrologists consider "the father of modern groundwater hydrology" in North America. The Meinzer Award includes a "traveling trophy," which consists of a year-long possession of a large silver Revere bowl, engraved with the names of former award winners, along with a smaller, "keeper" version of the bowl inscribed solely with the awardees name. Darcy Lecture Distinguished Series in Ground Water Science was established in 1986 to foster interest and excellence in groundwater science and technology. Today, the United States National Ground Water Research and Educational Foundation (NGWREF) sponsors the Henry Darcy Distinguished Lecture Series. The Darcy Lecturer receives support to spend a year traveling and lecturing at various institutions worldwide.]

Some scientists who contributed substantially to groundwater age discipline include

- A. N. Andrews for his work on noble gases dating.
- K. O. Münnich for his work on carbon-14 dating.
- F. M. Phillips for his work on chlorine-36 dating. F. M. Phillips won the Meinzer prize in 2001 partly for his contribution to groundwater dating and his publications in this arena. He was also Birdsall Dreiss Lecturer in 1994, discussing the use of fossil rat urine in chronology of groundwater.
- L. N. Plummer for his work on CFCs dating and also on the relationship between groundwater contamination and age. L. N. Plummer won the Meinzer prize in 1993 partly for his work on CFCs dating.
- E. Busenberg for his work on CFCs dating and also on relationship between groundwater contamination and age.
- P. G. Cook for his work on CFCs dating.
- D. K. Solomon for his work on noble gases in groundwater. D. K. Solomon, who runs the Noble Gas laboratory at the University of Utah, USA, was selected as the Darcy's Distinguished Lecturer in 2005 for his work on groundwater dating using noble gases.
- H. H. Loosli, for his long-lasting work on noble gas dating.
- B. E. Lehmann, for his long-lasting work on noble gas dating.
- T. Torgersen, for his work on helium-4 and argon-40 dating.
- H. W. Bentley for his work on chlorine-36 dating.
- T. M. L. Wigley for his work on correcting carbon-14 ages.
- S. N. Davis won the Meinzer Award in 1989 for publishing a report (Davis and Murphy, 1987—in the list of reports, i.e., Section 2.5.4) and co-authoring a book entitled *Groundwater Tracers* and published by NWWA Press (Davis et al., 1985).
- P. Maloszewski for his work on developing lumped parameter models to interpret groundwater age data.

- F. J. Pearson for his long-lasting work on various age-dating methods.
- F. Cornaton for his theoretical contributions on modeling groundwater ages and residence times distributions.
- J. Carrera for his conceptual work on simulating groundwater residence-time distributions.

2.4 NAMES FAMILIAR IN THE GROUNDWATER DATING BUSINESS

The names of the 50 scientists who have published considerably in the field of groundwater dating and their respective organizations are included in the list below. This list, although maybe not fair and complete, is to respect and appreciate the hard work of those mentioned. It may also serve as a database to help the junior researchers as well as everybody who needs, to quickly contact the appropriate and the nearest groundwater age expert.

1. A. G. Hunt, US Geological Survey
2. A. J. Love, CSIRO, Adelaide, Australia
3. A. Long, University of Arizona, USA
4. A. Zuber, Institute of Nuclear Physics, Crawcow, Poland
5. B. E. Lehmann, University of Bern, Switzerland
6. B. G. Katz, US Geological Survey, USA
7. C. M. Bethke, University of Illinois, USA
8. C. Sonntag, Heidelberg University, Germany
9. D. J. Goode, US Geological Survey
10. D. K. Solomon, University of Utah, USA
11. E. Busenberg, US Geological Survey, USA
12. E. Eriksson, University of Uppsala, Sweden
13. E. Mazor, Weizmann Institute, Israel
14. E. Modica, US Geological Survey, USA
15. F. Cornaton, University of Neuchâtel, Switzerland
16. F. J. Pearson, University of Texas, USA
17. F. M. Phillips, New Mexico Institute of Mining and Technology (NMT), USA
18. H. Oster, Heidelberg University, Germany
19. H. W. Bentley, University of Arizona, USA
20. H. H. Loosli, University of Bern, Switzerland
21. I. L. Kamensky, Kola Scientific Center of Russian Academy of Science, Russia
22. I. N. Tolstikhin, Cambridge University, UK

23. J. C. Vogel, Heidelberg University, Germany
24. J. Ch. Fontes, IAEA, Vienna (deceased)
25. J. Carrera, Technical University of Catalonia, Spain
26. J. K. Böhlke, US Geological Survey, USA
27. J. N. Andrews, University of Reading, University of Bath, UK (deceased)
28. J. T. Fabryka-Martin, Los Alamos National Laboratory, USA
29. K. O. Münnich, Heidelberg University, Germany
30. L. N. Plummer, US Geological Survey, USA
31. L. Sebol, Golder Associates, Canada
32. M. A. Tamers, Institute Venezolano de Investigaciones Cientificas, Caracas, Venezuela
33. M. C. Castro, University of Michigan, USA
34. M. E. Campana, University of New Mexico, USA
35. P. G. Cook, CSIRO, Adelaide, Australia
36. P. Maloszewski, GSF Institute for Hydrology, Germany
37. P. Schlosser, Columbia University, USA
38. R. J. Poreda, University of Rochester, USA
39. R. Kalin, Queens University, Belfast, UK
40. R. Purtschert, University of Bern, Switzerland
41. S. N. Davis, University of Arizona, USA
42. T. Iwatsuki, Tono Geoscience Center, Japan
43. T. M. Johnson, University of Illinois, USA
44. T. M. L. Wigley, University of Waterloo, Canada
45. T. Torgerson, University of Connecticut, USA
46. U. Fehn, University of Rochester, USA
47. W. E. Sanford, US Geological Survey, USA
48. W. F. Libby, University of Chicago, USA
49. W. M. Smethie, Columbia University, USA
50. Z. Szabo, US Geological Survey, USA

2.5 IMPORTANT PUBLICATIONS

Initially, groundwater dating publications dealt mostly with the theoretical aspects explaining the methods in order to increase the knowledge about the techniques. Later publications, in contrast, are the results of dating experiments on various aquifers worldwide and include a substantial discussion on various problems that might affect the accuracy of dates obtained. The publications dealing with age determination have been classified into books, Ph.D. and M.Sc. theses, full journals and important journal papers, and reports.

2.5.1 Book Chapters

A number of book chapters have been recently allocated to groundwater age, the oldest one dated 1991. The most closely related books to the present one that cover a wide variety of age-dating techniques are (1) *Applied Isotope Hydrogeology*: A case study in northern Switzerland by Pearson et al. in 1991, (2) *Environmental Tracers in Subsurface Hydrology* edited by Cook and Herczeg in 2000, (3) *Environmental Isotopes in Hydrogeology* written by Clark and Fritz in 1997, and (4) *Applied Chemical and Isotopic Groundwater Hydrology*, 3rd edition, by Mazor in 2004. Also, the reference by Wagner (1998), which has been published both in German and in English, is a very similar book to the present book, albeit devoted to age-dating young rocks and artefacts. The *Hand Book of Environmental Isotope Geochemistry* (Fritz and Fontes, 1980 and 1986) contains some scattered sections on groundwater age-dating but not much in the form of a full consolidated chapter. In addition, the reference by Davis and Bentley (1982), part of an edited book, may also be cautiously considered as a book chapter in groundwater age literature. Not yet published books such as that by Aggarwal et al. (2005) may also contain some sections on groundwater dating.

Book 1

Pearson, F. J., Balderer, W., Loosli, H. H., Lehmann, B. E., Matter, A., Peters, Tj, Schmassmann, H., and Gautschi, A. (Eds.), (1991) Applied isotope hydrogeology. A case study in northern Switzerland. Studies in environmental science, 43, Elsevier, Amsterdam: 439 p.

Chapter 4: Dating by Radionuclides	pp. 153–174
Chapter 5: Carbonate isotopes	pp. 175–238
Chapter 6: Isotopes formed by underground production	pp. 239–296
Section 7-3: Uranium and thorium series nuclides	pp. 336–373

Book 2

Cook, P.G., and Herczeg, A.L. (Eds), (2000) Environmental Tracers in Subsurface Hydrology. Kluwer Academic Publishers, Boston, 529 p.

Cook, P. G., and Böhlke, J. K.: Determining timescales for groundwater flow and solute transport	pp. 1–30
Kalin, R. M.,: Radiocarbon dating of groundwater systems	pp. 111–145
Loosli, H. H., Lehmann, B. E., and Semethie, W. M.: Noble gas radioisotopes: ^{37}Ar, ^{85}Kr, ^{39}Ar, ^{81}Kr	pp. 379–396
Solomon, D. K., and Cook, P. G.: ^3H and ^3He	pp. 397–424
Solomon, D. K.: ^4He in groundwater	pp. 425–440
Plummer, L. N., and Busenberg, E.: Chlorofluorocarbons	pp. 441–478
Phillips, F. M.: Chlorine-36	pp. 299–348

Book 3
Clark, I., and Fritz, P. (1997) Environmental Isotopes in Hydrogeology. Lewis Publishers, Boca Raton, Florida, 328 p.
Chapter 7: Identifying and dating modern groundwaters   pp. 171–196
Chapter 8: Age dating old groundwaters   pp. 197–244

Book 4
Mazor, E., (2004) Applied Chemical and Isotopic Groundwater Hydrology. 3rd edition, Marcel Dekker, Inc, New York, 453 p.
Part III: Isotopic tools of groundwater hydrology: Water identification and dating
Chapter 10: Tritium dating   pp. 210–230
Chapter 11: Radiocarbon dating   pp. 231–270
Chapter 12: Chlorine-36 dating   pp. 271–286
Chapter 13: The noble gases   pp. 287–311
Chapter 14: Helium-4 and Ar-40 long range dating   pp. 312–327

Book 5
Kendall, C., and McDonnell, J. J. (Eds) (1998) Isotope Tracers in Catchment Hydrology. Elsevier Science B.V., Amsterdam, 839 p.
Chapter 7: Isotopes in groundwater hydrology (partly related) pp. 203–246
Chapter 9: Dissolved gases in subsurface hydrology   pp. 291–318

Book 6
Alley, W. M. (Ed.) (1993) Regional Ground-Water Quality. Van Nostrand Reinhold, New York, 634 p.
Plummer, L. N., Michel, R. L., Thurman, E. M., and Glynn, P. D.: Environmental tracers for age-dating young groundwater   pp. 227–254

Book 7
Drever, J. I. (Ed.) (2003) Surface and ground water, weathering, and soils, Volume 5 of 10-volumes, Treaties on Geochemistry, Elsevier
Chapter 15: Phillips, F. M. and, Castro, M. C.: Ground water dating and residence time measurements   pp. 451–497

Book 8
Parnell, J. (Ed.) (1998). Dating and duration of fluid flow and fluid-rock interaction. Geological Society, London, Special publications, No. 144, 284 p.
Pinti, D. L., and Marty, B.: The origin of helium in deep sedimentary aquifers and the problem of dating very old groundwaters   pp. 53–68
Metcalfe, R. Hooker, P. J. Darling, W. G., and Milodovski, A. E.: Dating quaternary groundwater flow events: A review of available methods and their application   pp. 233–260

Book 9

Attendorn, H. G., and Bowen, R. N. C. (1997) Radioactive and Stable Isotope Geology. Chapman & Hall, London, 522 p.

Chapter 10: Tritium dating   pp. 268–271
Chapter 13: Cosmogenic radionuclides   pp. 347–366

Book 10

Dickin, A. P. (1995) Radiogenic Isotope Geology. Cambridge University Press, Cambridge, 490 p.

Chapter 14: Cosmogenic nuclides   pp. 360–396

Book 11

Taylor, R. E., Long A., and Kra, R.S. (Eds.) (1992) Radiocarbon After Four Decades. An Interdisciplinary Perspective. Springer, Berlin, 596 p.

Fontes, J. C.: Chemical and isotopic constraints on ^{14}C dating of groundwater   pp. 242–261

Book 12

Plummer, L. N., Busenberg, E., Cook, P. G., Solomon, D. K., and others (2004) Guidebook on the Use of Chlorofluorocarbons in Hydrology. IAEA (in press).

2.5.2 Ph.D. and M.Sc. Theses

Altogether 47 Ph.D. and M.Sc. theses have focused on the topic of groundwater dating (below here). The first one is that by F. J. Pearson in 1966 and the last before publication of this book is the one by L. Sebol in late 2004, a time span of about 40 years. Theses specified as "closely related" in the parenthesis, are those which are not exactly on the age-dating topic but are of immediate full relationship.

    Aeschbach-Hertig, W. (1994) Helium and tritium as tracer für physikalische prozesse in Seen, Ph.D. thesis, ETH, Zürich, (No. 10714). (closely related)

    Aravena, R. (1993) Carbon isotope geochemistry of a confined methanogenic groundwater system: The Allison aquifer complex. Ph.D. thesis, University of Waterloo, Canada. (closely related)

    Atakan, Y. (1972) Bomb tritium hydrology of a sand unconfined aquifer, Ph.D. thesis, University of Heidelberg, Germany, 102 p. (closely related)

    Bitner, M. J. (1983) The effects of dispersion and mixing on radionuclide dating of groundwater. M.Sc. thesis, University of Arizona, USA, 101 p.

    Brinkman, J. E. (1981) Water age dating of the Carrizo sand. M.Sc. thesis, University of Arizona, USA.

    Burbey, T. J. (1984) Three-dimensional numerical simulation of tritium and chlorine-36 migration. M.Sc. thesis, University of Nevada, USA. (closely related)

Cherry, A. (2000) Groundwater age with ^{3}H/^{3}He and CFCs. M.Sc. thesis, University of Ottawa, Canada.

Clark, I. D. (1987) Groundwater resources in the Sultanate of Oman: Origin, circulation times, recharge processes and paleoclimatology. Isotopic and geochemical approaches. Ph.D. thesis, Université de Paris-Sud, Orsay, France, 264 p.

Cornaton, F. (2004) Deterministic models of groundwater age, life expectancy and transit time distribution in advective-dispersive systems. Ph.D. thesis, University of Neuchatel, Switzerland, 147 p.

Cuttell, J. C. (1983) The application of uranium and thorium series isotopes to the study of certain British aquifers. Ph.D. thesis, University of Birmingham, UK, 269 p. (closely related)

Egboka, B. C. E. (1980) Bomb tritium as an indication of dispersion and recharge in shallow sand aquifers. Ph.D. thesis, University of Waterloo, Canada, 190 p. (closely related)

Elliot, T. (1990) Geochemical indicators of groundwater aging. Ph.D. thesis, University of Bath, UK.

Etcheverry, D. (2001) Une approche déterministe des distributions des temps de transit de l'eau souterraine par la théorie des réservoirs. Ph.D. thesis, University of Neuchatel, Switzerland, 118 p.

Fabryka-Martin, J. T. (1988) Production of radionuclides in the earth and their hydrogeologic significance, with emphasis on chlorine-36 and iodine-129. Ph.D. thesis, University of Arizona, USA, 400 p. (closely related)

Fabryka-Martin, J. T. (1984) Natural Iodine-129 as a groundwater tracer. M.Sc. thesis, University of Arizona, USA. (closely related)

Forster, M. (1984) C-14- und Ar-39-Gehalte in einem grundwasser des Saarlandes. Vergleich zweier grundwasser datierungsmethoden. Ph.D. thesis, University of Munich, Germany.

Fritzel, T. L. B. (1996) Numerical modeling of the distribution of ^{3}He and ^{4}He in groundwater flow systems with applications in determining the residence times of very old groundwaters. M.Sc. thesis, University of Illinois, Urbana-Champaign, USA, 109 p.

Fulda, C. (1998) Untersuchungen junger grundwässer mit Hilfe anthropogen erzeugter Spurenstoffe (In Deutsch). Ph.D. thesis, University of Heidelberg, Germany.

Goode, D. J. (1998) Ground-water age and atmospheric tracers: Simulation studies and analysis of field data from the Mirror Lake site, New Hampshire. Ph.D. thesis, Princeton University, USA, 194 p.

Groning, M. (1994) Noble gases and isotope tracer in the groundwater: Paleo-climatic changes and dynamics of regional groundwater systems (In Deutsch). Ph.D. thesis, University of Heidelberg, Germany, 136 p. (closely related)

Haran, R. (1997) Indicators of palaeowaters and groundwater mixing in the Lower Greensand of the London Basin. M. Phil. thesis, University of Reading, UK.

Heusser, D., and Langer, T. (2004) Geochemical groundwater evolution and age estimations for islands in the Okavango Delta. M.Sc. thesis. Swiss Federal Institute of Technology (ETH), Switzerland, 113 p.

Johnston, C. T. (1994) Geochemistry, isotopic composition, and age of groundwater from the Waterloo Moraine: Implications for groundwater protection and management. M.Sc. thesis, University of Waterloo, Canada.

Kilchmann, S. (2001) Typology of recent groundwaters from different aquifer environments based on geogenic tracer elements. Ph.D. thesis, Federal Institute of Technology, Lausanne, Switzerland.

Manning, A. H. (2002) Using noble gas tracers to investigate mountain-block recharge to an intermountain basin. Ph.D. thesis, University of Utah, USA. (closely related)

Mattle, N. (1999) Interpretation vpn tracer-daten in grundwässern mittels boxmodellen und numerischen strömungs-transportmodellen. Ph.D. thesis, University of Bern, Switzerland.

McGuire, K. J. (1999) Determining mean residence times of groundwater using oxygen-18 fluctuations in the Mid-Appalachians. M.Sc. thesis, The Pennsylvania State University, USA, 73 p.

Morse, B. S. (2002) Radiocarbon dating of groundwater using paleoclimatic constraints and dissolved organic carbon in the southern Great Basin, Nevada and California, M.Sc. thesis, University of Nevada, USA, 63 p.

Murphy, E. M. (1987) Carbon-14 measurements and characterizations of dissolved organic carbon in groundwater. Ph.D. thesis, University of Arizona, USA, 180 p. (closely related)

Osenbrück, K. (1996) Alter und dynamic tiefer grunwässer: Eine neue methode zur analyse der edelgase im perenwasservon gesteinen. Ph.D. thesis, University of Heidelberg, Germany.

Oster, H. (1994) Datierung von grundwasser mittels FCKW: Vorraussetzungen, Möglichkeiten and Grenzen. Ph.D. thesis, Universitüt Heidelberg, Germany, 121 p.

Patterson, L. J. (2003) ^{36}Cl and stable chlorine isotopes in the Nubian aquifer, Western desert, Egypt. M.Sc. thesis, University of Illinois at Chicago, USA.

Pearson, F. J. (1966) Groundwater ages and flow rates by the carbon-14 method. Ph.D. thesis, University of Texas, USA.

Pinti, D. L. (1993) Géochimi isotopique des gaz rares dans les huiles du Bassin Parisien. Implications sur la migration des huiles et la circulation hydrodynamique. Ph.D. thesis, Université Pierre et Marie Curie, France. (closely related)

Pracht, K. A. (2001) Flow and aquifer parameter evaluation using groundwater age-dating and geochemical tools: Missoula Aquifer, Western Montana. M.Sc. thesis, The University of Montana., USA.

Purdy, C. B. (1991) Isotopic and chemical tracer studies of ground water in the Aquia Formation, southern Maryland, including ^{36}Cl, ^{14}C, ^{18}O, and ^{3}H. Ph.D. thesis, University of Maryland, USA.

Purtschert, R. (1997) Multitracer-studien in der hydrologie; Anwendungen im glottal, am Wellenberg und in Vals. Ph.D. thesis, University of Bern, Switzerland. (closely related)

Rydell, H. S. (1968) Implications of uranium isotope distributions associated with the Floridan Aquifer. Ph.D. thesis, Florida State University, USA, 119 p. (closely related)

Sebol, L. (2004) Evaluating shallow groundwater age tracers: Br, CFCs, $^{3}H/^{3}He$, SF_6, and HCFCs/HFCs. Ph.D. thesis, University of Waterloo, Canada, 190 p.

Sebol, L. (2000) Determination of groundwater age using CFCs in three shallow aquifers in southern Ontario. M.Sc. thesis, University of Waterloo, Canada, 214 p.

Selaolo, E. T. (1998) Tracer studies and groundwater recharge assessment in the eastern fringe of the Botswana Kalahari—The Letlhaking-Botlhapatlou area. Ph.D. thesis, Free University of Amsterdam, The Netherlands, 229 p.

Shapiro, S. D. (1998) Evaluation of the ^{3}H-^{3}He dating technique in complex hydrologic environments. Ph.D. thesis, Columbia University, USA, 253 p.

Solomon, D. K. (1992) The use of tritium and helium isotopes to determine groundwater recharge to unconfined aquifers. Ph.D. thesis, University of Waterloo, Canada, 213 p. (closely related)

Swanick, G. (1982) The hydogeochemistry and age of groundwater in the Milk River aquifer, Alberta, Canada. M.Sc. thesis, University of Arizona, USA.

Thompson, G. M. (1976) Trichloromethane: A new hydrologic tool for tracing and dating groundwater. Ph.D. thesis, Indiana University, USA, 93 p.

Vitvar, T. (1998) Water residence time and runoff generation in a small prealpine catchment. Ph.D. thesis, ETH, Zurich, 111 p.

Wallick, E. I. (1973) Isotopic and chemical considerations in radiocarbon dating of groundwater within the arid Tucson basin aquifer, Arizona. Ph.D. thesis, University of Arizona, USA.

2.5.3 Journals

1. Applied Geochemistry The November 1997 issue of *Applied Geochemistry* (Vol. 12, No. 6, pp. 705–866), dedicated to the memory of John Napier Andrews (1930–1994), contains a full set of valuable papers dealing with many important aspects of groundwater dating using different tools. Also, the July-

August 1991 issue (Vol. 6, No. 4, pp. 367–472) of the same journal with the theme of "Dating very old groundwater, Milk River aquifer, Alberta, Canada" is wholly devoted to various age-dating techniques applied on Milk River aquifer. These two issues are recommended for consultation for those who want to get a broad view of topics covered under groundwater age.

2. Radiocarbon A journal allocated fully to carbon-14 related researches (www.radiocarbon.org). It is published by the Board of Regents, University of Arizona. Some of the *Radiocarbon* papers deal with groundwater age-dating. One very interesting article appearing in this journal is that by Geyh (2000), which discusses the historical trend in groundwater radiocarbon dating, especially evolution of analytical facilities. Some other groundwater-related *Radiocarbon* papers include Stuiver and Polach (1977), Yechieli et al. (2001), and Sivan et al. (2001). In addition to *Radiocarbon Journal*, in 1965, a conference entitled "6th International Conference, Radiocarbon and Tritium Dating" was held by the U.S. Atomic Energy Commission, Oak Ridge. The proceedings of this conference was published as U.S. Atomic Energy Commission Report, Conf-650652. It contains a number of papers dealing with groundwater dating. A number of other local workshops and small conferences such as that by Davis (1978) have been held on this subject.

3. Chemical Geology September 2001 issue of *Chemical Geology* (Vol. 179, Special issue, pp. 1–202), allocated to hydrochemistry of springs, contains an array of papers that deal with residence time of springs using almost all young groundwater dating methods. Anyone interested in the application of age-dating in the study of source, quality, and turnover time of spring waters is recommended to consult this reference.

4. Other important papers All groundwater age-related papers that have been published in the journals of *Nature* and *Science* are brief, but substantial, contributions to the field. These are cited throughout various parts of this book. Further, pre-1980 IAEA's hydrology symposiums and conferences include a number of very valuable, clearly written papers that cover initial developments and understanding of the dating techniques mainly for the tritium and carbon-14 methods (IAEA, 1967 Part III, 1970 Sessions 2 and 9, 1974, and 1976 Section III). The following, mostly, journal articles are considered as backbone papers for the specified method or topic they report on. These are, of course, in addition to the papers cited in Table 2.1 which are pioneering references for the related method.

The latest paper to include brief yet substantial information about groundwater dating topic is that by Glynn and Plummer (2005). In the *Water Encyclopedia* (Lehr and Keeley, 2005) there are 7 articles dealing with groundwater age (Divine and Humphrey, 2005; Divine and Thyne, 2005; Kazemi, 2005a,b,c; Priyadarshi, 2005; Testa, 2005). The trio-papers (Sturchio et al., 2004; Lehmann et al., 2003; Collon et al., 2000) cover the latest developments in ^{81}Kr dating techniques and contain, virtually, the entire knowledge that is presently available about this technique. Du et al. (2003) describe the Atom Trap Trace Analysis (ATTA), a recently invented technique to measure the concentration of ^{81}Kr in the water samples. The review paper by Phillips (1995) provides a good historical account of groundwater dating in shallow and deep aquifers. A large number of the applications of the groundwater age data are demonstrated by Solomon et al. (1995). The paper by Mazor and Nativ (1992) and subsequent related comments and reply by Torgersen (1994) and Mazor and Nativ (1994) are a good set to illustrate the problems involved in groundwater dating and also to explain the type of argument that may arise between scientists. Ferronskij et al. (1994) is a good reference to consult for the concentration of tritium in atmospheric precipitation in Russia. An interesting overview of the philosophy and history of time and dating, as well as time concepts in geology (oldest time), pedology (intermediate time period) and hydrology (youngest time) is presented by Back (1994). Fontes (1985) introduces and discusses, in a simple language, all the dating methods available until then. The two papers by Andrew et al. (1982 and 1989) include a very good short description of most age-dating methods as well as details about the Stripa granite aquifer in Sweden.

Cornaton and Perrochet (2005) introduce a generalized reservoir theory for the modeling of age and residence-time distributions in hydrodispersive systems with multiple inlets and outlets. Etcheverry and Perrochet (2000) provide a detailed and easy-to-understand description of the simulation of groundwater age distribution and the application of reservoir theory. In this regard, Goode (1996) is also a frontier work. Although not entirely on groundwater age, the long paper by Maloszewski and Zuber (1996) covers the mathematical aspects of isotopic dating and includes a description of some of the dating techniques.

2.5.4 Reports (Mainly by USGS)

Only reports with words like "age" and "residence time" in their title are cited here.

Bartolino, J. R. (1997) Chlorofluorocarbon and tritium age determination of ground-water recharge in the Ryan Flat subbasin, Trans-Pecos Texas. USGS WRI 96-4245, 29 p.

Böhlke, J. K., Plummer, L. N., Busenberg, E., Coplen, T. B., Shinn, E. A., and Schlosser, P. (1997) Origins, residence times, and nitrogen chemistry of marine ground waters beneath the Florida Keys and nearby offshore areas. USGS program on the south Florida ecosystem—Proc. of the technical symposium in Ft. Lauderdale, Florida, August 25–27, 1997, USGS OFR97-385, 6–7.

Busenberg, E., and Plummer, L. N. (1991) Chlorofluorocarbons (CCl_3F and CCl_2F_2): Use as an age-dating tool and hydrologic tracer in shallow groundwater systems. In: Proc. of USGS Toxic substances hydrology program, Technical meeting. Monterey, California, March 11–15, 1991, Mallard, G. L., and Aronson, D. A. (eds.) USGS WRI 91-4034, 542–5.

Busenberg, E., Plummer, L. N., and Bartholomay, R.C. (2001) Estimated age and source of the young fraction of ground water at the Idaho National Engineering and Environmental Laboratory. USGS WRI01-4265 (DOE/ID-22177), 144 p.

Busenberg, E., Weeks, E. P., Plummer, L. N., and Bartholomay, R. C. (1993) Age-dating ground water by use of chlorofluorocarbons (CCl_3F and CCl_2F2) and distribution of chlorofluorocarbons in the unsaturated zone, Snake River Plain Aquifer, Idaho National Engineering Laboratory, Idaho. USGSWRI 93-4054, 47 p.

Cox, S. E. (2003) Estimates of residence time and related variations in quality of ground water beneath submarine base Bangor and vicinity, Kitsap County, Washington. USGS WRI 03-4058.

Davis, S. N. (1978) Workshop on dating ground water. March 16–18, 1978, University of Arizona, Tucson, Arizona: Union Carbide Corporation-Nuclear division Y/OWI/Sub-78/55412. 138 p.

Davis, S. N., and Murphy, E. (1987) Dating ground water and the evaluation of repositories for radioactive waste: U.S. Nuclear Regulatory Commission, Report no. NUREG/CR-4912, 181 p.

Davisson, M. L. (2000) Discussion regarding sources and ages of groundwater in southeastern California. Lawrence Livermore National Laboratory UCRL-ID-138321, 10 p.

Davisson, M. L. (2000) Summary of age-dating analysis in the Fenner Basin, Eastern Mojave Desert, California, Lawrence Livermore National Laboratory UCRL-ID-139564, 6 p.

Focazio, M. J., Plummer, L. N., Böhlke, J. K., Busenberg, E., Bachman, L. J., and Powers, D. S. (1998) Preliminary estimates of residence times and apparent ages of ground water in the Chesapeake Bay watershed, and water-quality data from a survey of springs. USGS WRI 97-4225, 75 p.

Freethey, G. W., Spangler, L. E., and Monheiser, W. J. (1994) Determination of hydrologic properties needed to calculate average linear velocity and travel time of ground water in the principal aquifer underlying the southeastern part of Salt Lake Valley, Utah. USGS WRI92-4085, 30 p.

Goode, D. J. (1999) Age, double porosity, and simple reaction modifications for the MOC3D ground-water transport model. USGS WRI 99-4041.

Goode, D. J. (1999) Simulating contaminant attenuation, double-porosity exchange, and water age in aquifers using MOC3D, USGS FS086-99, 1 p.

Grove, D. B., Rubin, M., Hanshaw, B. B., and Beetem, W. A. (1969) Carbon-14 dates of groundwater from a Paleozoic carbonate aquifer, south central Nevada, USGS Prof. Paper, 650-C, pp. C215–C218.

Hayes, J. M., and Thompson, G. M. (1977) Trichlorofluoromethane in ground water—A possible indicator of ground-water age. Purdue University,

Water Resources Research Center, Technical report No. 90, NTIS Report PB 265 170, 25 p.

Hinkle, S. R. (1996) Age of ground water in basalt aquifers near Spring Creek National Fish Hatchery, Skamania County, Washington. USGS WRI 95-4272, 26 p.

Hinkle, S. R., and Synder, D. T. (1997) Comparison of chlorofluorocarbon-age dating with particle-tracking results of a regional groundwater flow model of the Portland basin, Oregon and Washington. USGS WRI 97-2483, 47 p.

Hinsby, K., Troldborg, L., Purtschert, R., and Corcho Alvarado, J. A. (2003) Integrated transient hydrological modelling of tracer transport and long-term groundwater/surface water interaction using four 30 year ^{3}H time series and groundwater dating for evaluation of groundwater flow dynamics and hydrochemical trends in groundwater and surface water. Report to the IAEA (TECDOC in preparation by the IAEA).

Hudson, G. B., Moran, J. E., and Eaton, G. F. (2002) Interpretation of ^{3}H/^{3}He groundwater ages and associated dissolved noble gas results from public water supply wells in the Los Angeles physiographic basin. Lawrence Livermore National Laboratory internal report, UCRL-AR-151447, 26 p.

Izbicki, J. A. (1996) Source, movement and age of ground water in a coastal California aquifer USGS FS126-96, 4 p.

Izbicki, J. A., and Michel, R. L. (2003) Movement and age of ground water in the western part of the Mojave Desert, southern California, USA, USGS WRI 03-4314.

Kauffman, L. J., Baehr, A. L., Ayers, M. A., and Stackelberg, P. E. (2001) Effects of land use and travel time on the distribution of nitrate in the Kirkwood-Cohansey aquifer system in southern New Jersey. USGS WRI 2001-4117, 49 p.

Kim, J. R., Artinger, R., Buckau, G., Kardinal, Ch., Geyer, S., Wolf, M., Halder, H. and Fritz, P. (1995) Grundwasser datierung mittels ^{14}C-Bestimmungen an gelösten Hu- rain- und Fulvinsiiuren. Final Report for BMFT project RCM 00895, TU-München, 221 p.

Knott, J. F., and Olimpio, J. C. (1986) Estimation of recharge rates to the sand and gravel aquifer using environmental tritium, Nantucket Island, Massachusetts. USGS WSP 2297, 26 p.

Lindsey, B. D., Phillips, S. W., Donnelly, C. A., Speiran, G. K., Plummer, L. N., Böhlke, J. K., Focazio, M. J., Burton, W. C., and Busenberg, E. (2003) Residence time and nitrate transport in groundwater discharging to streams in the Chesapeake Bay Watershed. USGS WRI 03-4035, 201 p.

Lopes, T. J., and Hoffmann, J. P. (1997) Geochemical analyses of groundwater ages, recharge rates, and hydraulic conductivity of the N aquifer, Black Mesa area, Arizona. USGS WRI 96-4190, 42 p.

McCartan, L., Weedman, S. D., Wingard, G. L., Edwards, L. E., Sugarman, P. J., Feigenson, M. D., Buursink, M. L., and Libarkin, J. C. (1995) Age and digenesis of the upper Floridan Aquifer and the intermediate aquifer system in southwestern Florida. USGS Bulletin 2122, 26 p.

McMahon, P. B., Böhlke, J. K., and Lehmann, T. M. (2004) Vertical gradients in water chemistry and age in the Southern High Plains Aquifer, Texas, 2002. USGS SIR2004-5053.

Moran, J. E., Hudson, G. B., and Eaton, G. F. (2002) A contamination vulnerability assessment for the Livermore-Amador and Niles Cone groundwater basins. Report to the California State Water Resources Control Board by the University of California. UCRL-AR-148831, 25 p + 5 pages appendices.

Nelms, D, L., Harlow, G. E., Jr., and Brockman, A. R. (2001) Apparent chlorofluorocarbon age of ground water of the shallow aquifer system, Naval Weapons Station, Yorktown, Virginia. USGS WRI 2001-4179, 51 p.

Nelms, D. L., and Harlow, G. E., Jr. (2003) Aquifer susceptibility in Virginia: Data on chemical and isotopic composition recharge temperature, and apparent age of water from wells and springs, 1998–2000, USGS OFR2003-246, 101 p.

Nelms, D. L., and Brockman, A. R. (1997) Hydrogeology of, quality and recharge ages of ground water in, Prince William County, Virginia, 1990–91: USGS WRI 97-4009, 58 p.

Phillips, S. W., Focazio, M. J., and Bachman, L. J. (1999) Discharge, nitrate load, and residence time of groundwater in the Chesapeake Bay Watershed. USGS FS-150-99, 6 p.

Plummer, L. N., and Busenberg, E. (1996) Chlorofluorocarbons (CFCs) as tracers and age-dating tools for young ground water: Selected field examples, in Stevens, P. R. and Nichols, T. J. (eds.), Joint U.S. Geological Survey, U.S. Nuclear Regulatory Commission workshop on "Research related to low-level radioactive waste disposal," May 4–6, 1993, Reston, VA: USGS95-4045, 65–71.

Plummer, L. N., and Friedman, L. C. (1999) Tracing and dating young ground water. USGS, FS-134-99, 4 p.

Plummer, L. N., and Mullin, A. (1997a) Tritium/helium dating of groundwater samples available through contact with Lamont-Doherty Earth Observatory of Columbia University, Palisades, New York. USGS technical memorandum 97-04, 1 p.

Plummer, L. N., and Mullin, A. (1997b) Supplementary information collection, processing, and analysis of groundwater samples for tritium-helium-3 dating. USGS technical memorandum 97-04S, 22 p.

Plummer, L. N., Dunkle, S. A., and Busenberg, E. (1993) Data on chlorofluorocarbons (CCl_3F and CCl_2F_2) as dating tools and hydrologic tracers in shallow ground water of the Delmarva Peninsula: USGS, OFR93-484, 56 p.

Rice, D. E., and Szabo, Z. (1997) Relation of ground-water flow paths and travel time to the distribution of radium and nitrate in current and former agricultural areas of the Kirkwood-Cohansey aquifer system, New Jersey coastal plain. USGS WRI96-4165-B, 41 p.

Robinson, J. L. (2004) Age and source of water in springs associated with the Jacksonville thrust fault complex, Calhoun County, Alabama, USGS SIR2004-5145.

Robinson, J. L., Carmichael, J. K., Halford, K. J., and Ladd, D. E. (1997) Hydrogeologic framework and simulation of ground-water flow and travel time in the shallow aquifer system in the area of Naval Support Activity Memphis, Millington, Tennessee. USGS WRI 97-4228.

Rowe, G. L., Shapiro, S. D., and Schlosser, P. (1999) Use of environmental tracers to evaluate ground-water age and water-quality trends in a buried-valley aquifer, Dayton area, southwestern Ohio. USGS WRI 99-4113, 81 p.

Rupert, M. G., and Plummer, L. N. (2004) Ground-water age and flow at the great sand dunes National Monument, south-central Colorado. USGS FS04-3051, 2 p.

Rupert, M. G., and Plummer, L. N. (2004) Ground-water flow direction, water quality, recharge sources, and age, Great Sand Dunes national monument, south-central Colorado, 2000–2001. USGS SIR2004-5027, 28 p.

Shapiro, A. M., Wood, W. W., Busenberg, E., Drenkard, S., Plummer, L. N., Torgersen, T., and Schlosser, P. (1996) Conceptual model for estimating regional fluid velocity in the bedrock of the Mirror Lake Area, Grafton County, New Hampshire. In: Mallard, G. E., and Aronson, D. A. (eds.) U.S. Geological Survey Toxic Substances Hydrology Program. In: Proc. of the technical meeting, Colorado Springs, Colorado, September 20–24, 1993: USGS WRI94-4015, 171–177.

Stewart, M., Trompetter, V., and van der Raaij, R. (2002) Age and source of Canterbury plains groundwater, Report No. U02/30, New Zealand Institute of Geological and Nuclear Sciences, 47 p.

Truini, M., and Longsworth, S. A. (2003) Hydrogeology of the D aquifer and movement and ages of ground water determined from geochemical and isotopic analyses, Black Mesa area, north-eastern Arizona. USGS WRI 03-4189, 38 p.

USGS (2002) Ground water age and its water-management implications, Cook Inlet Basin, Alaska. USGS FS-022-02, 4 p. http://water.usgs.gov/pubs/fs/fs-022-02/.

Weaver, J. M. C., and Talma, A. S. (1999) Field studies of chlorofluorocarbons (CFCs) as a groundwater dating tool in fractured rock aquifers. Pretoria: Water Research Commission, Report 731/1/99.

Yechieli, Y., Sivan, O., Lazar, B., Vengosh, A., Ronen, D., and Herut, B. (2001) Isotopic measurements and groundwater dating at the fresh-saline water interface region of the Mediterranean coastal plain aquifer of Israel. Report GSI/28/96. Israel Geological Society.

2.6 AQUIFERS SUBJECTED TO EXTENSIVE DATING STUDIES

In contrast to most aquifers, which have not been subjected to any level of dating, some aquifers have undergone extensive dating research. The latter group is listed here. In some cases, virtually all dating methods have been applied to a single aquifer.

Great Artesian Basin, GAB, Australia
Murray Basin, Australia
Milk River aquifer, Canada
Borden aquifer, Ontario, Canada
Strugen Fall, Canada
Kalahari desert, Africa
Illumedan Basin, Niger
Paris Basin, France
Northern Switzerland aquifers, Switzerland
Stripa granite aquifer, Sweden
Triassic sandstone aquifer, UK
Carizo sand aquifer, USA
Lucust Grove aquifer, USA
Snake River Plain aquifer of Idaho, USA
Tucson basin aquifer, USA
Groundwaters in the Chesapeake Bay Watershed, USA
Mojave River Basin, California, USA
Kirkwood-Cohnasey aquifer system in southern New Jersey Coastal Plain, USA
Deep groundwaters from the Tono area, central Japan

3

THE APPLICATIONS OF GROUNDWATER AGE DATA

Why do we need to know the age of groundwater? What use will it have? What we will learn when we have the age of a groundwater sample? What will we know when we have the age that we did not know before? The applications of age data are grouped into two categories, those that are general applications and those that are site-specific. Groundwater age data are discussed as a tool to evaluate the renewability of groundwater reservoirs, to constrain the parameters of groundwater flow and transport models, to study groundwater flow paths and vertical and horizontal flow velocities, to identify paleoclimate conditions (in combination with isotopes), to estimate groundwater recharge, to determine fracture and matrix properties and water velocities in fractured rock environments, to help study the trend of groundwater pollution, to identify past seawater level fluctuation, to manage groundwater-driven dryland salinity, to map susceptibility of groundwater systems to contamination, and to be used in many more hydrological applications such as mixing, groundwater–surface water interaction, and seawater intrusion.

3.1 RENEWABILITY OF THE GROUNDWATER RESERVOIRS

It is safe to argue that the most important and the unique application of groundwater age concept is its capacity to answer the question of renewabilty or replenishment of the groundwater resources. It is also safe to claim that

Groundwater Age, by Gholam A. Kazemi, Jay H. Lehr, and Pierre Perrochet
Copyright © 2006 John Wiley & Sons, Inc.

groundwater age is, as yet, the only sound and concrete piece of scientific evidence to show that groundwater resources are recharged by modern precipitation, or else, the extracted groundwaters were accumulated in the aquifers by slow infiltration processes that happened a very long time ago. This application is more highlighted in the arid zones where due to the scarcity and periodicity of the rainfall, the question of recharge (if any) often remains open (Payne, 1988). An important fraction of young water within an extracted water sample is an indication of an actively renewable reservoir; the opposite, i.e., a considerable amount of old water in the sample, depicts a poorly recharging reservoir and/or significant internal mixing processes (Cornaton, 2004). This is exactly what Libby (1953) in those early days suggested, "the determination of tritium in wells should reveal whether their ultimate supplies are closely connected to the rainfall." At a slightly larger scale, if there is a considerable percentage/volume of young water in a specific aquifer, we can be sure that the aquifer is being presently replenished by rainfall/snowfall recharges, and is sustainable. From a global perspective, it is important to establish what proportion of the world's freshwater resources is not being renewed at present, but was recharged instead during the Pleistocene (Metcalfe et al. 1998) or even earlier geologic times. Information about the age of groundwater is required if one is to confidently define the sustainability of groundwater resources of any particular well field. Zongyu et al. (2005, p. 485) writes, "Two fundamental questions related to the sustainability of the groundwater resources are (1) How old is this potable water supply? (2) What is the recharge of the deep confined aquifer? (are we 'mining' groundwater)?" By knowing the age of groundwater in our aquifers, we will be able to calculate the time that it takes for Mother Nature to replace the presently available groundwaters if we extract them. This will be a decisive tool enabling us to manage the precious groundwater resources in a sustainable manner.

To further explain the importance of the groundwater age concept, one should think of an important aquifer for which there is no solid information about renewability. Such an aquifer could be located anywhere in the world in an arid zone and in a less developed country, where no scientific data have been obtained to study the aquifer recharge processes. In such a case, it is believed that the most convincing and the quickest approach is to measure the age of groundwater. Techniques like measuring water-level fluctuations usually take months if not years to provide concrete and solid information. Age of groundwater in this hypothetical aquifer can quickly shed light on the renewability and sustainability issues, which now surround most aquifers worldwide. P. K. Aggarwal, Head, Isotope Hydrology Section, IAEA, has said in regard to the specific application of the groundwater age concept, "Estimates of renewable groundwater resources and an understanding of related hydrological processes are critically dependent upon determining the presence and age of modern groundwater" (Aggarwal, 2002, page 2257). In fact, one of the three new initiatives of the IAEA, to enhance the scientific understanding of non-renewable and renewable groundwater resources, is to map very old groundwaters and to develop a series of GIS-based global maps of nonrenewable groundwater resources (Gaye et al. 2005). We might conclude this section with the excerpt of an article from *The New York Times* (Broad, 2005) referring to the question of how often a particular body of water is renewed. "In the past, water engineers would address such unknowns by carefully measuring rain and the levels of rivers and other bodies of water for many decades—typically a half-century or more. It usually took that long to learn the subtleties of the local cycles. However, isotope hydrology can do it in days."

Example By measuring tritium and ^{14}C content of 17 groundwater samples, Vehagen et al. (1974) showed unambiguously for the first time that groundwaters in the northern Kalahari are directly, and in some cases rapidly, recharged by rain. The traditional scientific view was that the 6-m-thick sand overburden of the Kalahari desert prevented infiltrating rainwater from reaching phreatic water tables. Similarly, tritium content of water samples (0.04–12.9 TU) from shallow aquifer of Eastern Desert, near the intersection of Wadi El-Tarfa and the Nile River, Egypt, showed that the aquifer is at least partially maintained by modern recharge and the Nubian aquifer paleowater is not a significant component of this shallow aquifer (Sultan et al. 2000).

3.2 AN EFFECTIVE COMMUNICATION TOOL FOR SCIENTISTS AND MANAGERS—AND CURIOSITY TO LAYMEN AS WELL

One could also argue that the concept of groundwater age is the most easily understood scientific evidence when the groundwater professionals are to deal with resource managers and bureaucrats who often do not have a great deal of information about the hydrogeology and the groundwater hydraulics concepts. Effective communication between a resource manager and a groundwater professional can be made if the groundwater age data are available. It is easy to explain to resource managers what renewability means if we know the age of groundwater. Simply, tell "them" that this groundwater that "you" are extracting infiltrated into the subsurface environment 3,000 years ago. So if "you" use it now, it will take another 3,000 years to replace it. Therefore, "you" should take great care in using it and in preserving it for the future generations. Explain further that extracting very old, nonrenewable groundwater resources can have serious political and sociological implications (Clarke and Fritz, 1997) and should be viewed as a typical example of "groundwater mining." We refer again to the same article in *The New York Times* (Broad, 2005), where we read, "For instance, if the method reveals that the water in a well is young and recently derived from rain, villagers can pump away vigorously. But if it turns out to be very old—what scientists call fossil water—they need to move gingerly, taking care not to exhaust the water supply."

The groundwater age concept is also attractive for most laymen. Just have a look at this web posted question (in 2000) from Christchurch, New Zealand (www.madsci.org/posts/archives/nov2000/973783743.Es.q.html): "I heard somewhere that it is possible to tell the 'age' of groundwater (the year it was last on the surface of the earth) through a 'signature' of an isotope left on the water when it was last on the surface. I have a well/bore that is very deep and am curious to ascertain the 'age' of the water. Can anyone help? Thanks Geoff." Many nonscientists become immediately interested in the topic when they hear of such a thing as "age of groundwater". In fact, Payne (1988) argues that during the early days of isotope hydrology and isotope techniques, some of the most attractive work for hydrologists was the possibility of dating groundwater. In terms of scientific advancement, groundwater age-dating ranks high among other advances in the 20th century.

3.3 AGE MONITORING FOR THE PREVENTION OF OVEREXPLOITATION AND CONTAMINATION OF AQUIFERS

Increase in the population density often leads to an exponential increase in the demand on the aquifer. Once residences or industries are established, it is very difficult to limit their water supply. Overdevelopment can eventually lead to limited supply, with the greatest effects being to those districts farthest from the aquifers recharge zone (supply source). Since underutilized lands generally surround populated areas, housing and industrial development extends in directions reflecting the highest commercial yield. However, if the developing areas encroach into the recharge zone, new wells drilled to satisfy the eminent demand could create shortages if pumping exceeds recharge (The preceding lines are summarized from the Beta Analytic, Inc., Radiocarbon Dating Website.) By measuring the age of the water at certain time intervals within a district's well field (say once every five years), it would be possible to identify overexploitation before it happens. If the groundwater, being extracted, increases in age with time (becomes older and older), it means that a higher proportion of water is drawn from slow-moving storage. In contrast, if the age of groundwater being withdrawn decreases with time (becomes younger and younger), it means that a higher proportion of extracted water is derived from active present recharge. This shows that either the pumping rate has increased or the source water has changed (i.e., river recharge instead of rainfall recharge). This condition though does not imply groundwater mining, but it may not be a good sign in terms of contamination because eventually surface contaminants (if present) dissolved in very young waters (which may be contaminated) will reach the well field. Therefore, the best situation in terms of resource management is not to have sharp changes (either decrease or increase) in the age of groundwater that is being extracted. In fact, if the extracted water becomes too young, then there won't be enough time for natural purification and attenuation of pollutants like viruses or microbes to take place (this is explained in later sections). Hence, regular dating of the groundwater from well fields can provides a mechanism to monitor, understand, and control exploitation and contamination of the aquifer.

3.4 ESTIMATION OF THE RECHARGE RATE

This particular usage of groundwater age data is perhaps the most widely applied of all. An applied and comprehensive reference dealing with this topic is that by Xu and Beekman (2003), which is available freely in PDF format at http://unesdoc.unesco.org/images/0013/001324/132404e.pdf.

Figure 3.1 illustrates in a simple way the approach to calculate the recharge rate to a groundwater system by age data. The procedure is to have either

1. a minimum of two ages along the vertical line at the point of interest, i.e., age data must be obtained from a piezometer nest, which comprises at least two piezometers opened to the aquifer at different depths, or
2. the vertical position of the bomb peak tritium in the aquifer. It should be noted that the groundwater flow should consist of only one vertical component with negligible horizontal movement. The second approach may not be particularly accurate because of the difference between the flow rates in the saturated and unsaturated zones.

Example 1 A classic and a comprehensive example of calculating recharge rate by groundwater age data is the study by Solomon et al. (1993). They calculated the recharge rate to a shallow unconfined silty sands and clay lamellae aquifer, near Strugeon Falls, Ontario, Canada, by locating bomb peak tritium in 1986 and in 1991 in a set of piezometer nests as well as by measuring the age of groundwater along vertical profiles where the flow was vertical and little to no horizontal flow was present. Figure 3.2 depicts some results of their study that were used to calculate the recharge rate using the following two procedures:

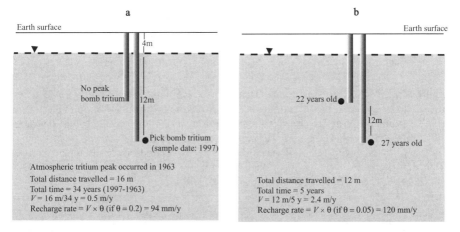

Figure 3.1. Determination of groundwater recharge rate: (a) by locating bomb peak tritium; (b) by measuring groundwater ages at two points along a vertical profile such as at a piezometer nest.

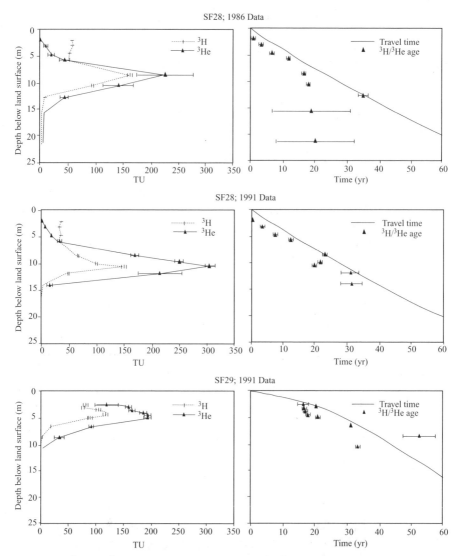

Figure 3.2. ^{3}H and ^{3}He measurements, calculated ^{3}H/^{3}He ages, and model travel times in a shallow aquifer near Sturgeon Falls, Ontario, Canada (from Solomon et al., 1993). Left diagrams illustrate ^{3}H and ^{3}He concentrations in vertical profile. Right diagrams show obtained ages based on ^{3}H/^{3}He method, and travel times obtained by modeling. The bars represent the range of variation of ages calculated or parameter measured. Note that the measurements have been made in two sets of piezometer nests (SF28 and SF29) in two different dates (1986 and 1991).

1. Recharge calculation based on the location of bomb peak tritium
   In one piezometer nest (Piezometer nest 28), the bomb peak tritium was at 8.7 ± 1 m depth in November 1986 and at 10.6 ± 0.5 m depth in July 1991. Assuming a porosity of 0.35, this yielded an annual recharge rate of 150 mm. The related calculations are

   $L = 10.6 - 8.7 = 1.9$ m T = July 1991–November 1986 = 4.6 years
   $\theta = 0.35$ $V = 1.9$ m / 4.6 years = 0.413 m/y
   $R = \theta \times V = 0.35 \times 0.413$ m/y = 145 mm/y (this was raised to 150 mm/yr because of some uncertainty about the exact position of the bomb peak).

   At Piezometer nest 29, the bomb peak tritium was located at 4.5-m depth below land surface. This gives a velocity of 0.16 m/yr, i.e., 4.5 m/28 yr (the peak tritium in the atmosphere was considered to have happened in 1963). Based on this velocity, the annual recharge rate at this piezometer nest was calculated to be 56 mm. The very low recharge rate at Piezometer nest 29 as compared to 28 was attributed to a discontinuous surface layer of fine-grained sediments that occurs near nest 29. [Note that the research by Robertson and Cherry (1989) has served as the background for recharge calculation through this study.]

2. Recharge calculation based on $^3H/^3He$ ages
   Solomon et al. (1993) also calculated the recharge rate to the same aquifer based on $^3H/^3He$ ages. Using the data presented in Figure 3.2, they obtained an age gradient of 2.15 yr/m or a flow velocity of 0.465 m/yr for the aquifer. This led to a recharge rate of 160 mm per year ($R = \theta \times V = 0.35 \times 0.38$ m/yr = 0.16 m/yr).

Example 2   Using the 4-m depth of bomb peak tritium, Dincer et al. (1974) estimated a mean recharge rate of 23 mm per year, about 35% of annual rainfall, for the period 1963–1972 in the Dahna Sand dune in Saudi Arabia. The parameters measured were

Depth of bomb peak tritium = 4 m
Year sample was taken = July 1972
Peak tritium in rainfall = January 1964
Specific yield (θ) = 0.05

The related calculations are

$L = 4$ m T = July 1972–January 1964 = 8.5 yr
$V = 4$ m / 8.5 years = 0.470 m/yr
$R = \theta \times V = 0.05 \times 0.470$ m/year = 23 mm/yr

More Examples   In an arid zone, based on ^4He and $^3H/^3He$ ages, the recharge rate to the eastern fringe of Botswana Kalahari aquifer was calculated at 2.2

to 3.1 mm per annum (Selaolo et al., 2000). A previous tritium study of the same region (Vogel et al., 1974) showed that the recharge rate to the Kalahari sand, Botswana, is 10 mm/yr and to Elandsfontein site, eastern Witwatersrand, South Africa, is 37–90 mm/year. Delcore (1986) analyzed the tritium content of 45 samples from an intermediate aquifer in southwestern Michigan, USA, from which he estimated the recharge rate to the aquifer at 176 to 351 mm per year. In a similar research, Tamers (1966) evaluated the recharge to a Venezuelan aquifer by radiocarbon ages. One more recent example to estimate the groundwater recharge using groundwater ages is the work by Cook and Robinson (2002) in Clare Valley, South Australia, where the recharge to a fractured rock aquifer was estimated at 60–75 mm per year. Similarly, Fette et al. (2005) used $^3H/^3He$ ages to calculate the average velocity of groundwater in an alluvial aquifer near Sion, Rhone river basin, Switzerland, at 1.7 km/yr.

3.5 CALCULATION OF THE GROUNDWATER FLOW VELOCITY

As shown in Figure 3.3, the velocity of groundwater flow can be calculated if we measure the age of groundwater at two separate points along a particular horizontal flow line. The ages should be measured at nearly the same depth and on the same flow line in order to avoid the effect of three-dimensional flows. The important point is that groundwater flow rates for aquifers can be gained from artificial (applied) tracer experiments as well, but age data offer

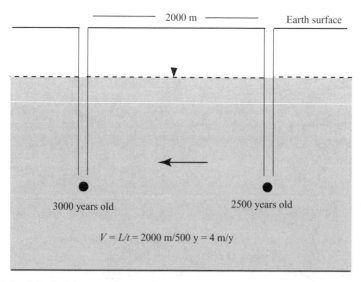

Figure 3.3. Calculating groundwater flow rate using ages of groundwater determined in two different boreholes. For this purpose, only distance between piezometers and age of groundwater in each piezometer are required.

the only realistic alternative if time scales of years or decades have to be taken into account (Zoellmann et al., 2001). Having obtained groundwater velocity, we can also back-calculate the hydraulic conductivity of the aquifer if we have an estimation of the effective porosity of the aquifer through $V = KI/\theta$ (the assumption is that the hydraulic gradient is easily obtainable). If there is no information about the porosity of the aquifer (i.e., we don't know the geology of the aquifer), the least we can do is estimate an upper and a lower range for the hydraulic conductivity by inserting theoretical minimum and maximum values of specific porosity into the equation.

Examples Pearson and White (1967) calculated the flow rate of groundwater in Carrizo Sand, Atascosa County, Texas, USA (Figure 3.4). They measured ^{14}C activities of 15 water samples in various parts of the aquifer, which yielded ages of 0 years at the outcrop recharging the aquifer to 27,000 years some 56 km downgradient. Based on these ages, the water velocities were calculated at 2.4 m/yr 16 km from the outcrop to 1.6 m/year 50 km from the outcrop. Another example is the work by Hanshaw et al. (1965), where they calculated the flow rate of groundwater in the principal artesian aquifer in central Florida (based on ^{14}C age of 5 samples) at 7 m/yr. In a recent and a fascinating study, the velocity of groundwater flow in different parts of the Nubian sandstone aquifer, southwestern Egypt, has been calculated at about 2 m/yr and 0.1 m/yr using groundwater ages obtained by ^{81}Kr and ^{36}Cl methods (Sturchio et al. 2004). These velocities are supported by both hydrogeological setting and some hydrodynamic models of the aquifer.

3.6 IDENTIFICATION OF THE GROUNDWATER FLOW PATHS

Groundwater flow paths in both vertical and horizontal directions can be determined by having ages that are increased along the inferred flow lines as shown in Figure 3.5. Accurate information about the groundwater flow path is needed in many water resources projects such as in construction of dams (determining different routes that water stored in the dam lake may escape), movement of plumes, mixing between different quality groundwaters, and study of surface water–groundwater interaction. Establishing flow directions in various parts of a deep, long regional aquifer is another example for use of groundwater age data. In such systems, however, very similar old ages in different parts of the aquifer may lead to the ambiguity in inferring the direction of groundwater flow. This shortcoming was demonstrated by Andre (2002) for the Infra-Molassic sands aquifer of the southern Aquitaine basin in France.

Example Figure 3.6 illustrates groundwater ages that have been measured along vertical and horizontal flow paths in Borden aquifer, Ontario, Canada, using Krypon-85 (Smethie et al., 1992). It is clearly demonstrated that groundwater ages are a reliable tool to identify the direction of groundwater flow in

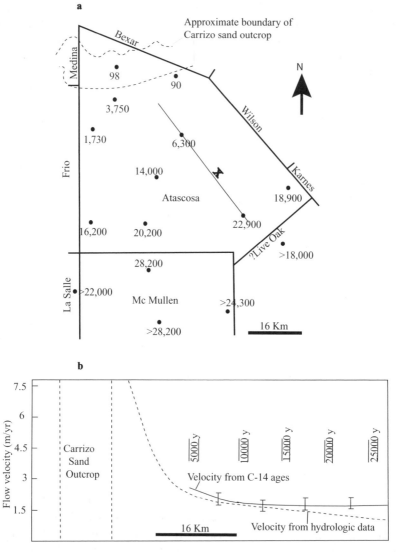

Figure 3.4. (a) Groundwater ages (year) in various parts of Carrizo sand, Atascosa, Texas, USA. Note the increase in groundwater age from the outcrop down dip. (b) Changes in groundwater flow rates down gradient of the Carrizo sand outcrop along line X in (a). The bars represent experimental uncertainty (modified after Pearson and White, 1967). The flow rate was calculated using two methods, age data and classical hydrological method.

GROUNDWATER AND CONTAMINANTS TRANSPORT THROUGH AQUITARDS

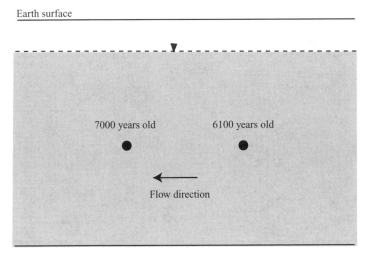

Figure 3.5. Identifying horizontal flow direction using groundwater ages. Groundwater age increases along the flow line.

a small-scale project. The study by Solomon et al. (1992) uses tritium and ^{3}He groundwater ages to calculate the recharge rate to this aquifer.

In contrast to the above example, various flow paths, lack of a dominant flow direction, and varying hydrological conditions all led to groundwaters ages that don't show a clear increasing trend with depth. An example of this is in Everglades National Park aquifers, USA, where groundwater ages vary significantly both in time and in different depths (Price et al., 2003). Figure 3.7 clearly depicts this mixture of ages in Everglades National Park aquifers, which were calculated using ^{3}H and ^{3}H/^{3}He dating methods. The varying hydrologic conditions were also reflected in the water-level hydrographs and potentiometric map of the same aquifer. In a similar study, Pint et al. (2003) demonstrated more than expected complex flow paths by simulating groundwater ages in the Allequash Basin, Wisconsin, USA. Here, the presence and interaction of a few lakes have led to a highly complex groundwater flow regime within a simple geological setting.

3.7 ASSESSING THE RATES OF GROUNDWATER AND CONTAMINANTS TRANSPORT THROUGH AQUITARDS

In studying the contamination migration, it is always crucial to identify the type of processes by which solutes are transported within various geological and hydrogeological entities. This has considerable implications with regard to the extent of contamination at a particular location within a particular time frame. Aquitards always act as a shield for aquifers. They may also restrict, or

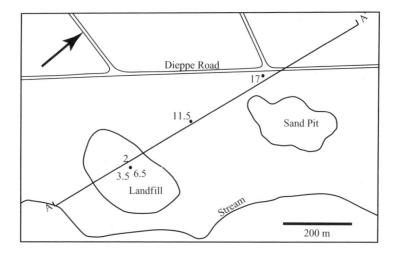
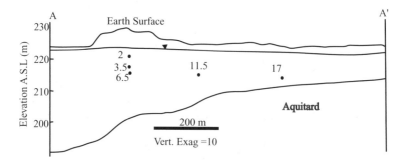

Figure 3.6. Identifying horizontal and vertical flow paths based on groundwater ages, Borden Aquifer, Ontario, Canada (modified after Smethie et al., 1992). The age of five groundwater samples from three different locations reveals the direction of groundwater flow. Increase in groundwater age with horizontal and vertical distances from the aquifer outcrop is also shown.

enhance, recharge to the underlying or overlying groundwater system. Therefore, it is important to understand hydrodynamic as well as geochemical behaviors of aquitards. The (aquitard) porewater ages have proved to provide useful information in this respect.

Example Wassenaar and Hendry (2000) measured the ^{14}C of dissolved inorganic content (DIC) from 13 aquitard pore water samples collected from piezometers ranging in depth from 1.5 to 37 m below ground surface (BG). The studied aquitard system is located 140 km to the south of Saskatoon, Canada, and comprises 80 m of uniform, plastic, clay-rich till deposited 18 to

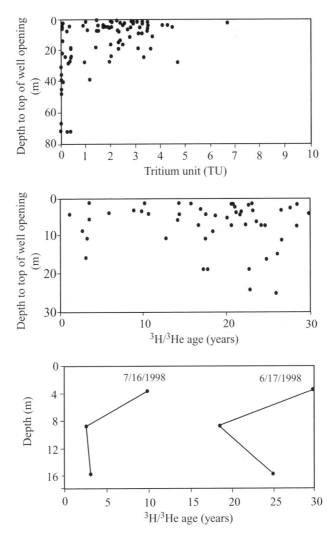

Figure 3.7. Groundwater ages do not show a clear trend with depth and with time in Everglade National Park, USA. Various hydrological conditions and presence of different flow paths are responsible for such a situation (Price et al., 2003).

38 ka BP. The results showed that ^{14}C activity ranges from 109 percent modern carbon (pmC) at 1.5 BG to 3 and 5 percent modern carbon at 29 and 37 m BG leading to age estimates of modern to 31,000 years, respectively. By this experiment, Wassenaar and Hendry (2000) showed that DIC migration through the aquitard is similar to ^{14}C transport and is mostly influenced by diffusion and radioactive decay.

3.8 CONSTRAINING THE PARAMETERS OF GROUNDWATER FLOW AND TRANSPORT MODELS (ESTIMATION OF LARGE-SCALE FLOW AND TRANSPORT PROPERTIES)

Groundwater flow models are increasingly employed as a powerful tool to help manage the groundwater resources. These models usually require an extensive volume of data and parameters, the quality of which is the key to the success of the modeling exercise. Estimation of some of these parameters such as hydraulic conductivity values, specific yields, and aquifer geometrics is always prone to some degree of error. Groundwater age data are the types of more precise data that can be incorporated into the models to complement the existing data, to eliminate some uncertainties, to serve as model-independent calibration data for groundwater transport models (Dörr et al., 1992), and to verify flow models especially those that predict the travel time of source water to wells.

Example 1 Reilly et al. (1994) used the CFCs and tritium dated ages as well as a numerical simulation technique to quantify groundwater flow rate, recharge ages of shallow groundwater, and mixing properties of the groundwater system at Locust Grove, Maryland, USA. They showed that the two methods together (age-dating method and simulation technique) enabled a coherent explanation of the flow paths and rates of movement, at the same time indicating weaknesses in the understanding of the system. They explain that "the ages from CFCs and tritium define the time of travel at specific points in the groundwater system, but without the numerical simulation the quantities of water and paths are unknown." They add that "the numerical simulations in the absence of the environmental tracer information provided non-unique and therefore, uncertain estimates of the quantities of water and flow paths within the groundwater system" (p. 433).

Example 2 In order to optimize the management of the deep groundwater systems in the city of Hamburg, Germany, a 3D groundwater flow model was developed by Taugs and Moosmann (2000). Groundwater flow velocities of 50 to 80 m/yr were calculated by the model, which were in large disagreement with ^{14}C ages (6,000–40,000 years) estimated velocities of 1–1.5 m/yr. This discrepancy was inferred to be due to drastically changing hydraulic conditions caused by increasing groundwater withdrawals since the beginning of the 20th century (present hydraulic conditions are much different than initial conditions). This observation led Taugs and Moosmann to conclude that a steady-state model covering the whole time period would be unsuitable for simulation, and thus a transient model was built. Similarly, in a study on Great Miami buried-valley aquifer in Southwestern Ohio, USA (Sheets et al., 1998), the discrepancy between $^{3}H/^{3}He$ ages and simulated travel times led the researchers to redefine and recalibrate the flow model of the aquifer to reduce the difference between the simulated and the measured hydraulic heads. A large number of case studies exploring the usages of groundwater age data in calibrating groundwater flow models can be found in IAEA (1994) and in Dassargues (2000).

3.9 IDENTIFICATION OF THE MIXING BETWEEN DIFFERENT END MEMBERS

The ages of those groundwaters that are mixtures of various components are helpful to calculate the mixing ratios and to identify the origin of each end member. For this, it is necessary to apply a minimum of 2 dating methods from different ranges; for instance, an old dating method and a young dating method (note that it is also possible to identify mixing by having the ratios of the three CFC-11, CFC-12, and CFC-13, which are all for dating young groundwaters; see Lindsey et al., 2003). The ages obtained from different dating methods are cross-plotted. On the plots, the best-case scenario would be to have a mixing line joining the oldest sample to the youngest sample. In such a situation, a simple two end-member equation such as Equation 3.1 can be applied.

$$A_s = A_y(x) + A_0(1-x) \tag{3.1}$$

where A_s is the age of sample in question, A_y is the age of youngest sample, A_0 is the age of oldest sample, x is the percentage of young fraction in the sample, and $(1-x)$ is the percentage of old fraction in the sample. Such simple mixing calculations allow hydrogeologists to study surface water–groundwater interaction and to investigate diffuse flow and conduit flow in fractured rocks aquifer. However, in more complicated cases, and where completely old and completely young samples do not exist or have not been collected, calculations become more complex and a larger number of dating methods and/or samples are required. A spreadsheet type of software called "Tracermodel1" has been developed by the USGS to determine hypothetical concentrations of atmospheric tracers in water samples with multiple ages and to permit plotting of ages and tracer concentrations in a variety of different combinations to facilitate interpretation of measurements (see http://water.usgs.gov/lab/shared/software/tracer_model.htm for more detail).

Examples Talma et al. (2000) measured CFCs (CFC-11, CFC-12 and CFC-113), ^3H, and ^{14}C in six boreholes in the Agter-Witzenberg valley of Table Mountain Group sandstone, 140 km north of Cape Town, South Africa. They plotted ages obtained from all methods against each other, produced 6 diagrams as presented in Figure 3.8, and compared the results with input curves of each tracers. Determination of a mixing line (dashed line) enabled them to calculate the percentage of old fraction in each sample, which were 0, 27%, 40%, 38%, 72%, and 90% for samples F, E, C, D, B, and A, respectively. They report that their findings are completely in agreement with the hydrogeolog-

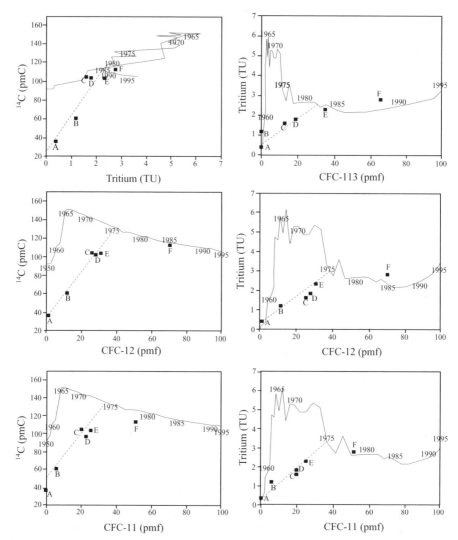

Figure 3.8. Cross plots of tritium, ^{14}C, and CFCs concentrations in six samples from the Agter-Witzenberg Valley, Table Mountain Group sandstone, South Africa. Curves on the diagrams depict atmospheric concentration of the tracers. For CFCs, pmf (percent modern Freon), an unusual unit, has been adapted, which is the ratio of atmospheric concentration of the species to the maximum historical concentration (in ppt, between 1990 and 1997) of the same species. The curve at ^{14}C-tritium plot shows the positive, albeit complex, correlation between historical concentrations of tritium and ^{14}C. Dashed lines show mixing trends (from Talma et al., 2000).

ical settings. For example, Sample F is from shallower depth at the recharge area of the aquifer, while Sample A is in the deeper parts and almost at the end of the flow line.

In an another study, Plummer et al. (2000) used ratios of CFCs ages to find that some rangeland water in the eastern River Plain (ESRP) aquifer, South Central Idaho, may contain between 5% to 30% young water (ages of less than or equal to 2–11.5 yr) mixed with old regional background water. The young water originates from the local precipitation. Further, the groundwater velocity along the top of the ESRP aquifer was also calculated by age data to be between 5–8 m/d. Some similar studies to this were undertaken by MacDonald et al. (2003) on Dumfries sandstone and breccia aquifer of Scotland and by Mattle (1999) and Heidinger et al. (1997) on two separate aquifers in Germany.

3.10 STUDY OF THE PRE-HOLOCENE (LATE PLEISTOCENE) CLIMATE

Evidence from groundwater systems may be used to help interpret the timing and nature of climatic changes during the Pleistocene era (Metcalfe et al., 1998). Groundwaters that are old or very old can be studied to identify the type of climate during which they entered the subsurface environment. To achieve this objective, first we have to know the age of groundwater. Then we have to study those parameters of groundwater that are climate indicators such as oxygen-18 isotope values (see Section 5.4 for more information). Combination of these two pieces of evidence will lead to an understanding of the pre-Holocene climate.

Examples Remenda et al. (1994), based on a detailed hydraulic age calculation, estimated the age of groundwater in three different aquifers/aquitards underlying Lake Agassiz in North America at the late Pleistocene era. They also measured the oxygen-18 composition of the same groundwaters, which was about −25 per mil. The depleted value of oxygen-18 was then considered to be associated with a recharge air temperature of −16°C. This is while the modern temperature at the Lake Agassiz area is 0°C. Therefore, with these two piece of information, i.e., a groundwater age of late Pleistocene and a recharge temperature of −16°C (based on depleted oxygen), it was logical to conclude that the climate was much colder during the Pleistocene era. If they did not know the age of the groundwater, it would have not been possible for Remenda and his colleagues to relate the depleted oxygen values to any particular geologic time table.

A similar study was conducted by Rademacher et al. (2002) on 11 springs in a high-elevation basin in central Sierra Nevada, USA, where the stable isotopes composition and age of spring waters were combined to understand the climatic and atmospheric circulation over short time intervals. Groundwater age data have shown that in Oman the average ground temperature during the Late Pleistocene (15,000–24,000 years BP) was about 6.5°C lower than present-day temperatures (Weyhenmeyer et al., 2000).

3.11 EVALUATION OF THE GROUNDWATER POLLUTION

In 1983, Jean-Charles Fontes (1937–1994), a leading isotope hydrologist and the co-editor of the famous *Handbook on Environmental Isotope Geochemistry* (Fritz and Fontes, 1980, 1986, 1989), made a provocative statement: "the concept of groundwater age has little significance." The reason for such a statement is unclear, but it could be due to the then-limited applications of groundwater age data. It could also be due to the inadequacy of worldwide analytical facilities, which made dating a difficult task and restricted to only a handful of countries. Fontes' statement was explored 20 years later by the delegates attending the 11th International Symposium on Isotope Hydrology and Integrated Water Resources Management, 19–23 May 2003, Vienna, Austria. The conclusion was that the use of multiple environmental tracers in groundwater systems could often help to redefine the interpretation of groundwater age, hydraulic concepts, and vulnerability of aquifers to contamination. However, because of some fundamental problems we explain in Chapter 8, the summary of the conference reads

> Thus, while the quote [Fontes' statement] is true, the "age" estimation must be translated into logical hydrological parameter. The isotopic tools for groundwater dating are important because shallow, young groundwaters provide drinking water supply in many parts of the world and also are the most vulnerable to contamination from anthropogenic activities. The Agency's [IAEA] recent commissioning of a $^3H/^3He$ facility endorses the development of improved isotope techniques for dating of young groundwater (IAEA, 2004a, p. 2).

Inference can be made from the above explanations that the focus of groundwater age discipline has now shifted toward contamination studies and the potential applications of age-dating technique to overcome the quality-related challenges facing the groundwater professionals. Contamination of groundwater resources has been the subject of intense investigations during the past four decades (Brown, 1964; Kimmel and Braids, 1974; McFarlane, 1984; Zubari, 1994; Michalak and Kitanidis, 2004). In this respect, shallow—often young—groundwater needs particular consideration because they are commonly used for drinking water and because they comprise much of the baseflow that discharges to streams. Various techniques including environmental isotopes have long been employed to study the pollution processes and the fate of contaminants in the groundwater systems (e.g., Wolterink, 1979). In this context, one could argue that the use of groundwater age data to assess groundwater contamination is the most recent, and perhaps the most precise and holistic, technique to deal with this issue. One other thing to be added is that the most comprehensive, 201-page USGS groundwater age report to date, i.e., Lindsey et al. (2003) and the associated nicely illustrative fact sheets (Phillips et al., 1999; Phillips and Lindsey, 2003) deal with groundwater-induced surface water contamination and why groundwater age data are required to quantify the lag time between reduction of nutrient sources in the watersheds and improvement of water quality downstream. This particular area of research, i.e., quantifying the lag time between land-use changes and water-quality degradation/improvement downstream using the groundwater age concept have attracted special attention over the past 10–15 years (e.g., Brawley et al., 2000; Wayland et al., 2002).

Most contaminants, especially human-induced ones, have entered aquifers recently, perhaps not earlier than 100 years ago. Further, groundwaters that were polluted 100 years ago or earlier have so far had enough time to be purified through natural processes because of their long contact with the subsurface environment. On the basis of these assumptions, the following conclusions can be drawn:

1. If a water sample is older than 100 years, then it should be pollution-free.
2. For old groundwaters, the contamination risk is low.
3. For young groundwaters, the contamination risk is high.
4. Along the age line the concentration of the contaminants should decrease, i.e., older groundwater should show less contamination (concentration of the contaminants should be less).
5. For a set of groundwater samples, if a contaminated groundwater sample is dated as young, then we can be sure that the dating exercise was most probably undertaken correctly and the age data obtained can be used for other purposes.
6. If there is a meaningful negative correlation between the age of groundwater and the concentration of the contaminants (the older the age, the lower the concentration of the contaminants), it would then be possible to predict the extent and timing of contamination plume, i.e., concentration reaches to what level and at what time. (a typical case study is the one by MacDonald et al. (2003) on the Dumfries aquifer of Scotland). Sometimes, it is possible to comment on the dinitrification reactions just by comparing the age and the nitrate concentration of groundwater (for example, in Pocomoke and East Mahantango Creeks, Chesapeake Bay watershed; see Lindsey et al., 2003). The procedure is to cross-correlate the age and the nitrate concentration of the groundwater samples. If there is an overall negative correlation, but some samples do not behave the same, then we can conclude that denitrification could be the cause of the decrease in the concentration of nitrate in those samples that show negative age-nitrate correlation. (It is good to note that denitrification can only be confidently proven if we have the

N isotopes as well. Also, during the 1970s and 1980s, there were higher nitrates inputs due to increased fertilizer use, but use declined in/since the 1990s, thus the above correlation may not hold for *very young* waters, i.e., water recharged during late 1990s or early 2000s.)

7. If there is a positive correlation between the age of groundwater and the concentration of contaminants (the older the age, the higher the concentration of the contaminants), it would then be logical to conclude that contaminants are gradually degraded over time (assuming that the contaminating source is no longer present, i.e., it has been removed).

In contrast to the preceding suggestions, there are cases where the water sample is dated as old and at the same time it is contaminated. Explanations for, and benefits associated with, this discrepancy are

1. The contaminants entered the aquifers via a route and time different than water molecules (e.g., DNAPL).
2. The dating exercise involves some errors (usually determining the contamination level of a groundwater sample is much less prone to error as compared to age-dating practice).
3. The contaminants have been produced on site within the aquifer itself sometime after water molecules entered the aquifer. For example, groundwater salinization due to deforestation in Australia is a recent phenomenon while the groundwater itself is of old age (see Section 3.15.4).

Examples Böhlke and Denver (1995) used CFCs and tritium age-dating along with chemical and isotopic techniques in an attempt to resolve and quantify the history and fate of nitrate contamination in two agricultural watersheds (Chesterville Branch and Morgan Creek) near Locust Grove, Atlantic Coastal Plain, Maryland, USA. Through this research, they were able to (1) provide a 40-year record (1945–1988) of flux of nitrate to groundwater (Figure 3.9), which increased in both watersheds by a factor of 3–6, most rapidly in the 1970s, (2) find a good correlation between the temporal increase in the groundwater nitrate concentration and increase in N fertilizer use in groundwatersheds, (3) show that denitrification has reduced the nitrate content of older groundwaters, and (4) demonstrate that groundwater resides for more than 20 years in the aquifers before being discharged into the streams.

The fate and transport of nitrate in a shallow alluvial aquifer in the Contentnea Creek subbasin of the Neuse River in the southeastern United States were studied using groundwater ages obtained by CFCs and tritium methods (Tesoriero et al., 2005). It was found that groundwaters with residence times of between 10–30 years has low nitrate and elevated excess N_2, while those with older residence times are low in both nitrate and excess N_2. This indicates that

CALCULATION OF THE TRAVEL TIME OF THE GROUNDWATER PLUME 71

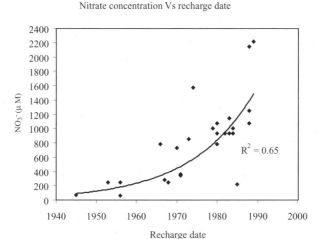

Figure 3.9. The correlation between nitrate concentration and groundwater age in two agricultural watersheds, Maryland, USA (data from Böhlke and Denver, 1995). Younger groundwaters are more contaminated, because active recharge carries contaminants into the aquifer.

1. denitrification has reduced the concentration of nitrate in younger groundwaters and has led to an increase in the excess N_2 concentration (note that in the denitrification process nitrate is reduced to N_2),
2. older groundwaters were free from, or low in, nitrate from the beginning, i.e., they were recharged prior to the excessive application of fertilizers. The usefulness of age data in such a study is that it enables us to find out whether low-nitrate groundwaters were originally low or else are low because of the denitrification process.

3.12 CALCULATION OF THE TRAVEL TIME OF THE GROUNDWATER PLUME TO THE POINTS OF INTEREST

By having the isochrone map of the aquifer, we would be able to answer the following:

1. When does the groundwater plume reach a drinking water production well?
2. When does the contaminated groundwater reach a surface water body like a lake or a river?
3. When does the contaminated groundwater reach the final discharge point such as an ocean or a sea?
4. When does the diluting effect of rainfall recharge appear in a specific point downstream?

If we know the age of groundwater, we can

1. predict the time required for a present-day polluting source (a leaking fuel station, for instance) to inflict damage on the water quality at the location of interest. For example, Fogg (2002) reports that nitrate groundwater contamination that began emerging in 1979' in the Salinas Valley, California, originated from the land-use practices of circa 1940s.
2. analyze that if we stop the polluting activities (e.g., discontinue applying excessive amount of fertilizers in agricultural lands), when we should expect to see the outcome of this measure. (This application of age data has been successfully demonstrated by Laier and Wiggers, 2004.) In general, we can predict the timing of the effect of land-use changes.
3. determine the likely sources of contaminants whose initial applications occurred during a specific time period. The age of the groundwater will help to determine if the sources of contamination can be traced to recent events, or if other unknown sources of contamination exist that would warrant further investigation. For example, if a particular fuel station commenced operation in an area 30 years ago, a contamination plume in the downgradient aquifer is, in the first instance, thought to result from leakage from underground storage tanks of the fuel station. However, if the age of groundwater samples from the contaminated aquifer is older than 30 years, then it is highly likely that the cause of contamination is not the fuel station. In contrast, if the age of contaminated groundwater is younger than the commencement date of the fuel station, then the likelihood that the contamination originated from the fuel station increases. In one simple sentence, an age of 30 years suggests that surface contaminant sources that have only existed for 20 years could not be present in the 30-year-old water (note that mixing of different age groundwaters could inflict a serious error in these types of interpretations).
4. estimate the time span required for self-purification of a polluted aquifer after removal of the pollutant.
5. determine whether there is enough time for natural purification of surface waters recharging aquifers (e.g., river water) whose quality should improve during the course of travel underground to make them suitable for domestic usages. For instance, Hofer et al. (2005) points out that in Switzerland, river water infiltrating into aquifers should spend a minimum of 10 days underground before being abstracted via production wells.

Example A typical example of the usefulness of groundwater age data is reported by Oudijk and Schmitt (2000) in the vicinity of Philadelphia, USA, where a 31.4-m-deep water well in a historical property was shown, in 1993, to be contaminated with trichloroethylene (TCE), cadmium, and lead. The contamination was suspected to be caused by an industrial site used for the remanufacturing of automobile engines, operating since the 1950s and located 550m upgradient of the well. The wastewater generated at the industrial site was initially discharged to a lagoon system (until the mid-1970s) and then to seepage pits until early 1982. Four lines of evidence showed that the lagoons and the seepage pits caused the groundwater contamination: (1) flow direction; (2) suite of contaminants; (3) groundwater migration rate, i.e., hydraulic age, and (4) groundwater $^3H/^3He$ age. With a groundwater/contaminant flow rate of 24.7m/yr, the hydraulic age of water entering the aquifer from the lagoons would be 22 years at the water well ($550m/24.7\,m\,yr^{-1} = 22\,yr$). The $^3H/^3He$ age of the well-water sample collected in 1999 was also determined at 24 years (^3H content of the well water was 11.3 TU). It was therefore concluded that the 24-year-old contaminants (water) entered groundwater in the mid-1970s, when the wastewater discharges (lagoon and seepage pits) were active (1999 − 24 = 1975). On the basis of this experiment, the owner of the remanufacturing facility accepted liability. As noted in Chapter 2, another typical and precise application of groundwater age data in pollution studies is that by Solomon et al. (1995).

3.13 MAPPING VULNERABILITY OF THE SHALLOW AQUIFERS

Construction of vulnerability map for the aquifers is a relatively modern exercise (e.g., Vrba and Zaporozec, 1994) that has received considerable attention during the past 10 years. For instance, a major research theme at the Centre of Hydrogeology, University of Neuchatel, has been to develop new methods to map the vulnerability of karstic systems to contamination (Doerfliger et al., 1999; Zwahlen, 2004, and many more). These maps can be used by resource protection agencies to focus prevention programs on areas of the greatest concern and to help prevent contamination of groundwater resources by identifying areas that are at greater risk of pollution. One of the earliest methods to evaluate vulnerability of the aquifers is DRASTIC, which considers 7 factors: *d*epth to water, net *r*echarge, *a*quifer media, *s*oil media, *t*opography, *i*mpact of vadose zone media, and hydraulic *c*onductivity of the aquifer (Aller et al., 1985). In this method, groundwater travel time (age) within the unsaturated zone has not been taken into account directly. However, some research heavily emphasizes the calculation of residence time of groundwater in the unsaturated zone and its role in the preparation of vulnerability maps. For instance, Hötling et al. (1995) classified the aquifers into five different categories (in terms of protection against pollution) ranging from very low to very high protection. The basis for this classification is the amount of time that the infiltrating water spends in the unsaturated zone. Aquifers with very high protection are those for which the travel time is over 25 years, in contrast to very low protection aquifers whose travel time is less than a year. More recently, a new technique has been presented to determine the vulnerability of an aquifer by calculating the transit time of the percolating water from surface to the groundwater table (Voigt et al., 2004). In this approach the retention time of the percolating water in the unsaturated zone is determined using a simple analytical model and incorporating the lithological characteristics of the vadose zone as well as groundwater recharge rate.

In a more practical sense, there is growing interest among regulatory agencies in using groundwater age dates to map susceptibility of groundwater systems to contamination (Weissmann et al., 2002, Manning et al., 2005). A number of recent works especially in the US have concentrated on the use of age data in this field (Harlow et al., 1999; Nelms et al., 1999, 2002; Nelms and Harlow, 2001). Through a series of major reports, the vulnerability of various groundwater basins in California have been assessed by measuring volatile organic compounds (VOCs) and groundwater ages (just a few examples include Moran et al., 2002a, 2004a, b).

The principal idea behind using age data for the construction of vulnerability maps is that the older the groundwater, the less chance for it to become anthropologically contaminated. Generally speaking, groundwaters with a high percentage of young water should be rated as highly susceptible, while those with a high fraction of old water should be regarded as insusceptible or less susceptible. Longer travel time of groundwater in the saturated and unsaturated zones provides a better opportunity for the degradation and sorption of contaminants. If the spatial distribution of aquifer groundwater ages is known, it would then be possible to subdivide the aquifer into different zones in terms of susceptibility. Groundwater vulnerability mapping based on the age data, if fine-tuned and tested successfully, would be a time-friendly method and could replace most of the currently available vulnerability assessment approaches, which require a large amount of data, a number of maps, and a lengthy development time. For example, Rupert (2001) demonstrates that some of the current methods such as DRASTIC can be improved.

Example Nelms et al. (2003) age-dated water from 171 wells and springs from both deep and shallow aquifers, and within various hydrogeological settings, across the Commonwealth of Virginia, USA, in order to use the results as a guide to classify aquifers in terms of susceptibility to contamination from near-surface sources. They employed a variety of dating methods including CFCs, SF_6, ^3H, $^3H/^3He$, and ^{14}C to increase the accuracy of the results. They found some fraction of young groundwater in the shallow wells (less than 33 m deep) and springs in the Coastal Plain shallow regional aquifer and wells and springs in the fractured-rock terrain. On this basis, these systems were rated as susceptible to near-surface contaminating sources. Another conclusion was that the ratio of the percentage and apparent age of the young fraction in binary mixtures of young and old waters indicates the relative degree of susceptibility, i.e., the larger the ratio, the higher the degree of susceptibility. In this respect, samples from the fractured-rock regional aquifer system such as Blue Ridge had the largest ratios.

Tritium-helium groundwater ages in samples from the public water supply wells in Salt Lake City, Utah, have shown that there is a clear correlation between the percentage of modern water fraction and both the occurrence and concentration of contaminants (Manning et al., 2005). This has been shown to have significant implications, despite the problem of mixing ages, for assessing the susceptibility of groundwater wells to contamination.

3.14 PERFORMANCE ASSESSMENTS FOR THE RADIOACTIVE WASTE DISPOSAL FACILITIES

In a number of countries, site characterization for disposal of radioactive wastes forms a substantial research theme (e.g., Arumugum, 1994; NEA, 1995; Wiles, 2002). In all of the investigations concerning the safety and feasibility of storing radioactive wastes underground, a high degree of certainty is needed to assure that the contaminants will not leach from the wastes. The likely maximum rates of future groundwater movements can be obtained from knowledge of the past history of flow events in the area under consideration. The fact that performance assessments for radioactive waste disposal generally consider time spans up to 1 to 2 million years, comparable to the duration of Quaternary (Metcalfe et al., 1998), emphasizes the importance of being able to deduce the timing and magnitudes of groundwater flow over a very long time scale to reflect the impacts of varying hydro-climatological conditions. Aquifer pumping tests, applied tracer tests, and water-level measurement studies all lack the capability to measure aquifer hydraulic properties, *and associated likely changes*, over very large temporal and spatial scales. For this reason, groundwater age data are of prime importance and can provide invaluable information on the hydraulic properties of the waste disposal facilities. Very old groundwater ages, in the order of millions, are the only scientific piece of evidence to show that a particular site has had no communication with the active hydrologic cycle for a very long period. Because of the time scale involved, the hydraulic properties inferred from age data (such as hydraulic conductivity or flow rate) include the historic variations in these parameters as well.

Example 1 A study was conducted by Marine (1979) to estimate groundwater velocities in order to demonstrate the safety of storing radioactive waste in the crystalline metamorphic bedrock at the Savanah River Plant (SRP), Atlantic Coastal Plain, USA. Horizontally, SRP is located 32 km from the bedrock outcrop, but vertically, it is more than 200 m above bedrock (i.e., bedrock lies more than 200 m below the land surface). Groundwater in the bedrock contains 6,000 mg/L of solutes and forms a 51-km-long flow line. It is separated from the overlying fresh alluvial surface aquifer by a layer of dense clay and sandy clay called saprolite. Using ^{4}He age-dating technique, Marine calculated the age of groundwater in the bedrock at 840,000 years, yielding a velocity of about 0.06 m/yr. The hydraulic conductivity values obtained by hydraulic tests, between 0.033 to 0.000012 m/yr, led to a much lower flow rate. Considering the length of the flow path (51 km) and the time scale (840,000 years), the results of this study were considered much more reliable for waste disposal purposes as compared to the packer tests, pumping tests, and artificial tracers test, which all require extensive time and space extrapolations.

Example 2 From 1980 until the early 1990s, NAGRA (the Swiss National cooperative for the storage of radioactive waste) undertook extensive investigations to evaluate the feasibility and safety of a repository for the disposal of high-level radioactive waste in northern Switzerland. As part of this study, 140 springs and wells were sampled for isotopic and age analysis, in addition to 60 water samples collected from very deep boreholes drilled specifically for this purpose. In fact, a major report, which took the form of an international book (Pearson et al., 1991), a number of smaller reports, and a few journal and conference papers are the result of the NAGRA hydrogeologic study. Pearson and his colleagues do not present a sharp conclusion about the suitability of the studied sites but suggest that waters in the bedrock in northern Switzerland are of deep origin, even though they may be present at shallow depth (Pearson et al., p. 375). Further, the concentrations of age-dating radionuclides in most samples were equivalent to water-rock equilibrium concentrations, indicating very old ages. It should be pointed out, however, that a site for disposal of radioactive waste in Switzerland was recommended by NAGRA in 1993 (Wellenberg), but the related formal, political, and legal process is still pending. A discussion about these issues is given by Loew (2004). Interested readers can refer to the NAGRA website for more information (http://www.nagra.ch).

Another example of dating groundwater for gaining insight into the process and impacts of nuclear waste disposal practices is the research by Marty et al., (2003) in the Paris Basin, which is described in detail in Chapter 6.

3.15 SITE-SPECIFIC APPLICATIONS

Some of the applications of groundwater age data have been undertaken on a site-specific basis. These may receive wider interest in the future. Comparatively, these types of applications have been demonstrated (or suggested) more recently. It is neither possible nor appropriate to include all these usages as the list is ever-growing and it will be inevitably incomplete regardless of its comprehensiveness at the present. However, it shows the usefulness of groundwater age discipline in the hydrological studies and in solving the water resources management problems. Apart from those applications described in detail ahead, some others include the evaluation of atmospheric acid deposition (Robertson et al., 1989; Busenberg and Plummer, 1996), estimating rates of geochemical and geo-microbial processes in the unsaturated zone (Plummer et al., 1990; Chappelle et al., 1987), estimating longitudinal dispersivitiy coefficient (Solomon et al., 1993), determining hydraulic conductivity and transmissivity of the aquifers (Phillips et al., 1989; Hanshaw and Back, 1974), finding the origin of methane in gas hydrate deposits (Fehn et al., 2003), identifying groundwater inflow into artificial lakes (Weise et al., 2001), evaluating ecosystem health (Chesnaux et al., 2005), quantitatively describing groundwater flow and hydrogeology (Boronina et al., 2005), and estimating volume of aquifer storage and location of recharge and extracting information on the rates of geochemical and microbiological processes in aquifers.

3.15.1 Identification of the Seawater-Level Fluctuations

Sea-level fluctuations during the recent geological past is an important topic both because it could signal an environmental problem and because it is a component in the global climate system (Kroonenberg, 2005). For instance, a full international conference entitled "International Conference on Rapid Sea Level Change: A Caspian Perspective" held in Rasht, Iran, in May 2005 was entirely devoted to the Caspian Sea fluctuations and associated impacts. One of the applications of groundwater age data is in identifying sea level and baseline changes during the Quaternary. Large discrepancies between hydraulic ages and isotopic ages can help reveal some information about the past fluctuations in erosion baseline and sea level. If hydraulic ages are much younger than isotopic ages, then it is highly likely that baseline was lower earlier, e.g., in the Pleistocene era. Previous lower baselines resulted in higher hydraulic gradients. Higher hydraulic gradients resulted in higher flow velocity and lower hydraulic ages. So, an isotopic age whose derivation is not affected by hydraulic gradient reflects the impact of earlier higher hydraulic gradients on the hydraulically calculated ages. It should immediately be added here that the younger hydraulic ages (compared to isotopic ages) in some deep confined aquifers have been attributed to the presence of "zones of imperceptibly slow moving groundwater" or stagnant groundwaters that are stored in an isolated part of the aquifer (Mazor and Native, 1992, 1994; Torgersen, 1994). Zones of imperceptibly slow-moving groundwater are much older than the overall groundwater surrounding them, and hence exert an elevating influence on the age of groundwater (they increase groundwater age). Hydraulic age calculations, however, are incapable of taking into account the presence of dead volumes of water, which slow down the groundwater movement in the nearby vicinity.

Example The application of groundwater age information to show shoreline fluctuation has been vividly shown by Love et al. (1994) for Gambier Embayment of Otway Basin in Southeast Australia. Gambier Embayment contains up to 5,000-m-thick mixed marine and terrestrial sediments forming two aquifers, unconfined Gambier aquifer and confined Dilwyn aquifer. By analyzing 52 water samples and extensive measurement of hydraulic parameters of the two aquifers, Love et al., (1994) calculated the groundwater age by the ^{14}C method at 12,800 years and by Darcy's law method (hydraulic age) at 49,000 years. The results from the two methods were hugely different. To explain these large differences, Love and his colleagues reviewed all possible sources of discrepancy such as initial ^{14}C activity, openness of the groundwater system, and the effect of isotopic exchange. They substantially counted for all these possibilities, but still a large difference remained for the two methods, i.e., ^{14}C ages and hydraulic ages. Their conclusion was that the discrepancy is due to increased hydraulic gradients during late Pleistocene-Holocene era caused by sea-level lowering. On page 177 they write,

> Because both aquifers are hydraulically connected to the sea, low sea level in the past would have resulted in increased groundwater velocities and a corresponding short groundwater residence time. The hydraulic calculated travel time is based on present relatively high sea-level. Much lower sea level recorded during the Pleistocene glaciations (as much as 150 m lower) have consequently increased hydraulic gradients, which corresponds to increased drainage rates. Because ^{14}C has a mean life time commensurate with that of the water residence time, it will record the variations in groundwater velocity over time. However, the Darcy calculation based on the present day water level will underestimate the average gradient because it represents a present day snapshot of the groundwater system and is not applicable when boundary conditions change significantly. In essence, the ^{14}C data integrate the effect of changing hydraulic heads on the groundwater flow regimes over the last 30,000 years. (Reprinted from Love et al., 1994, with permission from Elsevier)

3.15.2 Calculating the Timescale of Seawater Intrusion

In very recent research using ^3H and ^{14}C age data, Sivan et al., (2005) show that Mediterranean seawater penetrated the coastal aquifer of Israel between 15–40 years ago and moved inland to a distance of 50–100 m. Sivan and colleagues were able to classify chemical reactions associated with various stages of seawater intrusions and how groundwater and seawater mixed during the intrusion.

3.15.3 Disposal of Wastes into the Deep, Old Saline Groundwater Systems

Due to concern over environmental impacts of waste disposal systems, new sites and pertinent hydrogeologic conditions are continuously being evaluated. One main factor in evaluating a site for disposal of waste is its hydraulic connection with the surrounding environment. Deep old aquifers, which due to their high salinity values have no other use, can be considered as a possible disposal site (Freeze and Cherry, 1979). These aquifers represent little risk to the environment especially hydrosphere because their age suggests that they have been isolated from the hydrosphere for a very long time.

3.15.4 Management of the Dryland Salinity in Australia

In Australia, the clearing of native Eucalyptus vegetation and its replacement with shallow-rooted crops and pasture have resulted in reduced rates of rainfall interception and soil water usage by plants, and increased recharge. This has led to a rise in the water level of shallow unconfined aquifers and the potentiometric surface of the deep confined aquifers. Rising shallow groundwater dissolves the salts stored in the soil profile and carries them upward to the vicinity of ground surface where osmosis phenomenon is capable of conveying both water molecules and solutes to the surface. Water molecules are then evaporated, but solutes are left behind on the soil surface. The final and adverse outcome of this hydrological perturbation is widespread soil salinization in the nonirrigated agricultural land, locally known as dryland salinity, which in turn leads to stream salinization.

Different types of relationship between groundwater salinity and groundwater age in four shallow aquifers in the upper Macquarie Valley, New South Wales, Australia, shown in Figure 3.10, were used by Kazemi et al., (1998b) and Kazemi (1999) to investigate the time scale of dryland salinity process. On

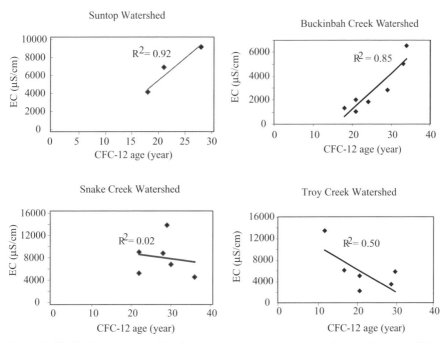

Figure 3.10. Various relationships between groundwater age and groundwater salinity in four salinized watersheds in New South Wales, Australia (after Kazemi, 1999). The difference in the degree of correlation is attributed to variation in the salinization stage of these watersheds, i.e., some watersheds underwent salinization after others. See Chapter 4 for more information.

the basis of these relationships, various watersheds are divided into three categories: (1) those that are at the early stages of salinization; (2) those already highly salinized; and (3) those that have passed their salinization peak and are starting to return to pristine, undisturbed conditions. Groundwater ages (1) show a positive correlation with groundwater salinity in the first type of watersheds, (2) show no correlation with groundwater salinity in the second type of watersheds, and (3) show a negative (reverse) correlation with groundwater salinity in the third category of aquifers (see Figure 3.10 and Chapter 4).

3.15.5 Hydrograph Separation

Rademacher et al., (2005) studied the relationship between the chemical and isotopic composition of groundwater and residence times to understand the temporal variability in stream hydrochemistry in Sagehen basin, California, USA. The mean residence time of groundwater feeding Sagehen Creek during base flow was estimated at 28 years, reducing to 15 years during the snowmelt periods. The residence-time data have shown that groundwater is not a single, well-mixed chemical component but rather is a variable parameter. This has implications for hydrograph separation because most current models of watershed hydrochemistry do not account for chemical and isotopic variability found within the groundwater reservoir.

4

AGE-DATING YOUNG GROUNDWATERS

This chapter describes the techniques used to date young groundwaters (0–60 years old) including ^{3}H, ^{3}H/^{3}He, ^{4}He, ^{85}Kr, CFCs, SF$_6$, and ^{36}Cl/Cl. Indirect and less used methods such as δ^{18}O are also briefly discussed. Rn-222 is discussed separately in Chapter 8, because the involved timescale is much shorter than the methods we describe here. The principle, characteristics, limitations, requirements, relevant formulas and equations, advantages, working examples, field sampling and laboratory measurements and associated costs of each method are spelled out (at the end of the chapter, advantages and disadvantages of all the dating methods are presented in Table 4.5). Some new CFCs data of the authors' unpublished works are reported. Interpretation of multi-tracer age data and the necessity for this approach are also discussed. This chapter is inevitably imbalanced. Some of the dating methods discussed here are well-established techniques with plenty of references. Some others, however, are quite new. For instance, there are a large number of references for tritium method, even in general groundwater textbooks. Also, with regard to CFCs and ^{3}H/^{3}He methods, substantial reviews by Plummer and Busenberg (2000) and Solomon and Cook (2000), respectively, do not leave much space for our maneuver. However, methods like SF$_6$ and ^{85}Kr are not sufficiently covered in the literature, because of their recentness. At the start of this chapter, there is a general discussion that applies to all the three chapters dealing with various age-dating techniques (Chapters 4 to 6). Hence, it is recommended for study before proceeding to the next two chapters of 5 and 6.

Groundwater Age, by Gholam A. Kazemi, Jay H. Lehr, and Pierre Perrochet
Copyright © 2006 John Wiley & Sons, Inc.

4.1 IMPORTANT POINTS

At the beginning of this chapter a number of important points, applicable to this chapter as well as the next two chapters, are discussed:

1. There is not much discussion on various, sometime lengthy, correction methods applied to some age-dating approaches because
   a. the corrections made most of the time do not change the dates obtained to a great extent, i.e., the corrected ages and the uncorrected ages are not much different. If the uncorrected and adjusted ages are vastly different, then there is some question on the validity of the correcting exercise as well. Wigley (1976) rightly emphasizes this problem by pointing out that the "largest adjustments (corrections) are those which are subject to greatest uncertainty."
   b. a variation of a few percentages or even higher does not usually change the overall outcome gained by applying age-dating techniques. For example, a 2,500-year-old sample is not really much different from a 3,000-year-old sample in terms of hydrogeological conclusions made on the base of its age. This statement has been strongly echoed by Freeze and Cherry (1979), where at the end of their discussion on the necessity to correct ^{14}C ages (p. 292) they dismiss the perception that extensive corrections may devalue the ^{14}C technique as a powerful hydrologic tool and write "detailed age estimates are often not necessary for a solution to a problem." It has also been suggested by many others that corrections will not affect the basic conclusions (e.g., Carrillo-Rivera et al., 1992).
   c. in most of studies, obtaining a relative age is satisfactory. For instance, the variations in age along a flow line to calculate flow rate or to determine the flow path, etc.
   d. there is some inherent uncertainties in the correction techniques. In the other words, the corrections need to be corrected or the correction, sometimes, leads to dates that are certainly wrong (Lloyd, 1981). Plummer and Sprinkle (2001) compared a number of adjusting models on 24 water samples and compared the results with NETPATH results. They found that all the previous adjusting models led, to some extent, to the ages that are different from NETPATH ages. Even two different assumptions in NETPATH led to ages that are in cases considerably different.
   e. it is safe to say that the ^{14}C dating method is the only technique that requires extensive corrections. However, the phenomenon of excess air and the calculation of subsurface production rate are also challenges in a number of groundwater dating studies.
   f. lengthy correction exercises are counterproductive and may detract the readers from the main content of the text and sometimes leads to disinterest and dissatisfaction.

g. as with all other nature-related sciences (biology, medical sciences, soil sciences, etc.), there are large, complex natural variations and many complicated phenomena that are never truly quantified and understood by scientists. Groundwater dating is no exception. In fact, it has been suggested very recently (Voss, 2005, p. 5) that "Hydrogeology is mostly a descriptive science that attempts to be as quantitative as possible regarding description but without the possibility (in many or most cases) of guaranteeing the accuracy of predictions."

h. as with all other disciplines dealing with groundwater, such as modeling, pumping tests, flow net analysis, etc., there is a large uncertainty involved in groundwater age-dating. Again excerpts from Voss (2005, p. 6) reads, "Hydrogeologic science is not well suited to quantitative prediction... some hydrogeologic problems cannot be solved."

2. It is more and more realized that a groundwater dating exercise should be undertaken using a number of tracers conjunctively, i.e., multitracer dating. Perhaps one could say a minimum of two dating techniques in two adjacent categories (e.g., young and old or old and very old) is required for a purely unknown aquifer, e.g., tritium and carbon-14, or carbon-14 and chlorine-36, etc. Application of two or more tracers for the same age range, i.e., SF_6 and CFCs are also recommended, especially for water table aquifers where ages are usually young. One major advantage of such an approach is that the difference between groundwater ages calculated by different methods can help to better understand some of the chemical or hydrodynamic processes that occur in the groundwater.

3. It is always a useful exercise to calculate the hydraulic age using simple Darcy equation especially for aquifers that are dated for the first time. This is an independent check on the accuracy of the dating results because it is based upon a different principle, which requires knowledge of the recharge area. It is also a worthy practice to check the dates obtained against the geological and hydrogeological settings of the aquifer such as age of the formation, aquifer depth, presence of confining units and flow system, chemistry of water, etc.

4. Radon-222 with a half-life of 3.8 days is often considered as a dating isotope. We have, however, refrained from describing it here in Chapter 4 and left it to the last chapter (Chapter 8) as a special or a potential method. The main reason for such differentiation is the much shorter half-life of Rn-222 as compared to other isotopes used in dating young groundwaters. If we wanted to include it here, then we should have added another chapter called "Age-Dating Very Young Groundwaters," which would have included methods like microbial and bacterial techniques as well.

5. Dating methods for young groundwaters are typically applicable to unconfined shallow aquifers only, while those methods for dating very

old groundwaters are often for deep, confined aquifers. It is, therefore, safe to argue that only dating methods for old groundwaters (Chapter 5) can be used for both confined and unconfined aquifers.

4.2 TRITIUM

The hydrogen atom is believed to have been created during the "Big Bang" by proton-electron combinations (Ottonello, 1997). It is the lightest and the most abundant element in the universe. At standard temperature and pressure, hydrogen is a colorless, odorless, nonmetallic, univalent, and highly flammable diatomic gas. It is present in water and in all organic compounds and living organisms. Tritium's commercial uses include medical diagnostics, sign illumination (especially exit signs), watch and drug industry, and many other small-scale applications. In the nuclear industry, the main consumer of tritium, tritium's primary function is to boost the yield of both fission and thermonuclear weapons.

Hydrogen has three isotopes: ^1H (common hydrogen or protium, H); ^2H (deuterium, D); and ^3H (tritium, T). Protium has one proton, deuterium has one proton and one neutron, and tritium has one proton and 2 neutrons. Deuterium (also referred to as heavy stable isotope of hydrogen) and protium are stable, but tritium is radioactive with a half-life of 12.43 years (some references give 12.3 years). In seawater, the relative abundances of hydrogen isotopes are H (99.984%), D (0.016%), and T (5×10^{-5}%) (Mazor, 2004). Tritium is the only environmental tracer that is part of the water molecule itself. All other tracers are solutes, i.e., they are dissolved in water. If tritium replaces one of the hydrogens in the water molecule, H_2O, the result would be what is called tritiated water, THO. Tritium content of a water sample is expressed as TU, where one TU represents one THO molecule in 10^{18} H_2O molecules. In simpler terms, one TU means one mg THO in 10^9 tones H_2O. This is equal to 6.686×10^{10} tritium atoms/kg and has an activity of 0.1181 Bq/kg (3.193 pCi/kg) or 7.1 disintegration per minute per liter of water. Tritium dating method is the first technique developed to date groundwater, and that was in 1957 by Begemann and Libby. However, as we explain in greater detail below, it is near its expiry date and we, therefore, limit our discussion to main points only. Tritium is also used in oceanography, meteorology, dispersion and diffusion studies, influx of leachate from the landfills, and much more.

4.2.1 Production of Tritium

Tritium is produced in at least five different ways:

1. By cosmic-ray bombardment of nitrogen and deuterium in the upper atmosphere (Testa, 2005):

$${}^{14}_{7}N + n \rightarrow {}^{12}_{6}C + {}^{3}_{1}H$$

$${}^{2}_{1}H + {}^{2}_{1}H \rightarrow {}^{3}_{1}H + {}^{1}_{1}H$$

The amount of this production is estimated at 0.5 ± 0.3 atoms/cm^2/s. Before the onset of atmospheric nuclear weapons tests, the global equilibrium tritium inventory was estimated at about 3.6 kg or 35 mega curies (some references give different figures). This leads to a rainfall tritium concentration of 3–6 TU in Europe and North America and 1–3 TU in South Australia (Solomon and Cook, 2000). Such estimations, however, are not quite accurate, and various references argue for a range of 0.5–20 TU pre-bomb rainfall. It is albeit clear that because of higher geomagnetic latitude, rainfall tritium content in the Northern Hemisphere is more than that in the Southern Hemisphere.

2. Through thermonuclear tests, which started in 1952, and, though banned, still form a significant source of tritium in the atmosphere and other spheres. In all textbooks, graphs are presented that illustrate temporal concentration of tritium in the rainfall as early as 1953 (in Ottawa, Canada, for example), reaching its peak in 1963 and 1964 in the Northern and the Southern Hemispheres, respectively, and declining since then to an almost natural level in the Southern Hemisphere. One such graph is illustrated in Figure 4.1, to show the concentration of tritium in the precipitation in Vienna, Austria. To better understand the effect of bombs on tritium level, we quote from an early reference (Barrett and Huebner, 1960, cited by Clebsch, 1961, p. C-124):

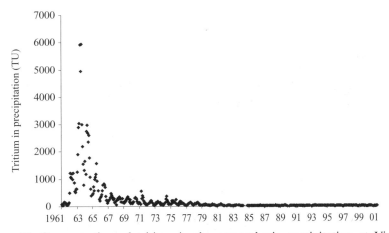

Figure 4.1. Concentration of tritium in the atmospheric precipitation at Vienna, Austria. [IAEA/WMO (2004). Global Network of Isotopes in Precipitation. The GNIP Database. Accessible at: http://isohis.iaea.org.]

The first increase in tritium concentration due to bomb testing resulted in a concentration of tritium in precipitation of 66 ± 1 T-units at Chicago on November 18, 1952. In the periods March–July 1954 and March–September 1956, tritium in Chicago rainfall exceeded about 100 T-units and between December 18, 1957, and July 29, 1958, the tritium level in Chicago rain and snow (and presumably in Nevada precipitation) ranged from a minimum of 54.0 T-units to a maximum of 2,160.0 T-units.

3. A newer source of tritium in the environment is the byproduct (or as a special product) of nuclear reactor operations (IAEA, 2004b). The amount of tritium produced, its behavior, its pathways to the environment, and its chemical form depend on the reactor design. The majority of tritium produced is in gaseous waste (e.g., reactor off-gases and dissolver off-gases at reprocessing plants), but it may be considerable in the liquid effluents from some nuclear facilities (IAEA, 2004b).
4. Watch industry wastes (e.g., in incinerator plants where old watches are burnt) release a huge amount of tritium to the environment. In Switzerland, for instance, such source of tritium is so high that using tritium as a tracer for dating becomes virtually impossible (H. Surbeck, University of Neuchatel, personal communication).
5. By neutron radiation of lithium in rocks (Andrews and Kay, 1982):

$$^6_3\text{Li} + n \rightarrow {}^3_1\text{H} + \alpha$$

The amount of such type of production becomes significant where lithium-and neutron-producing radio elements such as thorium and uranium are abundant. One example is granitic rocks, in which tritium production can lead to up to 2.5 TU in the fracture fluids (Andrews and Kay, 1982). For average Earth crustal rocks, however, rock-produced tritium is less than 0.2 TU (Lehmann et al., 1993).

For industrial, nuclear (main consumer), and commercial usages, tritium is produced through bombardment of lithium with neutrons. The demand for tritium is rather low. As an example, total U.S. tritium production since 1955 has been approximately 225 kg, an estimated 150 kg of which have decayed into ^{3}He, leaving a current inventory of approximately 75 kg of tritium (Albright et al., 1993).

4.2.2 Sampling, Analysis, and Reporting the Results

The volume of sample needed to measure tritium concentration depends on the method we choose for the analysis. All methods require a maximum of half a liter of sample, but one-liter bottles are preferred to allow repeated measurements if necessary. There is no special precaution except to avoid watches and compasses, which contain some tritium. Further, samples with a high amount of radon-222 should be degassed because ^{222}Rn is also an emitter of beta particles and can cause interference. In some cases, ^{222}Rn is removed by dissolving it in an immiscible organic solvent (which is added to the sample) and separating the water and the solvent afterward. Radon-222 is 10 times more soluble in the solvent than in the water, so all ^{222}Rn are dissolved in the solvent only. Another approach is to store sample for a few days to allow for the complete disintegration of ^{222}Rn atoms because of the short half-life of 3.8 days. Generally, four different methods with various detection limits are in practice for measuring ^{3}H content of water samples:

1. Liquid scintillation counting, LSC, or direct counting with a detection limit of about 8–10 TU involving distillation of the sample and counting the disintegrations. A liquid scintillation analyzer is shown in Figure 4.2. The sample volume required is 20 mL, but a larger volume should be collected. This method is generally employed as a first step to determine the feasibility of more precise tritium measurements by other techniques. It is applicable to high tritium environments such as leachate from landfills. The cost is about USD 150 per sample.
2. Low-level counting: This method involves electrolytic enrichment and radiometric measurement of ^{3}H concentration. The detection limit of this method is about 1 TU and needs a 250-mL sample. Cost is about USD 300 per sample.
3. Ultra low-level counting. This method uses the same principle as low-level counting, but includes double enrichment as well as extended counting time. The sample volume required is 500 mL and sample bottles should be flushed clean with argon. The detection limit of the method is about 0.1 TU, and the cost of analysis is about USD 450.
4. Helium-3 ingrowth method. This method also has a general detection limit of 0.1 TU. Some laboratories, The Lamont-Doherty Earth Observatory, for instance, claim a detection limit of up to 0.01 TU, and some references report a detection limit of as low as 0.005 TU (Bayer et al., 1989). Water samples for helium-ingrowth analysis are degassed on a vacuum line and stored in a sealed glass bottle for some long time, say a few months up to a year, after which the ^{3}He atoms generated by the decay of ^{3}H is measured by a mass spectrometer.

The result of tritium analysis of the water samples are usually reported in tritium units. They can be reported in activities as well, e.g., 1 Bq/L. Activities are convertible to TU by dividing to 0.1181.

4.2.3 Dating Groundwater by Tritium

Tritium, whether natural or artificially produced by thermonuclear tests, is incorporated in the meteoric water molecules by oxidation, becomes part of

Figure 4.2. Liquid scintillation analyzer at the Centre of Hydrogeology, University of Neuchatel, Switzerland. The detection limit of this instrument is about 8 TU. Computer, shown in the top figure, is connected to the analyzer for processing the analyses. A sample vial is shown in the bottom figure.

them, and moves within various compartments of the hydrologic cycle such as surface water, oceans, soil water, and groundwater. Atmospheric tritium, therefore, reaches groundwater at some stage. Theoretically, it should be possible to use tritium to date groundwater in two different approaches:

1. by analyzing the decay of natural tritium in the subsurface environment;
2. by locating the position of bomb pulse tritium underground or finding its signature in the water sample.

Dating Groundwater by Analyzing the Decay of Natural Tritium in the Subsurface Environment
This approach is, in theory, similar to dating groundwater by other radiotracers "Clocks" such as ^{14}C, but it may not be practically possible. This is because of the short half-life of tritium and because natural production of tritium in the atmosphere is low and does not accumulate to form an easily traceable quantity. In any case, tritium in the recharging water starts disintegrating into ^3He when it enters the subsurface environment:

$$^3_1H \rightarrow {^3_2He} + \beta$$

Therefore, with the passage of time, the concentration of tritium in the groundwater decreases according to the decay law.

$$C = C_0 \ln e^{-\lambda t} \quad \text{or} \quad ^3H = {^3H_0} \ln e^{-\lambda t}$$

To date a groundwater sample, we have to measure the concentration of tritium in the sample (C or ^3H), but we also have to find out what was the concentration of tritium in the recharge water, or the initial value (C_0 or $^3H_0$). Note that λ is the decay constant of tritium of 0.056 year^{-1}.

Example: Tritium content of a water sample from the Northern Hemisphere, collected in 1991, is 0.5 TU. What was the age of this sample if we assume a pre-bomb tritium concentration of 4 TU in the local rainfall?

$$0.5 = 4e^{-0.056t} \rightarrow t = 37 \text{ years}$$

The age of the sample, i.e., 37 years, shows that it was recharged in 1954 (1991 − 37 = 1954). We know that Northern Hemisphere rainfall tritium content was higher than 4 TU in 1954. Therefore, one of the two following scenarios may prevail:

a. The sample is a mixture of pre-1952 and post-1952 recharged waters.
b. The sample is pre-1952 recharge water, but in our calculation we assumed an incorrect initial value. An initial value (C_0) of 5 TU will increase the age of the sample to 41 years and will change the recharge date to 1950 (1991 − 41). In such a case, there will be no paradox in reporting the age of the sample as 41 years.

We should immediately add here that we will not be able to positively date such a sample if we do not find extra information (e.g., from other dating techniques) that could help us to reject one of the above-mentioned possibilities. For example, the presence of SF_6 in the sample is an indication that the sample consists of some post-1952 water component and scenario "a" is accepted, but the age determined would not be quite correct. We should also point out that a number of references (Mazor, 2004; Clark and Fritz, 1997) argue that a tritium level of 0.5 TU is an indication of pre-bomb age, but as we just showed it really depends on the assumed initial value.

Dating Groundwater by Locating the Position of Bomb Pulse Tritium Underground or Finding Its Signature in the Water Sample It is possible to date groundwater if we can (1) locate the position of the bomb-peak tritium or (2) find high tritium samples that we can confidently relate to the thermonuclear era. For (1) we need to measure the vertical concentration of tritium in a profile such as in a piezometer nest. The age of the groundwater would then be the "date of sampling—1963 (in Northern Hemisphere) or date of sampling—1964 (in Southern Hemisphere)." This approach, which will give us a quantitative age, works well at a flow divide where downward flow prevails; horizontal flow causes the loss of a distinct peak due to dispersion. For (2) we have to have water samples whose tritium concentration is over 10 TU. Such samples are certainly younger than 54 years (2006 − 1952 = 54), but their precise age cannot be determined. This occurs for the majority of cases.

At the end of this section we should point to the fact that although thermonuclear testing facilitated the practice of dating young groundwaters until about 1990, but it has prevented using naturally derived tritium as a dating tracer. This unfortunate situation will last for another 4 decades, until the impact of thermonuclear tests on the tritium in the environment is completely erased. By the year 2046, 99% of the 1963 bomb-peak tritium atoms will be decayed. We will be then able to again use naturally derived tritium as a dating tracer because we will have a near-precise initial value, which is the natural level of cosmogenic production. This statement will hold true if further bombs are not tested or exploded and if other significant widespread sources of tritium are not created. In such case, the dating range of the tritium methods is from 1.9 to 41 years if 10% and 90% of the initial atoms disintegrate into helium.

4.2.4 Advantages and Disadvantages

Advantages
1. Tritium is a well-established and a well-known method with plenty of references.
2. Laboratory facilities are worldwide and the cost of analysis is relatively small.
3. Tritium can still be regarded as a supplementary dating method.
4. It is the only tracer that is part of the water molecule.

Disadvantages
1. The method is approaching its expiry date.
2. Due to the strong latitudinal variation, it would be difficult to precisely determine the initial value even if the bomb-peak tritium effects on the environment are completely vanished.

4.2.5 Case Studies

The common perception is that tritium has been used in numerous case studies to date young groundwaters. This is not, however, the case and it is safe to argue that tritium studies leading to positive groundwater age determination are not ample. Basically, spatial and temporal variations in the input function have substantially lowered the dating capacity of tritium. Generally, it indicates whether groundwater was recharged before or after bomb peak of 1963–1964 (Smethie et al., 1992).

Kirkwood-Cohansey Aquifer System—Southern New Jersey Coastal Plain, USA The Kirkwood-Cohansey aquifer, New Jersey Coastal Plain, is an unconfined aquifer with a shallow unsaturated zone. Figure 4.3 shows the concentration of tritium in a vertical profile measured by Szabo et al. (1996) in 1991. Although the samples were collected from three different sites (albeit close), it is easy to locate the position of bomb-peak tritium at the depth of about 23 m. The age of the sample from this depth would, therefore, be 28 years (1991 − 1963 = 28). From the figure, it is interesting to note that the $^3H/^3He$ age of this sample was estimated at 24.4 ± 1 m, not much different from the tritium age. Further we can see that the $^3H/^3He$ ages increase with depth, while the same can be inferred from the tritium graph alone, i.e., samples shallower than 23 m are younger, and samples deeper than 23 m are older than 28 years. Tritium-helium ages are, however, more quantitative.

New Mexico Aquifers, USA One of the earliest studies using tritium as a dating tracer is on a few New Mexican aquifers by von Buttlar (1959). Through this study, groundwater samples collected in different months in 1956 and 1957 were analyzed for tritium content. The related data are presented in Table 4.1. As it is seen from the table, in the Alamogordo area, tritium content of the groundwater samples is much lower than that of the surface waters. Despite

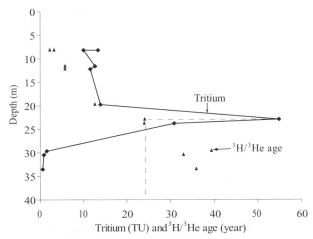

Figure 4.3. Vertical concentration of tritium and $^3H/^3He$ ages (filled triangles) in Kirkwood-Cohansey aquifer system, New Jersey Coastal Plain. The age of bomb peak tritium sample is 28 years, and its $^3H/^3He$ age is 24.3 years (raw data from Szabo et al., 1996). Sampling date is 1991.

TABLE 4.1. Tritium Concentration in Various Groundwaters, Two Surface Water, and One Ice Samples in New Mexico, USA

	Alamogordo area			Deming area	Artesia area
Sample	Nov. 1956	May 1957	Nov. 1957	Aug. 1957	June 1957
1	14.4 S	11.6 S	60.6 S	5.8	1.5*
2	5.8	1	6.5	8.6	1.7*
3	13.1 S	18.7 S	52 S	7.2	5.1*
4	3.2	4.6	11.6	9	6**
5	20.1	23.8	23.3	9.3	3.5 (April 1957)
6				4	4 (April 1957)
7				6.6	3.2 (May 1957)

S: Surface water sample. The source of Sample 5, Alamogordo area, is not accurately known.
* These samples are from one well but from different depths of 320, 73, and 27.5 m.
** Ice sample.
Source: von Buttlar, 1959.

the limited number of groundwater samples, their tritium content suggests that the aquifer did not receive substantial recharge from the then recent rainfalls or surface waters (i.e., ages of the samples were more than 5 years). However, interesting sharp increase in the tritium content of sample 4 from May to November 1957 indicates that some recharge had occurred during this period. Similarly, with regard to the Deming area, von Buttlar (1959) suggests that no appreciable post-bomb tritium is found in the samples. In contrast, Artesia area samples show a different picture. Their lower tritium content suggests ages of about 10 years or over. Further, samples taken from different depths (designated by *) clearly show that bomb tritium reached shallower part of the Artesia aquifer. For us, it may now be easier to interpret the data and conclude that groundwater samples in the Alamogordo (average 9 TU) and Deming (average 7.2 TU) areas were a mixture of pre- and post-bomb waters, while the Artesia area samples were all but one pre-bomb waters. However, we are still unable to precisely determine the age of the samples because of the uncertainty in the initial value.

4.3 $^3H/^3He$

The fading of the tritium dating method has led the scientists to revive an old technique, namely $^3H/^3He$, to replace it. As pointed out in Chapter 2, the $^3H/^3He$ dating method was proposed long ago by Tolstikhin and Kamensky in 1969, but it was not widely practiced or recognized until 1990 because of the difficulty in sampling and in measuring the ^3He content of the groundwater samples and because of the availability of the tritium method, which covers almost the same age range. The description about helium and its isotopes is given in full in Chapter 6, where we discuss helium-4 as a tool to age-date very old groundwaters.

4.3.1 Sources of ^{3}He

There are four sources for ^{3}He in the groundwater:

1. Atmospheric: The solution of atmospheric ^{3}He in percolating rain- and snow water. This is called atmospheric helium and includes the excess air component.
2. In situ: Production of helium from the fission of ^{6}Li by neutrons. This is called nucleogenic helium (in fact, ^{3}H is produced but it decays to ^{3}He).

$$^6_3Li + n \rightarrow {}^3_1H + \alpha \quad \rightarrow \quad {}^3_1H \rightarrow {}^3_2He + \beta$$

3. Tritium decay: A portion of ^{3}He in the aquifer is generated by the disintegration of tritium. This is called tritiogenic helium. The increase in the concentration of helium, which is directly related to the decay of helium, is shown in Figure 4.4.
4. Mantle: Mantle releases some ^{3}He, which may end up into the groundwater system.

All of the above sources have their own particular ^{3}He/^{4}He ratio and lead to specific ^{3}He/^{4}He ratios in groundwater. This is how heliums of different sources are differentiated. In young (and in most cases shallow) groundwaters, the contribution of nucleogenic and mantle helium is insignificant. However, it is still a difficult task to separate the atmospheric helium from the tritiogenic helium if there is considerable excess air component, though excess air can be quantified by measuring dissolved gases in water. Furthermore, if other sources of helium such as nucleogenic and mantle are significantly present, dating by ^{3}H/^{3}He becomes practically impossible.

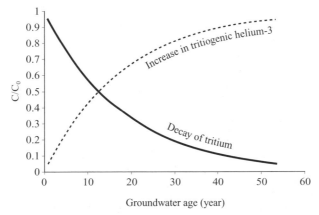

Figure 4.4. Decay of tritium atoms and increase in ^{3}He atoms in the groundwater system.

4.3.2 Sampling, Analysis, and Reporting the Results

To avoid contamination with air helium, samples for ^{3}He analysis are usually collected in 10-ml copper tubes sealed with stainless steel pinch-off clamps. Helium is then separated from all other gases dissolved in the sample and is analyzed by a mass spectrometer. The unit to report tritium concentration is TU and the unit to report ^{3}He concentration is also TU. One TU of ^{3}He represents one ^{3}He atom per 10^{18} hydrogen atom and is approximately equal to 2.487 pcm^3 ^{3}He per kg of water. In the other words, one TU of ^{3}H decays to produce 2.487 pcm^3 ^{3}He per kg of water at STP. Often in the literature, the concentration of ^{3}He in the groundwater samples is reported as %δ ^{3}He:

$$\%\delta^3\text{He} = \frac{R_{sample} - R_{at}}{R_{at}} \times 100$$

where R_{sample} is the ^{3}He/^{4}He ratio (units in cm^3 STP/g H$_2$O) of the water sample, R_{air} is the ^{3}He/^{4}He ratio in the atmosphere, which is 1.384×10^{-6}. Analytical precision for ^{3}He is about ±1%. The cost of analyzing of one groundwater sample for ^{3}H/^{3}He dating is about USD 600. Note must be made that ^{4}He is always simultaneously measured with ^{3}He, because sampling and sample preparation are the same (no further cost) and also because ^{4}He is needed to precisely determine the tritiogenic ^{3}He.

4.3.3 Dating Groundwater by ^{3}H/^{3}He

The tritium-helium method measures the relative abundance of tritium and ^{3}He in a groundwater sample. The amount of ^{3}He from the decay of tritium is measured along with the amount of tritium remaining in the water. That sum is equal to the amount of tritium that was present at the time of recharge, or the initial value. Mathematically we write:

$$^3\text{H}_0 = {^3\text{H}} + {^3\text{He}_{tri}} \quad \text{and} \quad ^3\text{H} = {^3\text{H}_0}\ln e^{-\lambda t}$$

Combining these two equations, we obtain:

$$^3\text{H} = (^3\text{H} + {^3\text{He}_{tri}})\ln e^{-\lambda t} \rightarrow \ln e^{-\lambda t} = \frac{^3\text{H}}{(^3\text{H} + {^3\text{He}_{tri}})} \rightarrow t = \frac{1}{\lambda}\ln\left(\frac{^3\text{He}_{tri}}{^3\text{H}} + 1\right) \quad (4.1)$$

It is seen from the equation that in order to measure the age of a groundwater sample, we simply need to measure its tritium and ^{3}He$_{tri}$ simultaneously. There is no need to have the initial value of ^{3}H, i.e., C_0 or ^{3}H$_0$.

Example: The concentration of tritium and tritiogenic helium in a groundwater sample are 53 TU and 24.6 TU, respectively. What is the age of this sample?

$$t = \frac{1}{\lambda}\ln\left(\frac{^3He_{tri}}{^3H}+1\right) \rightarrow t = \frac{1}{0.056}\ln\left(\frac{24.6}{53}+1\right) = 17.86(1.46) = 6.8 \text{ years}$$

The preceding example is a simple-case scenario, but in real world we have to

1. Measure the ^{4}He content of the sample to calculate ^{3}He/^{4}He ratio in order to determine the atmospheric component of helium.
2. Measure the excess air component of helium. This may require measuring some noble gases like neon, whose concentration in the groundwater is controlled solely by excess air and atmospheric dissolution.
3. Calculate the mantle and the nucleogenic components of helium if these are thought to be significant.

Jenkins (1987) proposed Equation 4.2 to calculate the tritiogenic helium (in TU) from the atmospheric helium:

$$^3He_{tri} = 4.021 \times 10^{14}\left[^4He_{sample}(R_{sample}-R_{atm})+{}^4He_{eq}R_{atm}(1-\alpha)\right] \quad (4.2)$$

where $^4He_{sample}$ is the ^{4}He content of the sample (cm^3 STP/g H$_2$O), R_{sample} is the ^{3}He/^{4}He ratio of the water sample, R_{atm} is the ^{3}He/^{4}He ratio of atmospheric helium, $^4He_{eq}$ is the ^{4}He concentration in solubility equilibrium with the atmosphere (cm^3 STP/g H$_2$O), and α is the effect of fractionation of ^{3}He and ^{4}He and is about 0.983.

Example: The concentration of ^{3}He and ^{4}He in a water sample are 20.64 × 10^{-14} and 5.27 × 10^{-8} cm^3 STP/g H$_2$O, respectively. What is the concentration of tritiogenic helium in this sample?

Known Parameters:

$$R_{atm} = 1.384 \times 10^{-6} \quad\quad {}^4He_{eq} = 4.75 \times 10^{-8} \text{ cm}^3 \text{ STP/g H}_2\text{O}$$
$$^3He_{eq} = 6.56 \times 10^{-14} \text{ cm}^3 \text{ STP/g H}_2\text{O}$$

We first calculate R_{sample} to be 3.92 × 10^{-6} and substitute it in the equation:

$$^3He_{tri} = 4.021 \times 10^{14}\left[5.27 \times 10^{-8}(3.92 \times 10^{-6} - 1.38 \times 10^{-6}) + 4.75 \times 10^{-8} \times 1.38 \times 10^{-6}(1-0.983)\right]$$
$$^3He_{tri} = 4.021 \times 0^{14}\left[5.27 \times 10^{-8}\ (2.54 \times 10^{-6}) + 6.57 \times 10^{-14}(0.017)\right]$$
$$^3He_{tri} = 4.021 \times 10^{14}\left[13.39 \times 10^{-14} + 11.5 \times 10^{-16}\right] = 54.1 \text{ TU}$$

There are more complex equations to calculate ^{3}He of each source, i.e., radiogenic, mantle, etc. Refer to Solomon and Cook (2000) or Kipfer et al. (2002) for such equations.

Points to Observe
1. Produced tritiogenic ^{3}He atoms must remain in the aquifer, i.e., they should not diffuse from the saturated into the unsaturated zone. These atoms are carried to the water table by diffusion against the downward movement of the natural groundwater flow; therefore, ^{3}He confinement is a function of the vertical flow velocity and diffusion coefficient (Schlosser et al., 1989). Significant loss of ^{3}He occurs if the vertical velocity of water infiltration through the unsaturated zone is less than 0.25–0.5 m/year.
2. Decay of the tritium commences in the unsaturated zone, but only those ^{3}He atoms produced in the saturated zone are preserved, i.e., ^{3}He atoms produced in the unsaturated zone escape to the atmosphere. Therefore, zero ^{3}H/^{3}He age is at the water table and in the case of a thick, unsaturated zone, ^{3}H/^{3}He ages will be much younger than the true ages (age as defined in Chapter 1). In case of a fluctuating water table, zero ^{3}H/^{3}He age is set at the seasonal low water table position (Cook and Solomon, 1997).
3. Excess air impact on the ^{3}H/^{3}He ages is –5 years per cm^3 kg^{-1} of excess air for very young water, reducing to –0.25 years per cm^3 kg^{-1} for water approximately 25 years old; however, recharge temperature's effect is less than 0.5 years per °C (Cook and Solomon, 1997).
4. Dispersion and mixing also have a profound effect on the ^{3}H/^{3}He dating results. We deal with these types of processes effects on the dating exercise in Chapter 7.
5. In ideal circumstances, the method is remarkably accurate for groundwaters up to 40 years old.

4.3.4 Advantages and Disadvantages

Advantages
1. In average situations (medium-thickness unsaturated zone, limited sources of helium, etc.), the resolution of this method is high.
2. Data collected can be used for both ^{3}H/^{3}He and tritium methods.
3. This method will be applicable for a long time, i.e., its effectiveness is not reduced in the future as is the case with some methods like CFCs, tritium, etc.
4. This method does not need the initial value, a parameter that is fundamental and problematic for many of the dating methods.

Disadvantages
1. Sampling and analysis are expensive and laboratory facilities are not available worldwide.
2. It is a difficult exercise to separate tritiogenic helium from the other heliums.

3. The method is excessively sensitive to excess air.
4. The ages obtained do not include the travel time in the unsaturated zone.

4.3.5 Case Studies

As stated before, since 1990, $^3H/^3He$ dating has received widespread application, particularly in combination with other techniques such as CFCs and ^{85}Kr. Some case studies include Schlosser et al. (1988) research on Liedern/Bocholt, West Germany; Ekwurzel et al. (1994) study of unconfined surficial aquifers in Atlantic coastal plain sediments of the Delmarva Peninsula, USA; Szabo et al. (1996) investigation of Kirkwood-Cohnasey aquifer system in southern New Jersey Coastal Plain; Nativ et al. (1997) research on the Antalya Plateau, Turkey; Johnston et al. (1998) study of Waterloo Moraine region; Aeschbach-Herrig et al. (1998) dating of a fractured rock aquifer northwest of New York City; Beyerle et al. (1999) research on the River Töss bank infiltration into Linsental aquifer in northeast Switzerland; Selaolo et al. (2000) work on East Kalahari, Botswana; Rademacher et al. (2001) age-dating of spring waters in Sagehen basin, Nevada County, California, USA; Plummer et al. (2001) study on Shenandoah National Park, Virginia, USA; Swanson et al. (2001) study of the springs of Nine springs watershed, south central Dane county, USA; Dowling et al. (2003) investigation of Ganges-Brahmaputra Flood plain, Bengal Basin; Price et al. (2003) study of surficial aquifer system beneath Everglades National Park, south Florida; Katz (2004) work on karstic springs of Florida, USA; and Corcho Alvarado et al. (2005) research on the Island of Funen, Denmark. Here, we describe one of the less usual cases, the Bengal Basin study.

Ganges-Brahmaputra Floodplain: BENGAL Basin More than 30 groundwater samples were analyzed for $^3H/^3He$ ages by Dowling et al. (2003) to assess the recharge rate and subsurface hydrology of the flood plain of the Ganges-Brahmaputra floodplain in the Bengal basin. The samples were collected along a 250-km length NW-SE transect as shown in Figure 4.5(a) and from various depths of up to 320 m [Figure 4.5(b)]. Two aquifer systems are present in the floodplain with complex hydrology but mainly of deltaic sediments. Salt water from the Bay of Bengal intrudes both upper and lower aquifers. Similarly, methane and ^4He from the deep natural gas reservoir diffuse into the lower aquifer. The main aim of the dating study was to prove whether the aquifers are under active recharge or not. Tritium-helium ages of over 100 years in the lower aquifer were confirmed by ^4He measurements, which also showed ages of over 1,000 years for the same samples. As shown in Figure 4.5(a), upper aquifer hydrodynamic pressure leads to the displacement of groundwater in the lower aquifer (arrows). Dowling et al. (2003) used $^3H/^3He$ ages in the upper aquifer to show that it receives active recharge. To compensate for this recharge, substantial volume of freshwater must be discharged into the Bay of Bengal. Further, age data calculations led to two different rates for downward

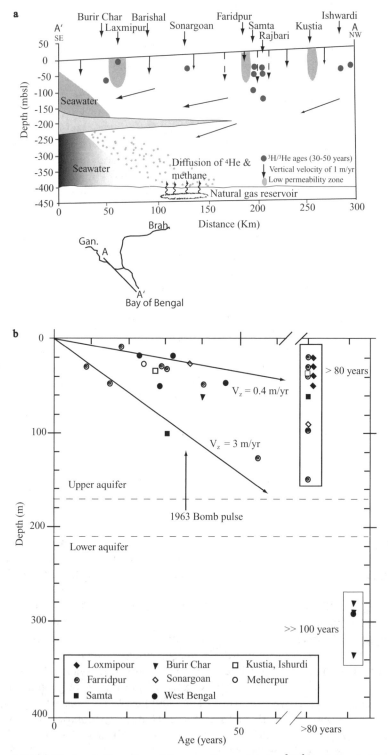

Figure 4.5. (a) Horizontal location of the sampling points for $^3H/^3He$ dating in Ganges-Brahmapura flood plain. (b) Vertical position of the sampling points and the ages (Dowling et al., 2003).

movement of the infiltrating rain water [Figure 4.5(b)], one with a velocity of 3 m/yr and one with 0.4 m/yr. The large difference in these velocities suggests that complex interfingerings of high- and low-conductivity layers exist within the subsurface, which is consistent with the floodplain settings. The recharge to the lower aquifer was estimated at 10–100 cm/yr with an average of 60 ± 20 cm/yr. The interesting finding of the study, however, was that bomb tritium was found in depths of up to 150 m.

4.4 HELIUM-4

We will discuss ^{4}He as a dating tracer for very old groundwaters in Chapter 6. Its main usage is in that area. However, some researchers (Solomon et al., 1996) have shown that ^{4}He can be used to date young groundwaters as well. Such a technique has not received any further attention as yet. Therefore, here we briefly deal with ^{4}He as the fundamental is the same as what we describe in detail in Chapter 6, where also general information about helium and its isotopes is provided.

Helium-4 enters groundwater from different sources including atmosphere, in situ production, inflow from the crust, inflow from the mantle, and release from the aquifer sediments. The basis of dating young groundwater by ^{4}He is the accumulation of such isotope in the groundwater as a result of the release from the sediments. Helium-4 in the sediments is within the crystal lattices and enters groundwater through solid-state diffusion. If we want to know the age of the groundwater, we should measure both the transfer rate of ^{4}He (from crystal lattices into groundwater) as well as the concentration of those ^{4}Hes (in the groundwater) that are from sediment release source. It is a very difficult task to compute each of these, especially the transfer rate. Solomon et al. (1996) suggest using laboratory experiments or calibrating groundwater ages from other methods to calculate the transfer rate, but they acknowledge that considerable uncertainties surround this approach. Further, Solomon (2000) argues that some sediment may not release their residual ^{4}He and this may affect the applicability of ^{4}He dating method in certain environments. The aquifer studied by Solomon et al. (1996) consists of fluvio-glacial (tillits) deposits containing pebbles of the Precambrian crystalline basement.

4.5 KRYPTON-85

Krypton is a colorless and inert gas belonging to the 8th group of the Mendeleyev table (the so-called noble gas or rare gases, (Ozima and Podosek, 2001), discovered by Sir William Ramsey and his student Morris William Travers in 1898. It is now realized that krypton forms some compounds such as KrF_2. The total inventory of krypton in the atmosphere, the main reservoir of such gas, is 4.516×10^{18} cm^3 STP (Verniani, 1966). Krypton is highly soluble in water, together with xenon. Typical abundances of atmospheric krypton in groundwater range from 7.61 to 12.57 × 10^{-8} cm^3 STP/g and from 2.26 to 3.80 × 10^{-8} in seawater (Ozima and Podosek, 2001).

Krypton has the atomic number and atomic weight of 36 and 83.8, respectively. It has 26 isotopes ranging in atomic number from 71 to 95. Six of these are stable: ^{78}Kr (0.35%), ^{80}Kr (2.28%), ^{82}Kr (11.58%), ^{83}Kr (11.49%), ^{84}Kr (57%), and ^{86}Kr (17.3%). All others are radioactive with half-lives of milliseconds to a few hours with the exception of ^{85}Kr and ^{81}Kr, whose half-lives are 10.76 and 229,000 ± 11,000 years, respectively. Krypton-81 has since long been proposed as a groundwater dating tool (see Chapter 6). Krypton-85 is used for dating young groundwaters. Krypton and argon are the only elements whose isotopes are used for dating different-range groundwater ages; ^{85}Kr for dating young groundwaters and ^{81}Kr for dating very old groundwaters; argon-39 for old groundwaters and argon-40 for very old groundwaters. Note that ^{4}He, ^{2}H, ^{18}O, and ^{36}Cl (and possibly ^{129}I) are also used for different age ranges, but these are the same isotope of the element for different age ranges. Because of the well-known sources and sinks of ^{85}Kr and its chemical stability, it is a good tracer for testing the global tropospheric transport models. In addition, many scientists are interested in the growth of ^{85}Kr in the atmosphere as a measure of the worldwide nuclear activity. As an example, it was possible to monitor the nuclear activities of the former Soviet Union by measuring ^{85}Kr content of the atmosphere (Clark and Fritz, 1997). Krypton-85 has also been used to investigate deep water mass formation in the ocean (Smethie et al., 1986).

4.5.1 Production of ^{85}Kr

Natural production of ^{85}Kr takes place in small amounts in the atmosphere by spallation and neutron activation of stable ^{84}Kr.

$$^{84}_{36}\text{Kr} + n \rightarrow {}^{85}_{36}\text{Kr} + \gamma$$

Manmade ^{85}Kr is formed when plutonium and uranium undergo fission. The main anthropogenic sources of ^{85}Kr to the atmosphere are, therefore, nuclear weapon testing and nuclear reactors used for both commercial energy production and weapons plutonium production. The nuclear weapons tests in 1945–1963 contributed about 5% of the total ^{85}Kr in the atmosphere (WMO, 2001). In nuclear reactors, most of the ^{85}Kr produced is retained in the fuel rods. When these fuel rods are reprocessed, the ^{85}Kr is released to the atmosphere. Most ^{85}Kr is produced in the northern hemisphere, specifically in North America, Western Europe, and the former Soviet Union. The atmospheric concentration of ^{85}Kr in Freiburg, Germany, is shown in Figure 4.6. In addition to Freiburg, recent stations in Poland, USA, Spain, South Africa, Austria, Australia, and Antarctica, measure the atmospheric concentration of ^{85}Kr (Weiss et al., 1992). The present background concentrations of ^{85}Kr in the atmosphere are about 1 Bq/m^3 and are doubling every 20 years (WMO, 2001).

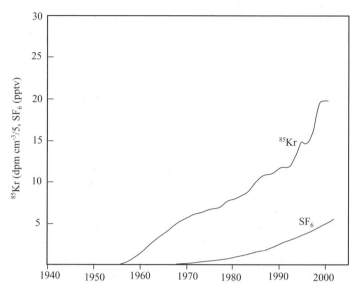

Figure 4.6. Atmospheric concentration of ^{85}Kr and SF_6 (Corcho Alvardo et al., 2005, reproduced with modification with the permission from Elsevier). Krypon-85 measurements are from Freiberg, Germany, and SF_6 data are average values for various coastal sites in the Northern Hemisphere.

4.5.2 Sampling and Analyzing Groundwater for ^{85}Kr

The main obstacle in preventing extensive use of ^{85}Kr as a dating tool is the extensive sample preparation and analysis. The volume of sample required is large (in excess of 100 liters) and depends on the krypton concentration of the water sample. The full procedure to analyze groundwater samples for ^{85}Kr includes the following various steps:

1. Degassing: Gases dissolved in the sample must be extracted from the liquid.
2. Separation of gases: Having done step 1, we will have a mixture of various gases. We have to separate each gas; in this case we have to separate ^{85}Kr from the rest. In a lot of cases, not all gases can be extracted from the water sample, and not whole krypton from the rest. To check the percentage of krypton extracted from the sample, it is compared with the amount of expected krypton in the sample based on the solubilities of krypton. Separation of krypton from other gases is carried by gas chromatography.
3. Counting: The separated krypton is place in a counter and its activity is measured using low-level gas proportional counting facilities. The counting time is about 2–3 days to one week per sample, and a few samples are counted simultaneously. One such equipment, presently installed at

the Underground Laboratory of Physics Institute, University of Bern, is shown in Figure 4.7. For more detail about the analytical facilities, see the website of the mentioned lab at http://www.climate.unibe.ch/~jcor/Group/Procedure_description.html.

Note must be made that other analytical techniques such as laser-based resonance ionization mass spectrometry, which may require 2–5 liters of the

Figure 4.7. Low-level counting facilities for ^{85}Kr and ^{39}Ar measurements at the underground laboratory, University of Bern. An inside view of a counter holder is shown in the top figure. This lab is the only one in the world to measure the ^{39}Ar content of groundwaters and one of only a few that measures ^{85}Kr.

sample, are under development to overcome the problem of large sample volume (Thonnard et al., 1997). Krypton-85 measurements of the groundwater samples are reported in disintegration per minute per cubic centimeter of krypton at standard pressure and temperature (dpm cm^{-3} Krypton STP). The cost of analysis is currently about 700 Swiss francs (approximately USD 550) per sample, not including the cost of sampling (R. Purtschert, University of Bern, personal communication, 2005).

4.5.3 Dating Groundwater with ^{85}Kr

Krypton-85 disintegrates by beta decay to stable ^{85}Rb. In this respect, it can be used as a "clock" or a radioactive tracer.

$$^{85}_{36}\text{Kr} \rightarrow {}^{85}_{37}\text{Rb} + \beta^-$$

However, its use in age-dating is more based on its nearly linear increasing concentration in the atmosphere as shown in Figure 4.6. Atmospheric ^{85}Kr is dissolved in rainwater and is carried out to the unsaturated and then saturated zones. The higher the concentration of ^{85}Kr is, the younger the groundwater age is. Also, due to the short half-life and minimal natural production in the Earth, the absence of ^{85}Kr verifies that groundwater is older than 1950. If ^{85}Kr is combined with an additional radioactive isotope with a similar half-life (such as ^{3}H), additional confidence in the results can be gained. Krypton-85 is commonly used with tritium because the two tracers have similar half-lives but completely different input functions. Tritium input has been decreasing since the 1960s, but ^{85}Kr concentration has been on the rise. Its short half-life and increasing concentrations in the atmosphere make ^{85}Kr a potential replacement for ^{3}H as tritium levels continue to decline.

To quantitatively date a groundwater sample, we should plot the sample on a chart using sampling date and the ^{85}Kr activity such as Figure 4.8(a). The chart is a logarithmic plot of ^{85}Kr in the atmosphere, on which some diagonal lines represent the radioactive decay of ^{85}Kr upon entering the groundwater system. A simpler chart, like Figure 4.8(b), can also be used to directly relate measured activity of the water sample to its age, but this needs upgrading every year, i.e., each year the scale of the horizontal axis needs to change.

Example: The concentration of ^{85}Kr in a water sample, collected in 1989, is 38.5 dpm cm^{-3} krypton STP. What is the age of this sample?

With the assumption that this sample was collected from a west European aquifer, we will first plot the sample on the chart in Figure 4.8(a). We then diagonally extend the sample position to reach the krypton concentration curve. The date corresponding to the point on the curve is then drawn, which is the recharge date of the sample. In this case, the recharge date is just after 1985 (say 1985.5), and the age of the sample is 3.5 years (1989 – 1985.5). Note that for samples from other regions, a relevant atmospheric krypton curve must be used.

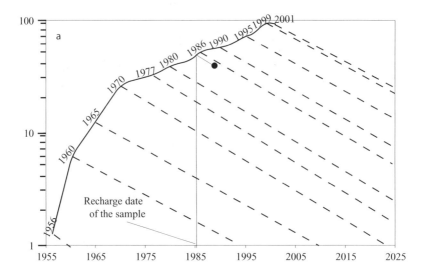
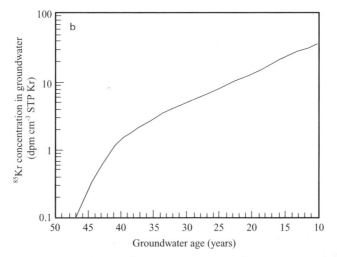

Figure 4.8. (a) Logarithmic plot of ^{85}Kr concentration at Freiberg, Germany, and how groundwater samples are dated. To draw the diagonal lines, atmospheric concentration of ^{85}Kr at any particular date is taken as C_0 in the decay equation, and C is taken as 1. The date corresponding to C (= 1) is then calculated. Lastly, C_0 date on the curve and C date on the horizontal axis are connected by diagonal lines. Black circle represents the position of the example sample on the chart, which is diagonally extended to reach the krypton curve. The recharge date is then identified, from where the sample age is calculated. (b) Relationship between groundwater age and ^{85}Kr activity for samples collected in 2004 based on the activity of ^{85}Kr in the atmosphere of the Northern Hemisphere (modified after Divine and Thyne, 2005).

4.5.4 Advantages and Disadvantages

Advantages
1. Because the ^{85}Kr method is based on an isotopic ratio (^{85}Kr/Kr), it is much less sensitive to degassing than the other dating methods for young groundwaters. This makes this method insensitive to excess air and recharge temperature problems. Also, in organically degraded aquifers where significant amounts of CH_4 and CO_2 may be produced, the generated gases do not change the ^{85}Kr/Kr ratio (Kipfer et al., 2002).
2. In anoxic environments and in CFCs and SF_6-contaminated areas, ^{85}Kr may prove to be superior because of low contamination possibility.
3. Some studies (Smethie et al., 1992) have shown that ^{85}Kr ages are insensitive to dispersion, which could be a substantial advantage of this method if it is confidently proved in future research.
4. The method continues to be applicable well into the foreseeable future because atmospheric concentration of ^{85}Kr is still on the rise.
5. Geochemical inertness of ^{85}Kr.

Disadvantages
1. The main disadvantages in using ^{85}Kr are large sample-size requirements and high costs due to the specialized measurement methods. There are only a few laboratory facilities worldwide.
2. In uranium-rich rocks aquifers, some proportion of ^{85}Kr is from the in situ production (Lehmann et al., 1993). For instance, the annual production rate of ^{85}Kr in Stripa granite is 1.4×10^{-2} atoms/cm^3 rock.year, 20 times higher than what is produced in the Milk River sandstone aquifer (Lehmann et al., 1997). This may mask the atmospheric component of ^{85}Kr or make it difficult to be distinguished.
3. In aquifers with thick unsaturated zones, there is a significant time lag for transport of ^{85}Kr to the saturated zone (Cook and Solomon, 1995). This leads to overestimation of groundwater ages.

4.5.5 Case Studies

A limited number of aquifers have so far been subjected to ^{85}Kr age-dating. A (partial) list of these aquifers includes three different aquifers in Poland (Rozanski and Florkowski, 1979), Borden aquifer, Ontario, Canada (Smethie et al., 1992), unconfined surficial aquifers in Atlantic coastal plain sediments of the Delmarva Peninsula, USA (Ekwurzel et al., 1994), an aquifer in Oberrheingraben near Offenburg, Germany (Heidinger et al., 1997), shallow unconfined aquifer, Rietholzbach catchment, Northern Switzerland (Vitvar, 1998a), quaternary multilayer aquifer of Singen, Germany (Bertleff et al., 1998, and Mattle, 1999), and quaternary sand aquifer of the Island of Funen, Odense, Denmark (Corcho Alvarado et al., 2005).

TABLE 4.2. Comparison of Groundwater Ages (in Years) from Various Tracers for Four Samples from the Delmarva Peninsula

Sample	^{85}Kr	CFC-12	CFC-11	^{3}H/^{3}He
KE Be 52	7.7	8.4	8.1	7.1
KE Be 61	14.4	21	20.8	17.4
KE Be 62	5.4	3.6	3.2	3.1
KE Be 163	5.7	9.2	9.3	7.8

Source: Ekwurzel et al., 1994.

Unconfined Surficial Aquifer, Delmarva Peninsula, USA Surficial aquifer, underlying 90% of the Delmarva Peninsula, is a shallow Pleistocene alluvial unconfined aquifer with a thin (average thickness of 2.5 m) unsaturated zone. It consists of four parts: well-drained northern part; poorly drained and low-hydraulic-gradient central part; fine-grained lowlands of northwestern shore; and poorly drained, shallow water table southern part. About 38 samples from this aquifer were dated by ^{3}H/^{3}He, CFC-11, and CFC-12 methods (Ekwurzel et al., 1994). However, only four of these samples were subjected to ^{85}Kr dating mainly because of the difficulty in sampling a large volume of water. The results are shown in Table 4.2. Groundwater ages from this study were used to calculate the recharge rate to the aquifer. The results showed considerable agreement with the previous recharge estimates. In addition, the good agreement between the ages from various tracers also led Ekwurzel and colleagues to conclude that some hydrogeological parameters such as mixing, dispersion, sorption-desorption processes, and gas loss to the atmosphere are at insignificant levels. It should be noted that all of the above dating methods are at their best when the thickness of the unsaturated zone is shallow, which was the case in the Surficial aquifer.

4.6 CFCs

The following paragraph, extracted mostly from a comprehensive paper by Badr et al. (1990), is a brief yet valuable introduction to chlorofluorocarbons (CFCs).

> Before 1930, the commonly used refrigerants were ammonia (R717), carbon dioxide (R-747), ethyl chloride (R-160), isobutane (R-600a), methyl chloride (R-40), methylene chloride (R-30), and sulfur dioxide (R-764). Due to either toxicity or fire hazards of the above, the need for a safer refrigerant became clear. The new refrigerant had to be non-flammable, with a normal boiling point of between –40 and 0°C, highly stable and low in toxicity. In response to these requirements, the CFCs family of refrigerants was developed. Dichloro-monofluoro-methane (R-21) was the first CFC developed with attractive thermodynamic properties for its use as a refrigerant, but due to its low toxicity,

stability, and inertness, it had never been in commercial use as a refrigerant. By 1974 approximately one million tones of CFCs were being produced worldwide. CFCs have been among the most useful chemical compounds ever developed [there is no known natural source of CFCs; Lovelock (1971)]. In addition to their uses as the working fluids in refrigerators and air-conditioning systems, they have also been employed, since 1950, as blowing agents for foams and plastics, aerosol propellants, solvents for cleaning precision and delicate electronic equipment as well as for dry cleaning, in sterilizers for surgical instruments and catheters, and as working media in Rankine-cycle engines for waste-heat recovery systems. A major cause of environmental concern for CFCs is their long atmospheric lifetime. The lifetime of CFC-11, CFC-12, and CFC-113 is 45, 100, and 85 years, respectively. (Montzka et al., 1999)

Hydrochlorofluorocarbons (HCFCs) are now replacing CFCs, and recent data suggest that the growth rate of CFCs concentration in the atmosphere has begun to decrease (Schoeberl, 1999). At a few stations worldwide, the concentration of CFCs in the atmosphere is measured regularly. The atmospheric concentrations for CFC-11, CFC-12, and CFC-113, those species that are used to date young groundwaters, are illustrated in Figure 4.9. Industrial use of CFC-12 started in the 1930s, CFC-11 in the 1950s, and CFC-113 in the 1970s (Plummer and Busenberg, 1999). Atmospheric concentrations of these species show little spatial variation, with only 10% variation observed between average concentrations in Ireland, Oregon, Barbados, Samoa, and Tasmania (source: Cunnold et al., 1994, in Cook and Solomon, 1997).

The use of CFCs in hydrology began when Thompson et al. (1974) discussed the usefulness of a special type of fluorocarbon known as freon 11 as a tracer in both ground- and surface water hydrology. They found a good agreement between CCl_3F data and hydrologic and tritium age data. Brown (1980) also examined the use of fluorocarbons as groundwater tracers through soil column

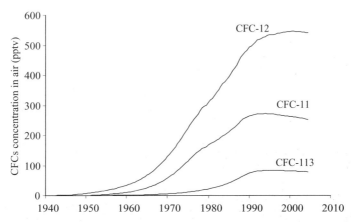

Figure 4.9. Concentration of CFCs in air at Niwot Ridge, Colorado, USA (data from Ed. Busenberg, pers. communication).

studies. There was no significant use of or development in CFCs dating techniques during the 1980s, possibly due to the availability of other methods like the tritium method and the low resolution of the available CFCs sampling and analytical techniques. From 1992, the application of CFCs in groundwater studies especially for age-dating and determination of rainfall recharge became important; e.g., Busenberg and Plummer (1992). A review of CFCs' sources, distribution, and applications in tracing and age-dating groundwater is given by Höhener et al. (2003).

4.6.1 Sampling and Analyzing Groundwater for CFCs

In sampling groundwater for CFCs analysis, due care must be exercised to avoid the sample from contacting air. The average CFC concentration in the air is about 300 times that of the groundwater. Therefore, any small contamination of air will have a very large impact on the concentration of CFCs in the groundwater sample. Several sampling methods are currently in use, including a sophisticated sampling apparatus like that developed by CSIRO (Commonwealth Scientific and Industrial Research Organization), Australia (Figure 4.10). A six-mm nylon tube attached to a stainless steel bailer is sent downhole to obtain the sample. Both nylon tube and bailer are initially flushed with ultrahigh-purity (UHP) nitrogen to remove any air from the system. The sample containers are 62-cm^3 borosilicate glass ampules, which are flushed, prior to filling, with UHP nitrogen for two minutes. After filling, the ampules are flame sealed using a propane oxygen torch. Another method of sampling is to use low-flow peristaltic pumps with Vitron tubing in the pump head connected to copper tubing downhole in the well. The third method is to use downhole Bennett pumps (a nitrogen gas bladder pump used by the USGS) to bring samples to a sample container. There are also three types of sample containers, including the glass ampule as described above (also used by the USGS), the copper tube method, and the new USGS glass bottle method. The glass ampule method gives the best precision followed closely by the new glass bottle technique and lastly by the copper tube technique. The latter is subject to air bubble problems due to bubbles sticking to the inside of the copper tube, resulting in the contamination of the sample (L. Sebol, personal communication, Golder Associates, Canada). The CFCs content of the water samples is measured using a purge-and-trap gas chromatography with an electron capture detector (ECD). For this, we need standard calibration gases that are usually provided by limited institutions. CFCs analyses are reported in pg/kg of water. The cost of analyzing groundwater for CFCs is about US $200 including rent of sampling equipment and consumable materials. Usually triplicate samples are needed to weed out bad data.

4.6.2 Dating Groundwater by CFCs

Atmospheric CFCs, dissolved in percolating rain- and snow water, reach the groundwater system after passing through the unsaturated zone. Groundwater

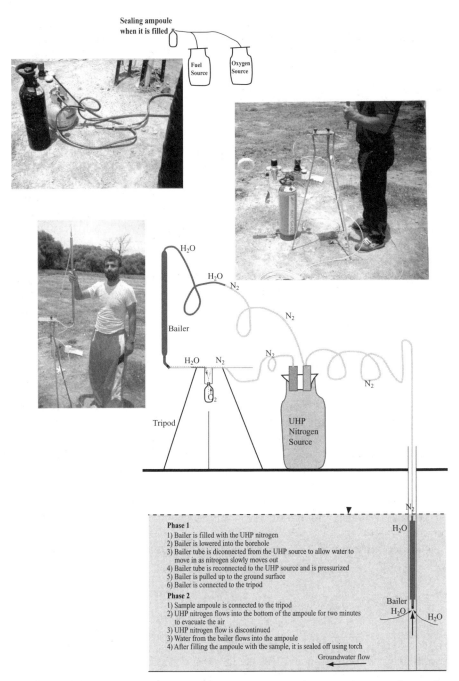

Figure 4.10. Apparatus (developed by CSIRO, Australia) to sample groundwater for CFCs analysis consists of bailer, tube connected to the bailer, UHP nitrogen source, tripod, ampoule, and torch. The diagram illustrates that sampling takes place through two phases.

CFCs ages are obtained by converting measured CFCs concentrations in the groundwater sample to equivalent air concentrations using known solubility relationships (Warner and Weiss, 1985, for CFC-11 and CFC-12, and Bu and Warner, 1995, for CFC-113) and the recharge temperature (Cook and Herczeg, 1998). Therefore, the first step is to measure the CFCs concentration of groundwater (usually less than 400 pg/kg). The second step is to determine the air temperature at which the rainfall recharge occurred. This is usually average summer temperature (for regions with summer-dominated rainfall), average winter temperature (for regions with winter-dominated rainfall), and mean annual air temperature (for regions with year-round rainfall). More accurate recharge temperatures can also be obtained by measuring the dissolved gases for each sample location. The mean air temperature does not allow for decadal shifts in temperatures; for example, air temperatures in North America have been rising since the 1960s. The third step is to calculate the solubility of the CFCs at the considered temperature (summer, winter or annual) with the use of the Warner and Weiss (1985) relationship. Solubilities are also available in a number of textbooks and websites. The final step is to use all three steps and the following formula to obtain the equivalent atmospheric concentration (or sometimes called "apparent atmospheric concentration") of CFCs:

$$\text{EAC} = \frac{\text{CFCs}_{gw}}{S \times MW} \quad \text{(adapted from Cook and Herczeg, 1998)} \quad (4.3)$$

where EAC is the equivalent atmospheric concentration, CFCs_{gw} is the concentration of CFCs in the groundwater, S is the solubility in $\text{mol kg}^{-1} \text{atm}^{-1}$, and MW is the molecular weight of CFCs with unit of g/mol (137 for CFC-11, 121 for CFC-12, and 187 for CFC-113). The EAC is then compared to the graphs of CFCs concentrations in the atmosphere (Figure 4.9) and the year that the recharge occurred is determined.

Example: The concentration of CFC-12 in a water sample collected in 1997 is 100 pg/kg. What is the age of this sample? The mean annual local air temperature is 10°C.

The solubility of CFC-12 at recharge temperature of 10°C is $5.48 \times 10^{-3} \text{mol kg}^{-1} \text{atm}^{-1}$.

$$\text{EAC} = 100/0.00548 \times 121 = \text{EAC} = \frac{100}{0.00548 \times 121} = 151 \text{ pptv}$$

By comparing the EAC obtained with graph in Figure 4.9, we see 151 pptv equals the concentration of CFC-12 in year 1972. Therefore, the age of the sample would be 25 years (1997 – 1972). In this sample, we have assumed negligible excess air and sampling location of North America. If the excess air component is appreciable, we have to correct the measured groundwater CFCs value and then apply the preceding equation. As mentioned earlier, we need the concentration of some noble gases such as argon or neon for this purpose. If excess air is present in the sample and we ignore it, we would calculate an age that is younger than the real age of the sample.

The second approach to more precisely date a groundwater sample is by using the ratios of various CFC species, i.e., CFC-12/CFC-11, CFC-12/CFC-113, etc. Graphs can be produced (from the data used for Figure 4.9) to show these ratios over time. Therefore, theoretically we have one particular value for each ratio in any specific date. For example, the atmospheric CFC-11/CFC-113 ratio in 1983 is about 6. To date groundwater through this approach, we have to calculate the CFCs ratios in the groundwater sample. Therefore, we need to analyze the sample for at least 2 species. Then, we relate the ratio in the groundwater sample to the same ratio in the atmosphere (the date the ratio in question was the same in the atmosphere). The issues with this approach are (1) the temporal variations in ratios are relatively small with the exception of CFC-12/CFC-113, (2) if one species is degraded, then this approach become invalid, and (3) different years have the same ratio, especially years immediately after 1945 and years after 1990. At the end of this section, it should be pointed out that "TRACER," an Excel-based program to calculate groundwater ages based on CFCs and tritium data, has been developed by Bayari (2002).

4.6.3 Limitations and Possible Sources of Error in CFCs Dating Techniques

Error in the Estimation of Recharge Temperature and Excess Air If the air temperature at the time of recharge is overestimated, then the solubility of CFCs that are calculated will be greater than the real values. This leads to overestimation of groundwater ages. It should be mentioned that the error associated with the estimation of recharge temperature is minimal for waters recharged before 1985 (Busenberg and Plummer, 1992).

Excess air is the air bubbles trapped during the recharge process. The CFCs content of these air bubbles is slowly dissolved in the groundwater, leading to a rise in the CFCs concentration of groundwater. This results in the underestimation of groundwater ages. However, excess air is not a serious problem for water recharged prior to 1990, but it is a source of considerable errors for post-1990 waters because of the decline in the air CFCs concentration since then (Plummer and Busenberg, 2000).

Thick Unsaturated Zone (More than 5 Meters) and the Associated Error
The assumption in the CFCs technique is that the concentration of CFCs immediately above the water table is the same as the CFCs concentration in the air. If there is a thick unsaturated zone and the soil water content is high, then the CFCs may be dissolved preferentially into the soil water. Consequently, the CFCs concentration above the water table differs from the CFCs concentration in the air. Furthermore, there will be a lag between the time that

rainwater, containing CFCs, leaves the Earth's surface and the time that it enters the groundwater. This phenomenon leads to overestimating the groundwater age.

Contamination During Sampling There is always a possibility that groundwater could be contaminated during sampling. Apart from air contamination, CFCs may enter the sample through sampling equipments.

Microbial Degradation of CFCs in the Aquifer This is perhaps the most common problem for using CFCs. Oremland et al. (1996) and a number of subsequent researchers have shown that the CFCs are degraded by microbial activity, especially in the soil zone, which is true for soils with high organic content. Earlier studies (Katz et al., 1995; Oster et al., 1996) suggested that CFC degradation is microbially driven in anoxic groundwaters associated with sulfate-reduction or methanogenesis. However, Hunt et al. (2004) have shown that degradation of CFCs could also occur by microbial reduction of iron. CFC-11 is the most degraded of CFCs. Degradation leads to an old bias in age, i.e., the measured ages are older than the real ages. If the CFCs are decomposed to a large extent, then it is difficult to obtain the true age of the groundwater.

Sorption and Desorption of CFCs from the Aquifer Matrix, Diffusion, and Dispersion Russell and Thompson (1983) showed that the CFCs are adsorbed to and released from the dry soil particles. This has two implications for the CFCs dating technique:

1. The rainwater infiltrating through the soil zone adsorbs some CFCs from soil particles (in addition to the CFCs gained from atmosphere) and hence it is enriched in CFCs.
2. Some of the CFCs content of infiltrating rainwater may be absorbed to the soil particles and hence it is depleted in CFCs. Due to the different sorptive properties of each of the three CFCs, sorption affects each one differently. CFC-113 is the most affected one.

In the saturated zone, the CFCs are adsorbed to the aquifer matrix. Some time later, the adsorbed CFCs may be released to the groundwater system. Diffusion and dispersion of CFCs in the groundwater lead to underestimation of CFCs concentration of groundwater and hence overestimation of the groundwater age.

Mixing of Water with Different Ages The problem of mixing of groundwaters from different sources and different aquifers that surrounds most of the hydrological calculations affect the CFCs technique, too. Groundwater in fractured-rock aquifers are the most problematic because of the large difference between the age of fractured water and the age of matrix water (Cook,

2003). This problem may present the biggest challenge to the CFCs dating technique but could be alleviated, to some extent, by comparing the CFCs ages with the hydraulic ages or ages obtained from other techniques. Similarly, Plummer and Busenberg (2000) describe the principles by which to separate the old and the young fractions of a groundwater sample by various ratios of different CFCs species and by ratio of SF_6 to CFCs. The reader is referred to page 462 of their publication for this purpose.

Uncertainty in the Estimation of Atmospheric Input Function There is always some uncertainty in the estimation of historical concentration of CFCs in the atmosphere at a particular site, especially if the site is far from CFCs measurement stations. This could lead to underestimation or overestimation of groundwater ages. However, the near-homogeneous concentration of CFCs in the atmosphere over long distances reduces the uncertainty.

Table 4.3 lists all the processes just described and others and their effects on the ages calculated by CFCs method.

4.6.4 Advantages and Disadvantages

Advantages
1. Presence of CFCs is a good indicator of post-1945 waters. CFC-113 indicates post-1965 water.
2. Input function is relatively well known because spatial variations in atmospheric CFCs concentration are relatively moderate.
3. It is possible to date the sample by EAC of one species and also by ratio of various species.
4. Concordant ages from various species may help to understand the geochemical processes in the aquifer.
5. Cost of analysis is cheap compared to all other methods.

Disadvantages
1. The method is losing its applicability (post-1990s).
2. Many parameters such as excess air, recharge temperature, degradation of CFCs, etc., can influence the accuracy of the ages.
3. Great care is needed for sampling, and large errors may be introduced if proper guidelines are not followed.

4.6.5 Case Studies

Shallow Aquifers in Eastern Australia We described in Chapter 3 that groundwater ages from four salinized watersheds, Buckinbah Creek Watershed (BCW), Suntop Watershed (STW), Snake Creek Watershed (SCW), Troy Creek Watershed (TCW), of Eastern Australia (Figure 4.11) were used by Kazemi (1999) to help understand and manage the problem of dry land salin-

TABLE 4.3. Summary of Processes That Can Modify CFCs Ages

Property	Environment most affected	Description of process	Effect on apparent age
Recharge temperature	Shallow water table	Temperature at the water table during recharge	
		Overestimated	Too young
		Underestimated	Too old
		±2°C, ≤1970, ±1 year or less	
		±2°C, 1970–1990, ±1–3 years	
		±2°C, >1990, >3 years	
Excess air	Rapid, focused recharge; fractured rock	Addition of air trapped and dissolved during recharge. Significant for post-1990 recharge	Too young
Recharge elevation	Mountain recharge	Water recharged at high altitude dissolves less CFCs because of lower barometric pressure	
		Overestimated	Too young
		Underestimated	Too old
		±100 m not important	
		±1000 m, <1987, ± few years	
		Significant for post-1990 recharge	
Thickness of unsaturated zone	Unsaturated zone >10 m	Air in deep unsaturated zone is older than that of the modern troposphere	
		0–10 m, error <2 years	Too old
		30 m, error 8–12 years	
Urban air	Eastern USA, Western Europe, urban areas	CFC mixing ratios in urban and industrial areas can exceed regional values	Too young
CFC contamination	Urban and industrial areas, sewage effluent	CFCs added to water from local anthropogenic sources, in addition to that of air–water equilibrium	Too young (impossibly young)
Microbial degradation	Anaerobic environments, sulphate reducing, methanogenic	No degradation in aerobic environments	No effect
		Sulphate reducing, and fermentation: CFC-11 and CFC-113 degraded, CFC-12 quasistable	CFC-11, CFC-113 too old
		Methanogenic: CFC-11 ≥ CFC-113 ≥ CFC-12	Too old
Sorption	Organic rich sediment, peat	Sorption of CFCs onto particulate organic carbon and mineral surfaces CFC-113 ≫ CFC-11 ≥ CFC-12	Too old

TABLE 4.3. Continued

Property	Environment most affected	Description of process	Effect on apparent age
Mixed waters	Production wells, fractured rock	Mixing of young and older water in water pumped from open intervals in wells	
		Apparent age of young fraction in mixture	Too old
		Apparent age of old fraction in mixture	Too young
Hydrodynamic dispersion	All groundwater environments	Generally small effect for CFCs 1975–1993	Too old
		<1975	Too young

Source: Plummer and Busenberg, 2000, with kind permission of Springer Science and Business Media.

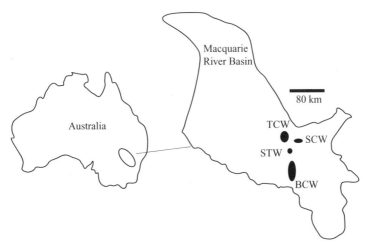

Figure 4.11. Locality map of the four salinized watersheds in eastern Australia that were subjected to groundwater dating by CFCs (Kazemi, 1999).

ity. Here, we present the relevant data in Table 4.4. To check the reliability of CFCs ages, ^{14}C measurements were also conducted on some of the samples from BCW (see also Section 5.3). As discussed, different relationships between age and salinity of groundwater were interpreted as an indication of the difference between salinization stages of these watersheds. Groundwater ages also showed that due to scattered local recharge, groundwater does not become older along the flow line. Further, a negative correlation between the age of groundwater and the maximum fluctuation in the water table in different piezometers in the BCW (Figure 4.12) shows that older groundwaters do not fluctuate as much as younger groundwaters. This is very good control on the accuracy of the ages and also a novel conclusion from the age data.

TABLE 4.4. Concentration of CFCs in Groundwater from Four Watersheds in New South Wales, Eastern Australia

Site	EC µS/Cm	CFC-12 Pg/kg	CFC-12 age (years)	CFC-11 Pg/kg	CFC-11 age (years)	DWT (m)	SD (m)
Buckinbah Creek Watershed							
P15	2,835	47	29	0	N.C.	1.58	3.25
P19	1,052	112	21	108	N.C.	1.18	2.3
P39	1,352	137	18	107	N.C.	3.27	4.18
P40	1,853	83	24	39	N.C.	0.54	5.47
P37	2,008	108	21	43	N.C.	3	6.72
P41	5,042	26	33	40	N.C.	0.55	4.68
P42	6,562	23	34	24	N.C.	0.88	5.18
Suntop Watershed							
P104	6,900	100	21	106	27	0.53	6.1
P106	6,860	N.M.	N.C.	401	10	0.5	5.8
P105	4,540	114	19	89	29	2.45	4.4
P110	4,150	N.M.	N.C.	42	33	−0.35	5
P108	9,160	48	28	27	35	2	4.7
P509	4,150	122	18	196	23	2.4	8.7
Troy Creek Watershed							
P120	5,780	41	30	51	32	0.68	4.08
P116	13,400	151	12	143	25	0.03	4.95
P130	5,010	103	21	99	28	3.8	6.96
P102	2,160	105	21	131	26	1.7	6.03
P101	6,100	135	17	186	N.C.	1.05	6
P111	1,500	48	29	45	N.C.	4.5	13.1
Snake Creek Watershed							
P65	5,230	93	22	150	25	0.13	3.65
P62	8,800	50	28	16	36	0.3	7.2
P63	9,010	91	22	92	28	0.65	4.4
P56	13,830	44	29	21	36	0.45	6.2
P57	6,800	40	30	28	N.C.	1.7	3.7
P58	4,530	18	36	0	N.C.	1.8	5.8

DWT: depth to water table; SD: sampling depth; N.C.: not calculated; N.M.: not measured.
Source: Kazemi, 1999.

Crystalline Bedrock Aquifer, Finnsjön, Sweden The concentrations of CFCs and tritium in the crystalline bedrock aquifer, Finnsjön, Sweden, were determined by Bockgard et al. (2004). The specific goal was to investigate the accuracy of CFC dating in such an environment where CFCs degradation due to anaerobic conditions and mixing may complicate the dating process. The

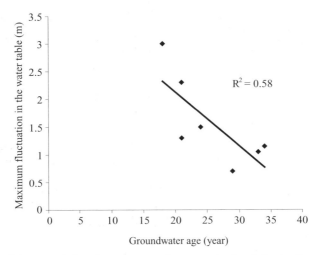

Figure 4.12. The correlation between maximum water-level fluctuation during a 3-year period and groundwater ages in various piezometers in the Buckinbah Creek Watershed, eastern Australia.

overall goal was to develop methods for quantitative estimates of groundwater recharge into Swedish fractured-rock aquifers. The studied aquifer, located 140 km north of Stockholm, consists of granodiorite, with a mean fracture frequency of 3 fractures/m. The depth to water table in the shallow local aquifer ranges from 0 to 3 m, but samples were collected from depths of up to 42 m. Altogether 10 samples from 4 boreholes, analyzed for CFCs and tritium, showed CFC-11 values of less than 0.07, with one exception of 4.6 pico mole L^{-1} (pmL^{-1}), CFC-12 values of 0 to 4.12 pmL^{-1}, and CFC-13 just above the detection limit in some samples. Tritium values range from below detection (2 samples) to as high as 18.7 TU. Bockgard et al. (2004), based on these values, calculated the percentage of old groundwater (CFC and tritium-free) in all samples, which ranged from 57% to 99%. The conclusions from this study were that (1) CFC-11 and CFC-113 degraded in the aquifer but CFC-12 did not and (2) mixing of old and young waters and complicated flow paths led to the ambiguity in age data with respect to the determination of flow path and flow velocity. It was possible to see an increasing trend in groundwater age with depth in only one borehole.

Various Studies Cook et al. (1996) used CFCs age data in County Musgrave on Eyre Peninsula and the Clare Valley, both in South Australia, to determine the recharge rate to the shallow groundwater. The results showed that the recharge rate to the aquifer was about 20–50 mm per year. Szabo et al. (1996) dated the shallow groundwater in the Southern New Jersey coastal plain using CFCs and ^{3}H/^{3}He. They found an excellent correlation between ages obtained from the two methods and also with those obtained by a groundwater flow model. The residence times of groundwater seepage to streams at the New Jersey Coastal Plain were determined by Modica et al. (1998) using CFCs data. They found a negative correlation between the age of groundwater and its nitrate concentration.

4.7 SF$_6$

Sulphur (or sulfur) hexafluoride, SF$_6$, is a colorless and odorless gas used in the electric power industry (to insulate high voltage lines, circuit breakers, and other equipment used in electricity transmission), in the semiconductor industry, in the production of magnesium and aluminum for degassing melts of reactive metals, in blood products, in running shoes, in the Princeton University Plasma Physics Laboratory, and as intraocular gas tamponades for a wide range of complicated vitreoretinal diseases, as well as in many other applications. It is the most potent greenhouse gas of all those defined under the Kyoto Protocol, with an estimated atmospheric lifetime of 1,935 to 3,200 years. Industrial production of SF$_6$ began in 1953 with the introduction of gas-filled high-voltage electrical switches. Concentrations of SF$_6$ in air have rapidly increased from a steady-state value of about 0.02 to 6 parts per trillion during the past 40 years. The rising concentration of SF$_6$ in the atmosphere at Niwot Ridge, Colorado, USA, is shown in Figure 4.13. SF$_6$ is primarily of anthropogenic origin but also occurs naturally in minerals, rocks, and volcanic and igneous fluids. The silicic igneous rocks appear to be the source of the SF$_6$ (Plummer and Busenberg, 2000). The solubility of SF$_6$ is low, the lowest in all environ-

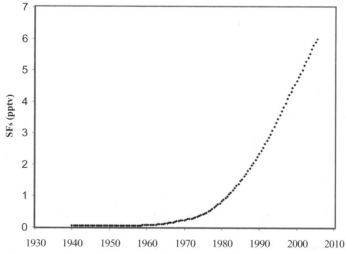

Figure 4.13. Concentration of SF$_6$ in air at Niwot Ridge, Colorado, USA (data from Ed. Busenberg, pers. communication).

mental gas tracers. It is 4.08×10^{-4} and $2.46 \times 10^{-4}\,\text{mol}\,\text{Kg}^{-1}\,\text{atm}^{-1}$ at 10 and 25°C, respectively (Cosgrove and Walkley, 1981). SF_6 is a conservative tracer for groundwater studies and behaves identically to bromide (Wilson and Mackay, 1993). In addition to its use in dating young groundwaters, SF_6 is applied (1) to estimate longitudinal dispersion coefficients in rivers (Clark et al., 1996), (2) in the study of groundwater nitrate pollution (Zoellmann et al., 2001), (3) to estimate gas exchange rate in streams (Wanninkhof et al., 1990) and (4) as a natural atmospheric tracer (Zahn et al., 1999).

4.7.1 Sampling and Analyzing Groundwater for SF_6

The precautions described for CFCs sampling are also applicable to sampling for SF_6. An analytical procedure has been developed by USGS's CFC laboratory for measuring concentrations of SF_6 in groundwaters to less than 0.01 femtomol per liter (fmol/L). Such a technique, which consists of an extensive sample preparation part, is shown diagrammatically in Figure 4.14. A gas chromatograph with an electron capture detector is used to analyze groundwater samples for SF_6 analysis. SF_6 analysis results are reported in fmol/L or pptv.

4.7.2 Dating Groundwater with SF_6

Atmospheric SF_6 is dissolved in rain and snow. Like other atmospheric-derived gas tracers, the amount of the dissolution is a function of the concentration of SF_6 in the air and the air temperature. The higher the atmospheric concentration, the higher the amount of the dissolution. Infiltrating rain- and snow water

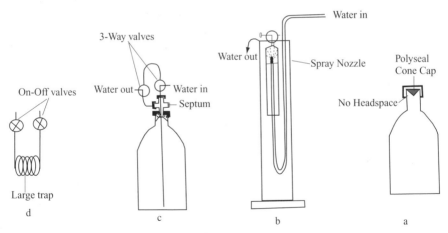

Figure 4.14. Diagram of apparatus for extracting SF_6 from groundwater: (a) bottle for collection of SF_6 sample, (b) apparatus for equilibrating a headspace with the gases present in groundwater, (c) apparatus for extracting SF_6 into a headspace, and (d) trap for the collection of SF_6 samples at the well site (redrawn from Busenberg and Plummer, 2000).

carry dissolved SF_6 into the unsaturated zone and finally to the saturated zone. Therefore, we can date a groundwater sample based on presence and concentration of its SF_6. Concentration above background level indicates post-1970 water. The higher the concentration, the younger the age. However, a number of points could detract from this simple principle:

1. At the time of recharge, groundwater should be in equilibrium with the atmospheric SF_6. This means that the inclusion of excess air could represent a source of error in the calculations.
2. If there are subsurface and onsite additions of natural or anthropogenic SF_6 (such as from volcanic rocks or near urban areas), then age calculations can include usually lengthy processes of computing the accumulation rates, etc. However, if the excess contribution in the vadose zone is measured, an average correction can be applied.

The dating range of this method is X-1970, where X is the date the groundwater is sampled for dating; e.g., in year 2010, the dating range of the method would be 0–40 years. However, due to low atmospheric concentrations during the early days of SF_6 production as well as the difference in various parts of the world, some may adapt year 1965 as the first year to count the age. The step-by-step procedure to calculate the age of a groundwater sample by SF_6 method is

1. Measure the concentration of SF_6 in the groundwater sample.
2. Calculate the recharge temperature by
   a. computing the mean annual air temperature, or
   b. measuring the concentration of noble gases, if possible Ne, N_2, Ar, and Kr and calculating the recharge temperature from them, or
   c. if neither a and b is available, use the annual average temperature of the groundwater.
3. Calculate the excess air component by having the recharge temperature from step 2.
4. Correct the measured concentration of SF_6 (step 1) for the excess air component; note that the corrected value is always less than the measured value unless there is no excess air.
5. Convert the corrected SF_6 concentration (step 4) to the equivalent of air concentration by using air temperature (step 2) and calculated solubilities of SF_6 at various temperatures using Henry's law in the same manner as is done for CFCs.
6. Finally, find in which year the concentration of SF_6 in the atmosphere equaled what you obtained from step 5. Use Figure 4.13 for this purpose for samples collected in North America; otherwise, it is best to use another global monitoring station (there are other stations in both the Northern and Southern Hemispheres with atmospheric SF_6 data avail-

able online at the NOAA website). Note that a robust graph has been (and can be) produced that combines steps 4 and 5 into 1 (Fulda and Kinzelbach, 2000). This graph shows the concentration of SF_6 in water with time since 1960.

Example: The concentration of SF_6 in a groundwater sample is 0.8 fmol/L, the recharge temperature is 10°C, and the sample was obtained in 2006. What is the age of this sample? Brief calculations are:

$$F = 1 + \text{excess air/solubility of } SF_6 \text{ at the recharge temperature} = 1 + 0.2 = 1.2$$

SF_6 concentration corrected for excess air would be 0.8/1.2 = 0.67 fmol/L. After calculating the EAC and referring to Figure 4.13, the recharge date of this sample would be 1983, or its age would be 2006 − 1983 = 23 years.

4.7.3 Advantages and Disadvantages

Advantages
1. The concentration of SF_6 in the atmosphere continues to rise and the method is, therefore, going to hold effective until this trend is stopped or reversed. This is in contrast to the CFCs method, which has lost some credibility due to the declining trend in the CFCs atmospheric concentration since the early 1990s approximately.
2. The atmospheric input is relatively well known, and the subsurface addition of SF_6 is thought to be insignificant.
3. The narrowness of the dating range is a plus in those situations where precise time scales are of interest.

Disadvantages
1. Groundwater ages obtained by the SF_6 method do not include the travel time of groundwater in the unsaturated zone. If this time is long, the ages obtained are substantially different from what is defined as groundwater age in Chapter 1. Although this deficiency may hold true for other methods, because the age range of this method is narrow (0–35 years), it becomes more pronounced.
2. As just stated, the age range of this method is narrow.
3. The method, being recently introduced, has to be proven by more case studies to see if subsurface or natural production of SF_6, microbiological degradation, and other unfriendly causes restrict and reduce the applicability of the method.
4. The main anthropogenic source of SF_6 is in the middle latitude of the Northern Hemisphere, and this may present some doubt on the applicability of the methods in the other parts of the world where measurements of atmospheric concentration of SF_6 may not be sufficient.

5. Low concentrations of SF_6 (due to its low solubility), in general, make the sampling and the analysis a delicate task. They also make the method very vulnerable to the excess air problem because solubility of SF_6 changes significantly with the temperature.
6. Researchers (Santella et al., 2002) have shown that concentration of SF_6 in air over large urban centers is up to twice that of SF_6 in clean air, and this anomaly may extend up to 100 km from the urban centers. This makes SF_6 dating of urban and near-urban groundwaters complicated, and most likely impossible, unless a correction is applied using measurements of vadose zone air SF_6 concentrations provided that the excess has been relatively stable over time.

4.7.4 Case Studies

SF_6 dating is the latest introduced groundwater dating method and, thus, case studies involving this technique are a handful as yet. Most relevant studies are summarized here.

Locust Grove, Maryland, USA The most comprehensive study of SF_6 in groundwater to date is by Busenberg and Plummer (2000) at Locust Grove, Maryland, part of the Atlantic coastal plain sand aquifer. Some 44 samples from 4 piezometer nests at various depths were analyzed to show that the concentration of SF_6 in groundwater ranges from 0.02 to 1.73 fmol/L, which corresponds to an age range of about 2.6–47 years. SF_6 ages were compared with CFC-12 and CFC-113 ages, which showed good correlation except for very young waters (less than 10 years old) and old waters (more than 30 years old). This discrepancy was then attributed to two fundamental inadequacies of these methods:

1. CFCs method is unable to date correctly groundwaters recharged after 1993 because the concentration of this gas in the atmosphere stopped rising after the Montreal Protocol, which imposed restriction on the production and use of CFCs.
2. The SF_6 method is not able to satisfactorily date groundwaters recharged prior to 1970 because of the low atmospheric concentration of SF_6 (Figure 4.13).

Busenberg and Plummer (2000) also show that at all four sampled piezometer nests, groundwater SF_6 ages increased with both depth and distance along the flow path (increasing from the recharge area).

Busenberg and Plummer's paper also includes a brief report on the SF_6 content of water from springs near the top of the Blue Ridge Mountains, Shenandoah National Park, Virginia (reported in full by Plummer et al., 2001); Eastern Snake River Plain aquifer, southeast Idaho; Middle Rio Grande Basin, New Mexico; and springs and groundwater from igneous and volcanic areas of various parts of the USA such as the Chesapeake Bay watershed. The number of analyzed samples reached 3,000, with some showing very high concentrations of SF_6—greater than air—water equilibrium and not explainable by a single atmospheric source. Take note of the extensive measurement of SF_6 concentration in groundwater in the Chesapeake Bay watershed by Lindsey et al. (2003), through which over 100 spring samples were analyzed, with the maximum recorded SF_6 concentration of 59.6 fmol/L.

Upper Limestone Aquifer, Stuttgart/Bad Cannstatt, Germany Eleven well-water samples from 80-m-thick upper limestone aquifer, in Stuttgart/Bad Cannstatt, Germany, the second mineral water system of Europe, were analyzed for SF_6 by Fulda and Kinzelbach (2000). Samples were collected along a transect extending from the recharge area of the aquifer to some 20 km downstream. The concentration of SF_6 declined from 2.2 fmol/L at the recharge area of the aquifer to virtually zero at the end of the transect. Groundwater ages were, therefore, estimated to range from 1 to more than 20 years, respectively. Starting from the recharge area, groundwater ages showed an exponential increase, as expected, except for one sample that was probably contaminated during sampling.

Tertiary Basalt Aquifer of Hessia, Germany Zoellmann et al. (2001) published the result of an analysis of SF_6 in six public and observation wells in Hessia, Germany, during various times from 1992 through 1995. SF_6 concentration ranges from 1.14 to 6.25 fmol/L. These data, in conjunction with tritium measurements, were modeled to study the transport processes in the saturated and unsaturated zones in order to predict nitrate level in wells for different land-use scenarios. Bauer et al. (2001) added some more analysis to the same SF_6 and ^3H results (by Zoellmann et al., 2001) and combined them with ^{85}Kr and CFC-113 to conclude that excess air plays a considerable role in the interpretation of gaseous tracers and to show that CFC-113 is retarded in the aquifer as opposed to SF_6 and ^{85}Kr.

Permian Sandstone Aquifer of Dumfries, Scotland The Permian sandstone (Locharbriggs Sandstone) and breccia (Doweel Breccia) aquifer of Dumfries produces $11 \times 10^6 m^3$ of water annually, which accounts for 11% of the groundwater abstracted in Scotland. MacDonald et al. (2003) analyzed 15 samples from this aquifer, collected from depths of 30 to 183 m, for SF_6 and CFCs. SF_6 values ranged from less than 0.1 to 2.1 fmol/L. On the basis of CFCs and SF_6 data, groundwater from the Dumfries aquifer was divided into old and modern components, older than 50 years (no detectable SF_6) and younger than 10 years (maximum SF_6 concentration), respectively. The percentage of modern portion ranges from 2 to 100% and occurs more in the Locharbriggs Sandstone. This study showed a positive correlation ($R^2 = 0.7$) between percentage of modern water in the sample and the concentration of nitrate. MacDonald and his colleagues also predicted that with the decline of old groundwater components in the aquifer, its nitrate concentration will rise.

Karstic Springs of Florida, USA Katz (2004) measured the concentration of SF_6 in 22 samples from 12 large karstic springs in Florida. The SF_6 values ranged from 1.55 to 4.55 pptv (four samples were discarded because of sampling contamination), corresponding to ages of 16.3 to 2.8 years, respectively. Katz's study showed that (1) CFC-113 ages agree with SF_6 values, but (2) both these tracers have sources other than atmosphere. This is why the ages obtained by these two methods were not concordant to $^3H/^3He$ ages, which, in fact, were found to be the most reliable.

Quaternary Sand Aquifer of the Island of Funen, Odense, Denmark The Quaternary sand aquifer, near the Island of Funen, Odense, Denmark, is a thin sandy aquifer confined by glacial tills from above and below. Four samples from three sites were analyzed for SF_6 by Corcho Alvarado et al. (2005) and showed SF_6 contents of below detection limit to 1.6 pptv. The aim of this study was to combine SF_6 analysis with a number of other dating tracers such as ^{85}Kr, ^3H, ^3He, CFCs, and ^{36}Cl to characterize and date the aquifer. The mean age of the samples (from all tracers) were determined to be 17, 22, 23 and 27 years.

4.8 $^{36}Cl/Cl$

The general information about chloride (chlorine), its isotopes, sources, and analytical facilities, is given in Chapter 6, where it is discussed in more depth for its application in dating very old groundwaters. Here, we aim to explain how ^{36}Cl can be used to identify and date young groundwaters, a less significant application of such tracer. Atmospheric ^{36}Cl is dissolved in rain- and snow water and enters the water cycle. Its concentration in the atmospheric precipitation is reversely proportional to the amount of rainfall—the higher the rainfall, the lower the concentration of ^{36}Cl. In contrast, ^{36}Cl concentration is directly proportional to the atmospheric reservoir of ^{36}Cl, itself, a function of natural and anthropogenic production rates. As rain and snow fall on the ground surface and become surface water or recharging groundwater, their ^{36}Cl content is further affected by the evapotranspiration. Such phenomenon enriches surface and recharge water in ^{36}Cl. Bentley et al. (1986a) suggested the following equation (4.4) to determine the concentration of ^{36}Cl in the surface and recharge waters:

$$[36_{Cl}] = \frac{F_{CL} \times 31,536,000}{P} \times \left(\frac{100}{100-E}\right) \quad (4.4)$$

where $[^{36}Cl]$ is the concentration of ^{36}Cl in atoms/L in the surface water or in the recharge water, F_{CL} is the fallout rate in atoms $m^{-2} s^{-1}$, P is the annual precipitation in mm, and E is the evapotranspiration in percentage (of precipitation). Note that the concentration of ^{36}Cl in recharge water is inversely proportional to rainfall amount while directly proportional to evapotranspiration.

Example: Natural fallout rate of ^{36}Cl at a location where annual precipitation and evapotranspiration loss are 300 mm and 80%, respectively, is 20 atoms m^{-2}s^{-1}. What would be the concentration of ^{36}Cl in the recharge water?

$$[36_{Cl}] = \frac{20 \text{ atoms m}^{-2}s^{-1} \times 31,536,000}{300} \times \left(\frac{100}{100-80}\right)$$
$$= 10,512,000, \text{ or } 10.5 \times 10^6 \text{ atoms/L}$$

In addition to natural production by cosmic activation of argon, ^{36}Cl was produced in large quantities from activation of seawater ^{35}Cl by thermonuclear tests during 1952–1958. The magnitude of this production, which is called bomb pulse ^{36}Cl, reached several times that of the natural production rate. The global average fallout rate for the integrated bomb peak was in the range of 2.8×10^{12} to 7.1×10^{12} atoms m^{-2}. This happened in a short period of time because, the estimated natural fallout rate is only 30.6 m^{-2}s^{-1} (Andrews et al., 1994), or 9.6×10^8 atoms m^{-2} year^{-1}. Investigation of a 100-m-long ice core drilled at Dye-3, Greenland, shows that ^{36}Cl in snowfall reached its maximum concentration in the late 1950s and early 1960s (Figure 4.15). Since thermonuclear explosions did not produce stable chloride ions, the ^{36}Cl/Cl ratio in the atmosphere increased exponentially. Such highly concentrated atmospheric precipitation with high ^{36}Cl/Cl ratio is a hydrologic marker that can be easily traced and located. Accordingly, a high level of ^{36}Cl/Cl in the groundwater indicates that recharge to groundwater has occurred since this period (i.e., between the late 1950s and early 1960s). Based on Bentley et al. (1982), the maximum post-1954 recorded ^{36}Cl in rainfall is 183×10^7 atoms/L at Long Island, New York (September 1957), in surface water is 167.5×10^7 atoms/L in Truckee River, Nevada (August 1958), and in groundwater is a water well in Arizona with 158×10^7 atoms/L (January 1979). This is while the maximum recorded ^{36}Cl in pre-1953 groundwater is 11.7×10^7 atoms/L and that is in the Tucson City well in Arizona (Bentley et al., 1982).

4.8.1 Dating Groundwater by the ^{36}Cl/Cl Ratio and Case Studies

The procedure to date groundwater by ^{36}Cl/Cl is partly similar to dating with ^{3}H. The job is to locate the position of the bomb-peak ^{36}Cl plume. We should start with studying the record of post-bomb ^{36}Cl concentration in the local air. Then, we must construct a vertical profile of ^{36}Cl concentration in the groundwater by measuring ^{36}Cl/Cl in samples from at least one piezometer nest. The best case and the lucky scenario would be to find a groundwater sample with a distinguishably high ^{36}Cl/Cl ratio, which reflects bomb ^{36}Cl. Finally, we can find the time lag between the peak ^{36}Cl/Cl ratio in the air and the peak ^{36}Cl/Cl in the groundwater (which is the sampling date). A major obstacle for this

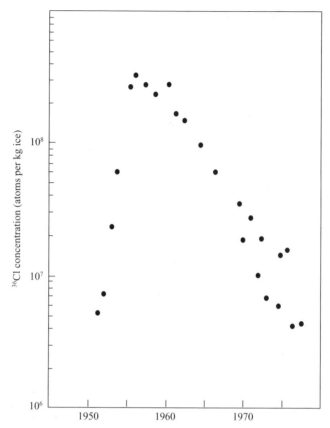

Figure 4.15. Chlorine-36 concentration in an ice core drilled at Dye-3, Greenland. Peak bomb concentration in late 1950s and early 1960s is evident. The age scale was determined by counting the seasonal variations in oxygen-18 content (Elmore et al., 1982, reproduced with permission from *Nature* publishing).

method is the recirculation of bomb ^{36}Cl in the environment. Initially, it was thought that the ^{36}Cl emanated from the thermonuclear tests was washed out from the atmosphere in a relatively short time after the release, probably before 1980. All studies, including measurements on the Dye 3 ice core, supported this theory. However, the latest research involving ^{36}Cl in young groundwaters by Corcho Alvarado et al. (2005) has shown that it is difficult to estimate the local ^{36}Cl/Cl input function. The uncertainty in estimating input function is attributed to the recirculation of the bomb ^{36}Cl in the various compartments of the environment. Such a problem has also been realized by Cornett et al. (1997), who found that ^{36}Cl concentrations in rivers and lakes in eastern North America are higher than those expected from natural production and fallout from the atmosphere, because of the continuous cycling of weapon ^{36}Cl through the biosphere. The other major problem that faces this method is the local scale of the explosions and the concern in extending such local inputs to areas far from the explosions. This is more appreciated if we bear in mind that considerable percentage of produced ^{36}Cl was washed out quickly before being able to spread. As a result of these two issues, and in spite of the fitness of ^{36}Cl/Cl for dating young groundwaters, not many case studies have been reported in the literature. Nevertheless, Phillips (2000) argues that availability of other dating methods such as CFCs and ^{3}H/^{3}He, which can cover the same age range, has led to the light use of the weapon ^{36}Cl in detailed tracing studies. This explanation may not hold true, because CFCs and ^{3}H/^{3}He methods came into widespread practice only in the 1990s, while there is not much dating exercise with ^{36}Cl/Cl even before the 1990s. Cook and Solomon (1997) describe two issues as the main problems for ^{36}Cl/Cl method: (1) the practical difficulty in preparing a detailed vertical concentration profile and (2) the weakening of the input signals by diffusion and dispersion processes. High costs of analysis and unavailability of analytical facilities could also form significant local problems. The best use of ^{36}Cl/Cl ratio is perhaps where it is used in conjunction with tritium and other young age tracers to better constrain the ages obtained. Such an example is the study by Corcho Alvarado et al. (2005), where ^{36}Cl is used in conjunction with CFCs, ^{3}H/^{3}He, SF$_6$, and ^{85}Kr.

Borden Aquifer Case Study, Canada Measurement of ^{36}Cl in Borden Aquifer, Canada, by Bentley et al. (1982) showed that the ^{36}Cl peak was found about 17m deep below ground surface (Figure 4.16). As can be seen from the figure, except for one sample, the tritium concentration of the samples is correlated with their ^{36}Cl values. With the atmospheric peak in 1955 (Figure 4.15) and sampling date in 1981, it has taken bomb ^{36}Cl about 26 years to reach a depth of 17m. This would give an average velocity of 0.65m/yr (17m/26yr). It should be pointed out, however, that Bentley and his colleagues refrained from reporting an age for the groundwater samples because of the uncertainty in the flow patterns.

Gambier Limestone, South Australia Herczeg et al. (1997) measured the ^{36}Cl/Cl ratio of 17 groundwater samples from the Gambier limestone aquifer of South Australia in order to assess the magnitude of small recharge sites to the aquifer. They measured ^{36}Cl/Cl ratios of up to 6 times greater than that of natural fallout in groundwater in the vicinity of the recharge sites where the water table was only less than 2m below the ground surface. This showed post-bomb recharge but also showed the sluggish nature of both the recharge process and groundwater movement (the facts that bomb ^{36}Cl moved less than 2m vertically and it was not flushed away from the recharge point by moving groundwater). It was, however, impossible to precisely date the groundwater samples because of the difference between the residence time of ^{36}Cl and that of groundwater, the uneven effect of the root zone on the concentration of ^{36}Cl and Cl$^-$, the recycling of weapon ^{36}Cl, and the contribution of atmospheric dusts to the soil and groundwater ^{36}Cl budget.

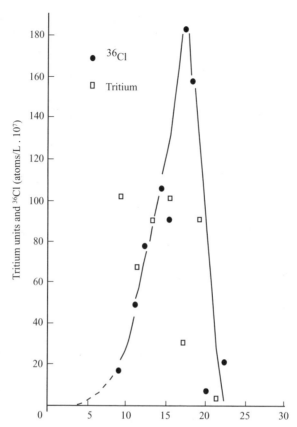

Figure 4.16. Vertical profile of ^{36}Cl and ^3H concentration in groundwater samples, Borden aquifer, Ontario, Canada (Bentley et al., 1982, reproduced with permission from *Nature* publishing).

Other Case Studies In the study of East Midland Triassic sandstone aquifer, UK, Andrews et al. (1994) found that the bomb peak of ^{36}Cl is entirely contained within the unconfined part of the aquifer and has not reached the deeper confined zones.

Murad and Krishnamurthy (2004) measured the ^{36}Cl content of groundwater samples along with other chemical constituents to show that recharge to the two studied aquifers in the eastern United Arab Emirates is post-1950. The Eastern Gravel Plain aquifer that occurs near the coastal area and the inland Ophiolite aquifer have $^{36}Cl/Cl$ values in the range of $13.3–17.4 \times 10^{-15}$ and $20–118 \times 10^{-15}$, respectively. These values indicate the presence of bomb ^{36}Cl.

A nationwide study conducted by Davis et al. (2001) has shown that bomb ^{36}Cl is still being discharged from some of the U.S. springs, which is a decisive indication of the residence time of these springs. The abstract (p. 3) of the work reads, "Chlorine-36 from testing of nuclear devices is still being flushed out of four of the spring systems that were sampled. Thus, more than 45 [2001 –

1952 = 49] years have passed since ^{36}Cl was introduced into the aquifers feeding the springs and the system, as of yet, have not been purged."

High (bomb) $^{36}Cl/Cl$ ratios of the water samples from the fracture fluids in the unsaturated zone at the Yucca Mountain tuffs enabled Levy et al. (2005) to show that faults have been a pathway for fast flow, which reached as much as 300 m into the mountains.

4.9 INDIRECT METHODS

As we also show in Chapter 5, some of the techniques to date groundwaters should be regarded as indirect because these do not lead to quantitative ages directly; rather they provide some clues to constrain the age of some groundwaters. Further, these approaches require information about parameters other than groundwater such as geology, climate, and atmospheric precipitation. We might be able to guess the age of a young groundwater system by shallowness of the depth, localness of the flow system, oxidizing nature of the water, activity of microbes and bacteria in the groundwater, etc. We briefly touch these noncommon approaches to date young or very young groundwaters in Chapter 8.

The use of the stable isotopes of the water molecule, especially $\delta^{18}O$ to estimate the age of groundwater, is a well-established technique, though it is, in fact, limited to calculating the residence time of groundwater only. This takes place by measuring the stable isotope content of spring waters and of precipitation over a long period of time. Identification of paleowaters based on the stable isotopes of water is discussed in Chapter 5.

4.9.1 Stable Isotopes of Water

The stable isotope content of atmospheric precipitation, rain or snow, is determined by many factors including air temperature, elevation, geographical position, etc. Summer precipitations are generally isotopically heavier when compared with the precipitation that occurs in winter. Also, storms have isotopic signatures different from the long rainfall extended over a period of time. Further, we could see a temporal increase in the heavy isotope component of rainfall as a result of global warming and climate change. In principle, it should be possible to estimate the age and especially the residence time of groundwater if the weighted half-year (winter and summer) average concentration of stable isotope in precipitation differs strongly. This difference is carried out to the groundwater system, where it will be preserved for a relatively short time and short space, and later recovered by water sampling. The amplitude of the variations (how much winter and summer rains are different in terms of isotopic composition) is a key factor in the application of such technique in the estimation of the residence time of groundwater. Also, recognizable differences may exist in the oxygen-18 content of precipitation in different years, say annual or bi-annual, etc. If such differences are preserved and manifested in the groundwater system, we should be able to determine the residence times up to a few years. In summary, determination of the residence time of groundwater is based on the preservation and recognition of temporal changes in the stable isotope content of the precipitation.

Burgman (1987) proposed Equation 4.5 for calculating the residence time of springs from the amplitude of the seasonal variations of isotopic content of spring waters and rain waters.

$$T = 1/2\pi \times (1-A)^{0.5}/A \qquad (4.5)$$

where T is the residence time, A is the ratio of the amplitude of seasonal variation in the spring water to the same for rain water (which is of course always less than unity), i.e.,

$$A = \text{variation in spring water}/\text{variation in rainwater}$$

Example: The seasonal amplitude in the stable isotope content of local precipitation is 3.29% and that of the discharging spring water is 0.26%. What is the residence time of this spring?

$$A = 0.26/3.29 = 0.079$$
$$T = 1/2\pi \times (1-A)^{0.5}/A \rightarrow T = 0.159 \times (1-0.079)^{0.5}/0.079$$
$$= 0.159 \times 0.9597/0.079 = 2 \text{ years}$$

The main task in using the preceding equation is the accurate determination of the seasonal amplitude, because short-term and irregular variations may be imposed on the seasonal variation. Also, it is a difficult task to determine which spring water sample is related to which rainfall event. Long-term sampling and analysis are required to overcome this issue.

Maloszewski et al. (1983) proposed a similar equation (4.6) to calculate the mean transit time of recharge water in soil:

$$T = c\left[(A/B)^2 - 1\right]^{0.5} \qquad (4.6)$$

where c is the radial frequency of annual fluctuations, or 0.017214 rad/day, and A and B are the amplitudes of rainfall and soil water isotopic variations, respectively.

Lately, a number of researchers have concentrated on the calculation of the residence time using stable isotope contents of rain- and spring waters (Vitvar, 1998a; Frederickson and Criss et al., 1999; Rademacher et al., 2002; Winston and Criss, 2004; McGuire et al., 2005). Equation 4.7 has been developed for this purpose.

$$\delta^{18}O_{flow} = \frac{\sum \delta i P_i e^{-t_i/\tau}}{\sum P_i e^{-t_i/\tau}} \quad \text{(Frederickson and Criss, 1999)} \quad (4.7)$$

where $\delta^{18}O_{fllo}$ is the oxygen-18 content of the springwater sample, δ_i and P_i are the $\delta^{18}O$ value and amount for a given rain event, t_i is the time interval between the storm and the spring sample, and τ is the aquifer residence time. This equation should be solved for each rainfall and spring sampling event and, therefore, large calculations are needed. For long residence times (τ), the calculated $\delta^{18}O_{fllo}$ will be very close to a volumetrically averaged isotopic value of the rainfall; for too short residence times, only the most recent rain event will have any significance (Frederickson and Criss, 1999). A main difficulty with using such an equation is the large volume of data required. Time series analysis of springwater samples, precipitation, and sometimes soil water for stable isotopes are required. Also, in groundwater systems where annual recharge is small compared to the reservoir volume, and where interflow and delayed return flow are significant, the preceding equation fails to provide a reasonable estimation of residence time.

4.9.2 Case Study

Bluegrass Spring, Meramec River Basin, Eastern Missouri, USA Bluegrass spring, with an average discharge rate of 8.5 L/s (ranges from 1.4 to 280 L/s) and average temperature of 13.3°C, issues from the base of the Plattin limestone, Meramec River basin, eastern Missouri. Numerous water samples from this spring were analyzed for oxygen-18 contents from 1996 to 2003 and bimonthly local precipitation samples for a longer time (Winston and Criss, 2004; Figure 4.17 here). The average oxygen-18 content of Bluegrass spring based on over 5 years of measurements is −7.0‰, identical to the average value of the local precipitation. The difference is that precipitation values range from −1.0 to −16‰, while spring values range from −5.0 to −9.0‰ only. Therefore, the isotopic content of Bluegrass springwater is a subdued reflection of the isotopic content of the precipitation. Figure 4.17 clearly demonstrates the fluctuating nature of the springwater as far as the oxygen-18 values are concerned. Using Equation 4.7 and fluctuation observed for precipitation and spring waters, Winston and Criss obtained a residence time of 2 years for the Bluegrass spring.

TABLE 4.5. Advantages and Disadvantages of Various Methods

Method	Advantages	Disadvantages
^{3}H	Well-established and known method	The method has reached its expiry date
	Can be used as an additional tracer for checking other methods	Difficulty in knowing precise initial value even if bomb tritiums are completely disintegrated
	It is revivable in 4 decades from now	
^{3}H/^{3}He	Initial value is not required	Expensive sampling and analysis
	High-resolution ages	Limited worldwide laboratories
	The method will not expire	Difficulty in differentiating between various sources of helium-3
	Data collected can be used for other purposes	Travel time in the unsaturated zone is not counted
		Excessively sensitive to excess air
^{4}He	Could be a supporting method for other tracers	Its validity still needs to be approved
^{85}Kr	Insensitivity to degassing problems	High sampling and analysis cost and unavailability of analytical facilities
	Limited subsurface contaminating sources	
	Insensitivity to dispersion	Subsurface production in U-rich rocks
	The method will not expire soon	
CFCs	Simplicity and low cost of analysis	The method continues to lose its applicability
	Good results if sorption and degradation processes are absent	Degradation of CFCs
	Good indicator of post-1945 waters	CFCs contaminating sources (not atmospheric)
		Sampling delicacy
SF$_6$	The method will continue to be applicable into the future	Not well known yet
	Well-known input function	It may not be applicable in some regions due to local production of SF$_6$
	Short but precise age range	Narrow age range
		Difficult to apply in and close to urban centers
^{36}Cl/Cl	Good indicator of post-bomb waters	Not many case studies available
		Qualitative nature
	The method will not expire soon	
Stable isotopes of water	Sampling and analysis are routine	It needs ages from other methods for calibration
	Can be used as a check for other methods	It requires extensive data sets and calculations

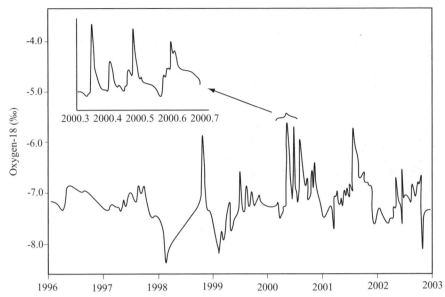

Figure 4.17. Temporal variations in the oxygen-18 content of Bluegrass Spring water, Meramec river basin, eastern Missouri (modified after Winston and Criss, 2004).

5

AGE-DATING OLD GROUNDWATERS

The methods to age-date old groundwaters (60–50,000 years old) include mostly ^{14}C, but less used and indirect methods such as ^{32}Si, ^{39}Ar, ^{18}O, ^{2}H, and conservative and reactive tracers are also explained. In addition to presenting some new data from salinized watersheds in Australia, various aspects of ^{14}C dating such as initial value problem, ^{14}C modifying geochemical reactions, half-life variations, and Sanford correction for diffusion in fractured-rock and aquifer-aquitard environments are discussed. The principle, characteristics, limitations, requirements, correction factors applicable to each method, relevant formulas and equations, advantages, working examples, field sampling and laboratory measurements, and associated costs of each method are also spelled out (at the end of the chapter, advantages and disadvantages of all the dating methods are presented in Table 5.3). With regard to ^{14}C, however, there are at least three major book chapters (Clark and Fritz, 1997; Kalin, 2000; Mazor, 2004) that discuss the topic in detail. This has caused us to refrain from going deep into all aspects of the method because it would inevitably be a repetition of the previous books. We have therefore limited our discussion to major aspects.

Silicon-32 ($t_{0.5} \sim 140$ years) and argon-39 ($t_{0.5} = 269$ years) have been classified as old groundwaters-dating methods because their half-lives are much longer than those of isotopes for dating young groundwaters. Also, it was intended to have a correspondence between human average life expectancy and the age of young groundwaters. Silicon-32 and ^{39}Ar dating methods have

---

Groundwater Age, by Gholam A. Kazemi, Jay H. Lehr, and Pierre Perrochet
Copyright © 2006 John Wiley & Sons, Inc.

been applied to very limited extent, and the related publications are only a few. As a consequence, not much emphasis is needed to be placed here on these methods. Another point worthy of note is that in a number of references the ^{4}He method is referred to as a dating method that could cover old groundwater age ranges (say from about 1,000 years and longer), in addition to both young and very old groundwaters. Practically, however, this has not been satisfactorily demonstrated as yet and we have therefore discussed the ^{4}He method in the context of young and very old groundwaters only, i.e., Chapters 4 and 6.

5.1 SILICON-32

Silicon, discovered by Jons Berzelius in 1822, has the relative atomic weight of 28 and 3 stable isotopes: ^{28}Si (92.27%), ^{29}Si (4.68%), and ^{30}Si (3.05%). It also has a number of radioactive isotopes such as ^{26}Si, ^{27}Si, ^{31}Si, and ^{32}Si; but only ^{32}Si is of hydrologic and geoscientific interest. Despite being the second most abundant element (after oxygen) in the Earth's crust, forming 25.7% of the Earth's crust by weight, silicon isotopes are not widely used in geosciences because the nature of Si compounds is such that the δ^{30}Si variations are small and the fractionation processes are limited (Attendorn and Bowen, 1997). The cosmic abundance of silicon, however, is only 0.05% by weight (Cameron, 1973, cited by Attendorn and Bowen, 1997).

The half-life of silicon-32 was and still is under intense investigation and doubt; estimates range from 60 to 700 years (Alburger et al., 1986). The most recent estimation, which is the weighted average of a number of earlier measurements, puts it at 140 ± 6 years (Morgenstern et al., 1996). However, one could argue that this is not quite satisfactory because the latest related report, i.e., Lal (2001), tends to adapt the 150-year estimate.

5.1.1 Production of ^{32}Si

Silicon-32 is produced in the upper atmosphere by cosmic-ray spallation of argon according to the reaction:

$$^{40}_{18}\text{Ar} + \text{P} \rightarrow\ ^{32}_{14}\text{Si} + \text{P} + 2\alpha$$

The production rate of ^{32}Si, like that of other cosmogenic radionuclides, shows latitude and seasonal variations, anticorrelation to solar activity, and highest natural production in Antarctica (Morgenstern et al., 1996). Silicon-32 has also been produced by thermonuclear tests, but this was washed out from the atmosphere quickly within a few years afterwards. So far, there is no evidence to suggest underground or other source of production for ^{32}Si. In the laboratory, silicon-32 is prepared by spallation from chlorine in a beam of protons at 340–420 MeV by a reaction of the type:

$$^{37}_{17}Cl + P \rightarrow {}^{32}_{14}Si + 2P + \alpha \qquad \text{Schink (1968)}$$

Silicon-32 disintegrates to ^{32}P through a beta emission reaction; the half-life of ^{32}P is 14.3 days:

$$^{32}_{14}Si \rightarrow {}^{32}_{15}P + \beta^- + \bar{\nu}$$

5.1.2 Sampling and Analyzing Groundwater for ^{32}Si

The activity of ^{32}Si in the water samples is usually low, and hence a sample volume of between 5 to 20 m^3 is required to achieve satisfactory results. Due to the low activity of ^{32}Si, its measurement in ground- and surface water samples is undertaken by counting its daughter product (^{32}P) through a five-step procedure described ahead (from Lal et al., 1970, and Morgenstern, 1995). The full description of the method can be found in Gellermann et al. (1988).

1. scavenging of dissolved silicon by a ferric hydroxide precipitation from the water samples (this process is undertaken in the field and only the precipitate and 30 liters of water sample are transported to the lab),
2. radiochemically pure silica is then extracted from the ferric hydroxide scavenge slime,
3. ^{32}Si decays to ^{32}P in the pure silica extract and reaches a secular equilibrium within 1–3 months. A precise amount of carrier phosphate is added two weeks before counting to make the total $Mg_2P_2O_7$ content of the sample 62 mg,
4. radiochemically pure sample of ^{32}P is then re-extracted (milked), and
5. finally, the milked ^{32}P activity is counted over a period of about 40 days. [Note that *milking* is a term used to describe counting of daughter products in a given sample because these products are counted one by one upon their creation, just like milking a cow by which milk is drained slowly.]

Presently, the Institute of Geological and Nuclear Sciences (INGS), New Zealand (http://www.gns.cri.nz/) is the primary laboratory where ^{32}Si activity in water and sediment samples is measured. It has developed the accelerator mass spectrometry, AMS, technique, which offers the major advantage of a reduction in the sample size by at least 2 orders of magnitude (Morgenstern et al., 1996). Because AMS techniques perform analysis on a small amount of silicate, great care must be exercised in handling the sample and undertaking the analysis. So far, the results of ^{32}Si analyses have largely been reported as the activity, i.e., dpm/m^3 or milliBq/m^3. Morgenstern et al. (1996) report that AMS measurement yields the results both in activity and in ^{32}Si/Si ratio. This ratio varies from 10^{-15} to 10^{-13}, depending on the type and ^{32}Si content of water samples. Rainwater samples show higher ratios, while ground, surface, and ocean water samples contain the lowest ratio. If we have the ^{32}Si/Si ratio and the concentration of silicon, it would be possible for us to calculate the ^{32}Si values in atoms/liter of water (see Section 5.2.2 for more detail).

5.1.3 Dating Groundwater by ^{32}Si

Since 1961, ^{32}Si activity in atmospheric precipitation has been measured at three laboratories including Physical Research Laboratory, Ahmedabad, India; University of Copenhagen, Denmark; and Bergakademie Freiberg, Germany (Morgenstern et al., 1996). The activities are generally less than 10 milliBq/m^3 except largely for bomb-produced ^{32}Si during 1963–1965. Atmospheric silicon-32 is dissolved in rain- and snow water in the form of silicic acid. It is then deposited on the soil surface where it either infiltrates into the subsurface environment or is absorbed on the soil particles. [In contrast to most hydrologic tracers, atmospheric ^{32}Si reaches the ground surface through dry precipitation (attached to aerosols) as well.] Morgenstern et al. (1995) have shown that about 50% of the deposited ^{32}Si is retained on the soil surface, and the remaining is transported to the subsurface groundwater system. That portion of silicon-32 that reaches groundwater decays to ^{32}P, and this forms the basis of groundwater dating. Using the basic law of isotopic decay, the age of a groundwater sample can be determined by having an initial ^{32}Si value of the recharged water and the given sample ^{32}Si concentration:

$$^{32}\text{Si} = {}^{32}\text{Si}_0 e^{-\lambda t}$$

Theoretically, the silicon-32 dating method can date groundwaters whose ages range from 21 to 465 years, if 10% and 90%, respectively, of the original silicon-32 atoms have decayed. This is, however, an invalid suggestion since the effectiveness of the method is under serious doubt. The reasons for such a statement are explained as we proceed.

Example: If the ^{32}Si activity of a groundwater sample is 2 milliBq/m^3, what would be its age? Assuming the half-life of 140 years, and an initial value of 5 milliBq/m^3, the age of the sample would be 185 years:

$$2 = 5e^{-0.00495t} \rightarrow t = 185 \text{ years}$$

It should be pointed out that both ^{32}Si and ^{39}Ar (Section 5.2) dating methods were initially developed for ocean and glacier studies and were employed in the field of groundwater hydrology thereafter. Silicone-32 is used for dating of sediments, ice cores, and marine siliceous biota (e.g., glass sponges). It is also useful for study of oceanic and atmospheric circulation (exchange processes between stratosphere and troposphere). Attention has recently been drawn to the potential application of silicon-32 for evaluating the recharge processes in semi-arid regions (Einloth et al., 2001, 2003).

5.14 Advantages and Disadvantages

Despite being proposed in 1966, ^{32}Si has not received widespread attention and application as a groundwater dating tracer. Professor Devendra Lal, currently at Scripps Institute of Oceanography, California, USA, and formerly from Tata Institute of Fundamental Research, Bombay, India, who has been behind the idea of using ^{32}Si as a tracer in various fields including groundwater dating since the late 1950s (Lal et al., 1959, 1970, 1976, 1779; Nijamparkar et al., 1966; Lal and Peters, 1967; Lal, 1999, 2001) concludes in his latest publications (Lal, 2001) that ^{32}Si is not a suitable method for dating groundwater, though it is useful for calculating groundwater infiltration rates. In a personal communication with the senior author, Lal writes, "Silicon-32 exchanges with the soil and some ^{32}Si is lost as the water percolates down. So, it is not suitable for age determination, but it is an ideal tracer for measuring ground water infiltration rates (since ^{32}Si is lost in transit and the rate of loss gives the infiltration rate)."

Silicon-32 may be used for dating if one or more of the following conditions are met (Morgenstern, 2000):

1. the surface water containing ^{32}Si recharges directly into the aquifer without passing through an unsaturated zone,
2. the initial value of ^{32}Si to the aquifer can be measured directly from shallow wells in the recharge area, or
3. ^{32}Si losses in the unsaturated zone are well understood.

Considering the above points and the fact that the geochemistry of silicon is complex (Wagner, 1998), there is very little chance of revitalizing the silicon-32 dating method. The sample volume required is also large, and not many institutions worldwide have the laboratory facilities to analyze water samples for ^{32}Si activity. In spite of all the shortcomings, the currently accepted half-life of 140 years makes it the ideal tracer to cover the age gap between ^{14}C and young dating methods. Lack of subsurface production and precisely known atmospheric values of activity are the other advantages. Also, the recent attempts by IGNS to analyze water samples by high-resolution mass spectrometer requiring only one cubic meter of water sample may help scientists to undertake more case studies and gain some insight into the processes and overcome shortcomings of this method.

5.1.5 Case Studies

Case Study in Germany Silicon-32 concentration along with ^{14}C and ^{3}H contents of 26 samples from a variety of phreatic aquifers in Germany whose lithology ranges from sand to karstic limestone were measured by Fröhlich et al. (1987) during 1977–1985. The ^{32}Si values ranged from 0.1 to 5.2 milliBq/m^3, with the exception of one sample, which contained 10.5 milliBq/m^3 of silicon-

32. The results from this study gave mixed signals. The positive outcome was that the activity of ^{32}Si showed negative correlation with groundwater ages calculated by tritium, suggesting decay of ^{32}Si through time. On the other hand, some young samples with appreciable tritium content were ^{32}Si-free. The conclusion from this study was that ^{32}Si can be used in phreatic sand and limestone aquifers as an additional tracer, but it is not applicable in aquifers with strong geochemical reactions in the unsaturated zone.

A Joint Case Study in Estonia and Germany From 1985 until 1989, seven well-water samples from Devonian and Silurian marine carbonate aquifers in western and northern Estonia and three karst spring samples from central Germany were analyzed for their ^{32}Si content by Morgenstern et al. (1995). Tritium ages and ^{14}C content of the samples were also measured. Silicon-32 activities ranged from 0.1 to 2.4 milliBq/m^3, showing some decrease to a depth of 50 m, below which samples contained only minor, undetectable ^{32}Si activity. Tritium ages were only partially in agreement with ^{32}Si activities indicating that the reduction in ^{32}Si concentration is via both radioactive decay and other geochemical processes such as sorption. The conclusion from this research was that (1) ^{32}Si is strongly sorbed during infiltration of rainwater by soil layers and limestone, (2) ^{32}Si cannot yield meaningful groundwater ages in limestone aquifers, and (3) ^{32}Si in some favorable circumstances may serve as an additional guide for the identification of very young groundwaters.

Case Study in India Lal et al. (1970) measured the ^{32}Si activity of 24 groundwater samples from 6 different hydrological regimes in arid–semi-arid regions of India, which ranged from 0.2 to 5 milliBq/m^3. Considering the then-accepted half-life of 500 years for ^{32}Si, Lal and his colleagues obtained ^{32}Si ages of modern to 2,100 years for the samples, generally younger than the corresponding ^{14}C ages. At that time, the results of the two methods were regarded as in good agreement, and thus satisfactory. Nevertheless, if the presently accepted half-life of 140 years is applied to those ^{32}Si results, they will become "unacceptably too" younger than ^{14}C ages.

Case Study in New Zealand Measurement of ^{32}Si in a confined gravel aquifer in the coastal alluvial basin of the Lower Hutt Valley, New Zealand, by Morgenstern and Ditchburn (1997) showed a 1.2 milliBq/m^3 decrease in the ^{32}Si content of groundwater from recharge area (4.5 milliBq/m^3) to 4-km downgradient (3.3 milliBq/m^3). This study though demonstrated that ^{32}Si content of groundwater in the Hutt Valley originates from the atmospheric sources; it did not yield any concrete age result.

5.2 ARGON-39

Argon (Greek *argos*, meaning inactive), a noble gas with the relative atomic mass of 39.948 g/mol, was discovered by Lord Rayleigh and Sir William

Ramsay in 1894. It is inert, colorless, and odorless and makes up about 0.934% of the Earth's atmosphere, the third abundant gas after nitrogen and oxygen. Argon has 3 stable and 12 radioactive isotopes ranging in atomic mass from 32 to 46. All radioactive isotopes, but ^{39}Ar, have short half-lives of milliseconds to days (Attendorn and Bowen, 1997). The most abundant isotopes of argon in the terrestrial atmosphere are ^{40}Ar (99.6%), ^{36}Ar (0.337%), and ^{38}Ar (0.063%). However, argons-37, -39, and -40 are the most extensively studied isotopes of argon for their various applications in the fields of environment, planetary, and earth sciences. Argon-39 has a half-life of 269 years, and its activity in the troposphere, measured for the first time by Loosli and Oeschger in 1968, is 0.107 ± 0.012 dpm/L of argon (Loosli and Oeschger, 1979). (1 disintegration per minute, dpm, = 0.45 picocourie = 0.017 Bq.) The decay constant, λ, of argon-39 is 0.002577 year^{-1}. The solubility of argon in water in equilibrium with the atmosphere at 10°C and 1 atmosphere is 3.9×10^{-4} cm^3 STP/mL water. This gives rise to a ^{39}Ar concentration of 8.5 atoms/mol (Lehmann et al., 1993).

5.2.1 Production and Sources of ^{39}Ar

Production of ^{39}Ar takes place:
1. In the atmosphere,
    a. By cosmic-ray activity, primarily with ^{40}Ar with a rate of 0.1 dpm/L

    $$^{40}_{18}Ar + n \rightarrow {^{39}_{18}}Ar + 2n$$

    Such a reaction does not occur underground because it requires neutrons with energy in excess of 12.8 MeV, and
    b. Through nuclear weapon testing at a rate of less than 0.005 dpm/L
2. In the subsurface environment:
    a. Through irradiation of ^{39}K with fast neutrons in a neutron-capture, proton-emission reaction,

    $$^{39}_{19}K + n \rightarrow {^{39}_{18}}Ar + P$$

    b. Via a less important reaction of ^{39}K,

    $$^{39}_{19}K + \bar{\mu} \rightarrow {^{39}_{18}}Ar + \nu_\mu \quad \text{(Florkowski and Rozanski, 1986)}$$

    c. By neutron activation of ^{38}Ar, and

    $$^{38}_{18}Ar + n \rightarrow {^{39}_{18}}Ar + \gamma$$

    d. Via alpha emission by calcium-42 (Zito and Davis, 1981)

    $$^{42}_{20}Ca + n \rightarrow {^{39}_{18}}Ar + \alpha$$

5.2.2 Sampling and Analyzing Groundwaters for ^{39}Ar

Sampling groundwater for ^{39}Ar analysis requires a huge volume of water, and this is the major drawback of this technique, especially where groundwater flow rate into boreholes and piezometers is slow. The amount of water required depends on the concentration of argon in the sample, the degassing efficiency, and the amount of the argon required (about 2 liters of argon is usually needed). It is 15 m^3 (15,000 L) on average (Loosli et al., 1989) with a minimum of 10 m^3 (Pearson et al., 1991). Initially, all gases present in the water sample are separated (degassing) by pumping water through an evacuated cylinder of acrylic glass from where the gas is extracted with a vacuum pump followed by two high-pressure compressors. Different gases in the extracted gas such as argon, krypton, etc. are separated using distillation, gas chromatography, and chemical procedures. Once argon gas is separated through the lengthy process, its argon-39 activity is measured by low-level proportional radioactive counting. This technique requires ^{39}Ar-background-free environment and careful operation. A reference (Forster and Maier, 1987) indicates that about two decades ago, the University of Munich in Germany started a program to analyze water samples for ^{39}Ar, but this apparently did not progress. At present, the primary and the sole place to measure ^{39}Ar in groundwater is the Physics Institute, University of Bern, in Switzerland (see also Section 4.5). Its web address (http://www.climate.unibe.ch/~jcor/Group/Procedure_description.html) contains an illustrative description of the analysis procedure. This website includes the latest development in the ^{39}Ar measuring protocol and reports that progress is underway to use less amount of water samples to perform ^{39}Ar analysis. The paper by Forster et al. (1992b) is another reference to consult for a description of the analytical technique for ^{39}Ar analysis of water samples. The cost of ^{39}Ar analysis was initially estimated to be about 3–5 times that of ^{14}C analysis (Oeschger et al., 1974). It is now some 2,700 Swiss francs for a combination of ^{39}Ar and ^{85}Kr, which are often measured together (R. Purtschert, University of Bern, personal communication, 2005). The cost of sampling and sample preparation (degassing, etc.) should be added to this. We should add that a new method to measure the ^{39}Ar content of an ocean water sample is currently being tested at the Argonne Tandem Linac Accelerator System (ATLAS), USA. This method is an AMS-based technique and uses an electron cyclotron resonance positive ion source equipped with a special quartz liner (Collon et al., 2004).

Argon-39 results of analysis are reported in four ways:

1. Activity: is the direct measurement of activity of ^{39}Ar in the sample, i.e., 0.15 dpm/L of water (or per liter of argon) or 0.0025 bq/L. If the reported activity is the activity in water, its quantity (value) is much less compared to when it is reported in argon.
2. Atoms in groundwater, i.e., atoms/L or atoms/mL. Activity values can be converted to atoms/mL if we have the concentration of argon in ground-

water. The formula for this, which needs careful attention in harmonizing the units if used, is

$$^{39}\text{Ar (atoms/L water)} = \text{Activity (dpm/L of argon)} \times \text{Ar (cm}^3\text{ Ar/cm}^3\text{ water)} \times 204{,}020{,}779$$

Note that argon-39's half-life in seconds divided by (ln 2 × 60) equals 204,020,779 [8,483,184,000/41.58 = 204,020,779]. The coefficient of 60 is used to convert dpm to Bq.

3. Percent modern argon (pma): This unit is the ratio of the ^{39}Ar activity in the sample to the argon-39 activity in the atmosphere and it is expressed as percentage, e.g., 76% modern argon. 100 pma is equal to 1.78×10^{-6} Bq/cm^3 of Ar (1.05×10^{-4} dpm), or 1.67×10^{-2} Bq/cm^3 of air.
4. ^{39}Ar/^{40}Ar ratio. This ratio expresses the number of ^{39}Ar atoms to the number of argon-40 atoms in the sample. In the atmosphere, this ratio is 8.1×10^{-16}, which means the number of argon-40 atoms is almost 10^{15}–10^{16} times that of ^{39}Ar.

Example: The ^{39}Ar activity of a water sample is 0.09 dpm/L of argon, and the concentration of argon is 0.0004 cm^3 Ar/cm^3 water. What is the concentration of ^{39}Ar in pma, the ^{39}Ar/^{40}Ar ratio, and the concentration of argon-39 in atoms/L of water?

The percent modern argon of this sample would be 0.09/0.107 = 84 pma.
The ^{39}Ar/^{40}Ar ratio of this sample would be $0.84 \times 8.1 \times 10^{-16} = 6.8 \times 10^{-16}$.
The concentration of ^{39}Ar in atoms/L would be $0.09 \times 0.0004 \times 204{,}020{,}779$ = 7,345 atoms/L.

5.2.3 Age-Dating Groundwater by ^{39}Ar

Atmospheric ^{39}Ar is dissolved in the rain- and snow water molecules and reaches the Earth's surface. Upon entering the subsurface environment, the communication between ^{39}Ar content of rainwater (now groundwater) molecules and the atmospheric ^{39}Ar stops. From this point onward, ^{39}Ar decays to ^{39}K by a beta emission as

$$^{39}_{18}\text{Ar} \rightarrow\, ^{39}_{19}\text{K} + \beta^- + \bar{\nu}$$

Therefore, as time goes by, the number of ^{39}Ar atoms in the groundwater decreases, albeit if there is no subsurface addition of ^{39}Ar. The age of the groundwater can then be determined using the simple decay equation:

$$^{39}\text{Ar} = \,^{39}\text{Ar}_0 e^{-\lambda t}$$

What we measure by analyzing water samples is ^{39}Ar; ^{39}Ar$_0$ is the initial concentration of ^{39}Ar in the recharging rainwater (usually can be taken as 0.107 dpm/L because the atmospheric production is well known and constant), λ is the decay constant for ^{39}Ar (0.002577 year^{-1}), and t is the age of the given groundwater sample in year. Theoretically, ^{39}Ar dating method can be used to date groundwaters in the age range of 41 to 894 years, if 10% and 90%, respectively, of the original atoms have decayed. This is, however, a rough suggestion and the actual dating range and the effectiveness of the method depend on the accurate estimation of subsurface production of ^{39}Ar as well as on the precision and detection limit of the analytical facilities, which are not available widely. In addition, it is becoming more evident that the application of this method should be limited to sedimentary and carbonate terrains where subsurface production of ^{39}Ar is limited.

Example: Argon-39 activity in a groundwater sample is 0.08 dpm/L; what is the age of this sample? Assuming no underground production and using the decay equation:

$$0.08 \, \text{dpm/L} = 0.107 \, \text{dpm/L} \, e^{-0.002577 t} \rightarrow t = 113 \text{ years}$$

(In this example, the initial value has been taken as equal to the atmospheric activity of ^{39}Ar.)

To properly calculate the age of groundwater, the in situ (underground) production rate of ^{39}Ar should be known. This is done through the following procedure:

1. The elemental composition of the aquifer lithology, especially its U and Th contents, should be determined. Also, distribution of the fractures and porosity of the rock must be known.
2. Having obtained the above data and information, the in situ flux of neutrons needs to be calculated. This can be determined experimentally or it can be calculated from the neutron production rate and the neutron absorbing capacity of the rock (Andrews et al., 1989).
3. Important reactions responsible for the production of ^{39}Ar and their characteristics such as reaction cross sections should be identified.
4. Production rate of the related radionuclides, i.e., ^{39}K, etc. should be calculated.
5. Production rate of ^{39}Ar is, finally, calculated.

It should be emphasized, however, that it is a difficult exercise to calculate the subsurface production of ^{39}Ar because cross sections of the ^{39}K reaction (as the dominant reaction) increase with increasing neutron energy resulting in the significant influence of fast neutrons (Lehmann and Purtschert, 1997). Moreover, the cross sections of some of the ^{39}Ar-producing reactions are not accurately known (Andrews et al., 1989).

5.2.4 Advantages and Disadvantages

Advantages
1. Covering the age gap between ^{14}C method and young groundwater dating methods.
2. Relatively precisely known initial value of ^{39}Ar, a problem that surrounds a number of other dating techniques.
3. Limited amount of subsurface production of ^{39}Ar within the sedimentary and carbonate rock aquifers and limited number of reactions involving ^{39}Ar.
4. Measurement of ^{39}Ar in groundwater samples sheds some light on the underground production rate of other nuclides such as ^{85}Kr, ^{222}Rn, etc., which are required in other dating methods. Due to the well-known atmospheric inputs (item 2 above), underground production of ^{39}Ar can be recognized and quantified. It is also possible to quantify ^{39}Ar subsurface production rate by measuring ^{37}Ar concentration in the groundwater. The latter has a half-life of just over a month and, therefore, in a groundwater sample, it is wholly of subsurface origin, because atmospheric inputs are totally decayed by the time that the sample is collected and analyzed.

Disadvantages
1. The large volume of water sample required.
2. The long time needed to degas the samples and perform the analysis.
3. Low values of ^{39}Ar activity and the possibility of introducing error in the analysis due to background values.
4. High cost of the equipments for low-level and proportional radioactivity counting.
5. Underground production of ^{39}Ar especially in groundwaters from crystalline formations with higher U and Th concentrations, which can completely mask the atmospheric ^{39}Ar inputs (Lehmann and Purtschert, 1997; Loosli et al., 2000).

The limitations of the ^{39}Ar method, some of which are now ratified, are better described by Forster et al. (1992a). It is, however, pointed out that despite its almost unsuccessful history in dating groundwater, $^{40}Ar/^{39}Ar$ dating is a well-established method (Merrihue and Turner, 1966; Mitchell, 1968) for dating rocks especially for dating small whole-rock samples of lunar material, particularly fine-grained mare basalts (Dickin, 1995). In all isotope geology textbooks, a chapter or a subchapter is allocated to ^{40}Ar-^{39}Ar dating technique. The $^{37}Ar/^{39}Ar$ activity ratio in the meteorites with a known orbit can be used to determine the spatial gradient of the cosmic radiation in the solar system (Florkowski and Rozanski, 1986). Argon-39 is also used for ice cores and ocean water dating; the practice started before groundwater dating.

5.2.5 Case Studies

The list of aquifers for which ^{39}Ar measurements have been conducted is given ahead. Not all these studies resulted in the age determination of the concerned aquifer.

- Zurzach thermal spring, Switzerland (Oeschger et al., 1974)
- Karstic aquifers near Ingolstadt; Thermal springs, Bad Gögging, Baden and Wildbad (Germany); Thermal springs, Zurzach and Lostorf (Switzerland) and Triassic sandstone aquifer, Lincoln, UK (Loosli, 1983)
- Bunter sandstone aquifer, Saar, Germany (Forster et al., 1984)
- Granitic Stripa groundwater, Sweden; Granitic Leuggern groundwater, Switzerland (Loosli et al., 1989)
- Northern Switzerland and adjacent region, NAGRA program (Pearson et al., 1991)
- Bocholt aquifer, Germany (Forster et al., 1992b)
- Milk River aquifer, Canada (Lehmann and Purtschert, 1997)
- Confined sedimentary aquifer, Glatt Valley aquifer, Switzerland (Beyerle et al., 1998)
- Fontainebleau sands aquifer of the Paris Basin, France (Corcho Alvarado et al., in press)

Most of the above case studies have been summarized in the relevant published textbooks. The case of granitic Stripa aquifer, located 200 km N-NW of Stockholm, Sweden, is discussed here. Stripa granite has been under intense international investigation since 1977 as part of a nuclear waste repository study. (The entire August 1989 issue of *Geochimica et Cosmochimica Acta* was allocated to report part of this international investigation.) Before that, Stripa granite's nearby meta-sedimentary rocks were home to a number of hematite ore bodies, mined for about 400 years (Andrews et al., 1982). Groundwater in Stripa granite occurs in two systems, a shallow system with a depth of less than 100 m below ground surface and a deep system with a depth of up to 900 m below ground surface. Three groundwater samples from the latter system were analyzed for argon-39 yielding very high activities of 0.36, 0.83, and 1.71 dpm/L of argon (Loosli et al., 1989). These values correspond to 51, 120, and 240 atoms/mL of ^{39}Ar. Such high values led the researchers to devise methods to calculate the underground production rate of ^{39}Ar. It also resulted in no age estimate except to report that the atmospheric argon-39 was entirely disintegrated and, therefore, the age of the Stripa groundwater is several hundred years, i.e., 2–3 half-lives of ^{39}Ar. Andrews et al. (1989) calculated the subsurface production rate of ^{39}Ar in Stripa granite at 1.3 atoms/(cm^3 rock.year), and Lehmann et al. (1993) estimated that the concentration of ^{39}Ar in groundwater would be about 1.3×10^4 atoms/cm^3 if all matrix-produced ^{39}Ar atoms escape into the groundwater. These studies suggest that quantifying the underground production rate of ^{39}Ar is a huge challenge that may never be satisfactorily achieved.

5.3 CARBON-14

Carbon occurs in all forms of organic life and is the basis of organic chemistry. Carbon's different forms include one of the softest (graphite) and one of the hardest (diamond) substances. The major economic use of carbon is in the form of hydrocarbons, most notably the fossil fuels, methane gas and petroleum. The most prominent oxide of carbon is carbon dioxide, CO_2, which is a minor component of the Earth's atmosphere and is produced and used by living things. Carbon monoxide, CO, is formed by incomplete combustion and is a dangerous colorless and odorless gas. The paths that carbon follows in the environment are called the *carbon cycle*. For example, plants draw carbon dioxide out of the environment and use it to build biomass. Some of this biomass is eaten by animals, where some of it is exhaled as carbon dioxide.

The atomic weight and atomic number of carbon are 12 and 6, respectively. Carbon has 15 isotopes ranging in mass from 8 to 20, but only three of these occur naturally. There are two stable carbon isotopes: carbon-12, or ^{12}C (98.89%), and carbon-13, or ^{13}C (1.11%). In 1961, the International Union of Pure and Applied Chemistry adopted the isotope carbon-12 as the basis for atomic weights. Carbon-14, or radiocarbon, the naturally occurring radioactive isotope of carbon, was discovered by Martin Kamen and Sam Ruben in 1940, when they bombarded graphite with protons in the cyclotron at the Lawrence Radiation Laboratory, University of California, Berkeley. There is only one ^{14}C atom for every trillion (10^{12}) ^{12}C atoms.

5.3.1 Production of ^{14}C

Sources of ^{14}C and how they enter rain- and groundwater molecules are shown in Figure 5.1 and are described as follows:

1. As a result of cosmic radiation, a small number of atmospheric nitrogen nuclei are continuously transformed by neutron bombardment into radioactive nuclei of ^{14}C. A neutron knocks a proton out of nitrogen and takes its place.

$$^{14}_{7}N + n \rightarrow {}^{14}_{6}C + P$$

The amount of cosmic rays penetrating the Earth's atmosphere varies with the sun's activity and the Earth's transit through magnetic clouds in the Milky Way Galaxy. This governs the amount of ^{14}C produced and the half-life used to date various materials. The average annual natural production rate of ^{14}C, which is temporally and spatially variable, is

CARBON-14

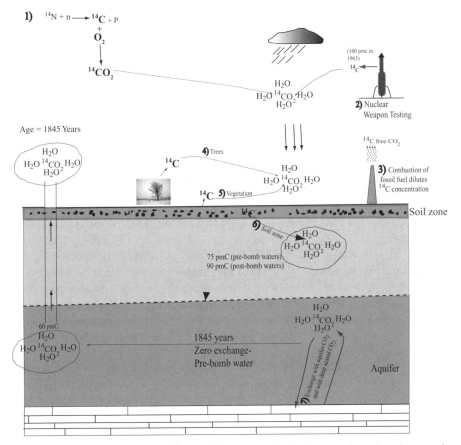

Figure 5.1. Various sources of ^{14}C and how they mix and enter the subsurface environment. An example of dating a pre-bomb groundwater sample is also illustrated.

7.5 kg, and the total mass of ^{14}C on Earth is about 75 tons (Wagner, 1998). However, recent research by Lal et al. (2005) has shown that the cosmic-ray production rate of ^{14}C at Summit, Greenland, was close to its estimated long-term average production rate, except during three periods; 8,500–9,500 and 27,000–32,000 year B.P., when the production rate was higher by about a factor of 2 and during 12,000–16,000 year B.P., when production rate was lower by a factor of about 1.5. The ^{14}C formed is rapidly oxidized to $^{14}CO_2$ and enters the Earth's plant and animal life through photosynthesis and the food chain. The rapidity of the dispersal of ^{14}C into the atmosphere has been demonstrated by measurements of radioactive carbon produced from thermonuclear bomb testing. ^{14}C also enters the Earth's oceans in an atmospheric exchange and as dissolved carbonate.

2. A large amount of ^{14}C was produced during the thermonuclear testing in 1950s and 1960s enhancing the concentration of ^{14}C in the atmosphere to double the natural level.
3. In the Earth, in situ production of hypogenic ^{14}C by neutron activation of nitrogen or neutron capture by ^{17}O is insignificant (Zito et al., 1980). Considerable neutron fluxes are required for this purpose. In exceptional circumstances such as in uranium ore deposits, hypogenic ^{14}C production may become appreciable.

5.3.2 Sampling, Analysis, and Reporting the Results

An earlier method of measuring ^{14}C of groundwater samples depended on the radioactive decay counting, which requires a large volume of the water sample. As ^{14}C decays back to ^{14}N, it emits a weak beta particle (b), or electron. For the counting method, which is still in use, at least 2 g of carbon should be present in the water sample. The major carbon-containing species in the groundwater is bicarbonate. We should, therefore, determine the volume of sample required for ^{14}C analysis by determining the sample's bicarbonate concentration. This is done in the field prior to sampling for ^{14}C. For example, if the bicarbonate concentration of a groundwater is 205 mg/L, it contains 40.3 mg of carbon in each liter. Therefore, we need to collect a sample volume of 50 L to obtain 2 g carbon. The carbon content of the groundwater sample can be precipitated as $BaCO_3$ in the field and only the residue be sent to the laboratory for analysis. This reduces the volume of sample and the cost of shipment, but increases time in the field and requires experienced staff. The problem of a large sample volume was overcome with the development of the accelerator mass spectrometry, which literally extracts and counts the ^{14}C atoms in the sample and at the same time determines the amount of the stable isotopes ^{12}C and ^{13}C. The AMS injects negative carbon ions from the analyte into a nuclear accelerator. The negative ions accelerate toward the positive potential where they pass through a thin carbon film or tube filled with low-pressure gas. Nitrogen ions are so unstable that they self-destruct before getting to the accelerator terminal leaving the ^{14}C alone to be counted. The molecules dissociate into their component atoms and the kinetic energy they had accumulated is distributed among the separate atoms all different from the ^{14}C. Accelerating the ions to high energy also has the advantage of allowing well-established nuclear physics techniques to detect the individual ^{14}C ions. As a consequence, a measurement that may last 12 hours and require several grams of sample using decay counting may take only 30 minutes and consume only a few milligrams using AMS. The cost of analysis of water sample using AMS is about USD 500, which includes the preparation cost as well, i.e., if the sample is delivered to the laboratory in the form of water. Finally, in sampling for ^{14}C analysis, care must be taken to avoid a long temporal gap between sampling and analysis. Delay may lead to equilibrium between air and the sample.

Carbon-14 analyses are reported in percent modern Carbon, pmC. The modern activity of ^{14}C is set by convention as 13.56 decays per minute per gram of carbon. The zero year for this activity is 1950 AD. Therefore, all samples that have an activity lower than this are pre-1950 AD, and samples greater than this value are younger (Cook and Herczeg, 2000). This value is also considered to have an activity of 100% modern carbon (pmC). After 1950, atmospheric thermonuclear testing doubled the amount of radiocarbon in the atmosphere leading to initial values of up to 200 pmC, therefore making the determination of modern activity of ^{14}C a big problem. Further complication on the modern ^{14}C activity stems from the burning of fossil fuels, which has the reverse effect of the thermonuclear testing.

5.3.3 Groundwater Dating by ^{14}C

Background Carbon-14 quickly became a favorite dating technique for archaeologists because living organisms continually exchange ^{14}C with the atmosphere until they die when the exchange ceases. Accurate archaeological dating depends on the assumption that the atmosphere today contains the same amount of ^{14}C as the atmosphere at the time of the organism's death. We actually can calculate the amount of ^{14}C that was in the atmosphere for the past 3,000 years by analysis of the molecular structure of tree rings whose ages are quite certain, but beyond 3,000 years we are considerably less certain. Ordinary ^{14}C is found in atmospheric carbon dioxide (CO_2) taken up by plants which are eaten by animals. It was perhaps the first isotope used for age-dating once living artifacts and materials by Willard F. Libby and his colleagues at the University of Chicago in 1946. They developed a technique designed to measure the ratio of ^{14}C to ^{12}C. The method made it possible to estimate the time that has elapsed since something is no longer exchanging carbon with the atmosphere. A bone or leaf or tree or even a piece of furniture contains some of it. The ratio in air remains relatively constant over short periods as it does in living things constantly exchanging with the air around it. Once a plant dies, no carbon is further exchanged and the continuous decay and changing ^{14}C–^{12}C ratios give us a type of clock.

By measuring the reduced ratio of ^{14}C atoms to regular carbon atoms in an artifact that was once alive, we can determine how long the ^{14}C has been decaying, or how long it was since the living substance died. A piece of wood used in the construction of a spear could also be dated. Carbon-14 dating is particularly useful to archaeologists for establishing the age of objects found in the remains of ancient human settlements. The most famous example is the so-called Shroud of Turin, which was claimed to be the burial cloth of Christ, but ^{14}C analysis gave a date of around 1300 BC. Libby and his team initially tested the radiocarbon method on samples from prehistoric Egypt. They chose samples whose age could be independently determined. A sample of acacia wood from the tomb of the pharoah Zosar was obtained and dated. Libby reasoned that since the half-life of ^{14}C was 5,730 years, they should obtain a ^{14}C concentration of about 50% of that which is found in living wood. His values on this and other woods that were dendochronologically aged proved to be within 10% of his experimental calculation, which convinced him of the efficacy of his method. Science published his paper (co-authored by his colleague, Dr. Arnold), "Age determination by radiocarbon content: Checks with samples of known age" in 1949. Within a few years radiocarbon dating laboratories began to spring up and numbered nearly two dozen by 1960.

Dating Groundwater   Atmospheric ^{14}C is dissolved in the percolating rainwater, as shown in Figure 5.1 and reaches the groundwater table. In groundwater, ^{14}C starts decaying to nitrogen:

$$^{14}_{6}C \rightarrow {}^{14}_{7}N + \beta^{-}$$

If no further ^{14}C exchange occurs, measurement of the remaining ^{14}C atoms can be used to date groundwater following the first-order kinetic rate law for decay:

$$C = C_0 e^{-\lambda t}$$

where C_0 is the activity assuming no decay occurs (initial activity or activity at $t = 0$), and C is the observed or measured activity of the sample. The groundwater datable by ^{14}C ranges in age from 870 to 19,000 years if 10% and 90%, respectively, of the original atoms are assumed to have decayed. However, ages up to 40,000 years or longer have been reported by this method. This can be achieved if the right conditions are met. AMS technique has also helped to increase the dating range of ^{14}C method by enabling lower detection limits.

Inherent Difficulties Associated with the ^{14}C Dating Method
A. **Initial value problem**: The ^{14}C content of the rainwater is modified through a number of processes numbered from 1 to 6 in Figure 5.1 and listed below. This makes dating by ^{14}C a delicate exercise. The final outcome of these processes together is an unknown and difficult to calculate initial value (C_0), which could vary from less than 75 pmC to over 200 pmC. The ideal situation for determining initial value, C_0, is to be able to measure the ^{14}C concentration of a young but tritium-free pre-bomb water in the upgradient area of the portion of the aquifer we want to date. The classical approach to determine the initial value is by tree rings for almost 7,000 years, with no way to accurately determine it prior to 7,000 years ago (Priyadarshi, 2005).
   1. Cosmic-ray production of ^{14}C.
   2. Thermonuclear testing production of ^{14}C. In order to cope with such problems, some references suggest (Hiscock, 2005) that any groundwater with a ^{14}C age of up to 500 years can be considered as young.

3. Fossil fuel burning: Since the industrial revolution of the early 19th century, large amounts of fossil fuels have been burned, causing an increase of about 10% in the concentration of atmospheric CO_2 (Mazor, 2004). This added CO_2 is devoid of ^{14}C and has therefore lowered the $^{14}C/^{12}C$ ratio in the air by about 10%.

4. and 5. Plants and trees influence: Atmospheric ^{14}C is incorporated into vegetation by photosynthesis and later released in the soil by decay and root respiration. Bacterial degradation of organic litter, as well as organic substances released by living plants, pumps enormous amounts of CO_2 into soils (Clark and Fritz, 1997). Plants, trees, and animals utilize carbon in food chains taking up ^{14}C throughout their lives, they exist in equilibrium with the ^{14}C in the atmosphere until they die and cease the metabolic function of carbon uptake.

6. Soil zone influence: Although rainwater contains some CO_2 containing ^{14}C from the atmosphere, it is the soil zone that gives recharging groundwater its radiocarbon signal. Carbon-14 in groundwater comes from the solution of carbon dioxide in the soil zone. Once the water moves below the water table into the saturated zone, this solution ceases and a constant rate of decay can be expected.

B. **Carbon-14 modifying geochemical reactions**: In addition to the above processes, which affect the initial value of ^{14}C and make the interpretation of ^{14}C measurements a challenge, a number of geochemical processes also affect the dating results. Domenico and Schwartz (1990) refer to the work of Reardon and Fritz (1978) in listing geochemical processes in groundwater that can alter the ^{14}C quantities in groundwater in ways not fully reliant on radioactive decay. These include

1. The congruent dissolution of carbonate minerals, which adds carbon without ^{14}C activity to the groundwater, which results in a lower ^{14}C ratio for the sample.

2. The dissolution of carbonate or other calcium-containing minerals accompanied by the precipitation of calcite, which could remove ^{14}C.

3. The addition of dead carbon from other sources such as the oxidation of old organic matter, sulfate reduction, and methanogenesis can reduce the ^{14}C activity of the sample.

4. Possible isotopic exchange involving CO_2 and carbonate minerals could lower the ^{14}C activity, though this process is negligible at normal groundwater temperatures. It becomes significant in geothermal waters, however.

C. **The impact of diffusion**: Sanford (1997) developed a set of equations to correct the ages obtained by the ^{14}C method for the effect of diffusion. Sanford's equation regards molecular diffusion as a significant process affecting the transport of ^{14}C in the subsurface when occurring either from a permeable aquifer into a confining layer or from a fracture (flow zone) into a rock matrix (stagnant zone). The ^{14}C age can be corrected for diffusion using Equation 5.1:

$$t_c = t_u \left(\frac{k}{k + k'} \right) \tag{5.1}$$

where t_c is the corrected age, t_u is the uncorrected age, k is the decay constant of ^{14}C, and k' is the diffusive rate constant. Depending on the thickness of flow and stagnant zones, the ratio of k'/k varies and ranges from 0.0001 to 1,000. A curve has been produced by Sanford (p. 359) to calculate the k'/k for various environments. If the k'/k ratio is more than 1, the corrected age would be less than the uncorrected age. However, if the k'/k ratio is much less than 1, the corrected age would be only slightly different from the uncorrected age.

D. **Half-life**: To obtain true ages of groundwater samples, two adjustments must be done on the ages obtained. The first is to correct the half-life used for dating from 5,730 years to Libby's 5,568 years. This causes insignificant modification; for example, a 17,000-year-old age is reduced to 16,519 years (17,000 × 5,568/5,730 = 16,519 years). The second adjustment is to take into account the secular variations in ^{14}C production and to calibrate the ages to years before 1950 (a standard calibration curve is needed to do this). For younger groundwaters, these two adjustments do not result in substantial changes, but for older groundwaters changes become appreciable.

How to Deal with the Complexities One can interpret ages from a single sample or use ion and isotopic data from many samples using mass balance modeling to sort out the major inputs and outputs of carbon. The simplest way is to account only for the most important process affecting ^{14}C activity, which is the congruent dissolution of calcite. Domenico and Schwartz give the reaction between water containing CO_2 and calcite as

$$CO_2 + H_2O + CaCO_3(s) = Ca^{+2} + 2HCO_3^-$$

At equilibrium according to this reaction, half the bicarbonate would be generated from a source containing ^{14}C (CO_2), and the other half would be generated from a dead source like calcite.

A number of complex computer codes have been developed to account for various mineral dissolution and isotopic exchange reactions. Fontes and Garnier (1979) developed a simple one, while Wigley and others (1978) take into account an arbitrary number of carbon sources such as the dissolution of carbonate minerals and oxidation of organic matter, and sinks that include mineral precipitation, CO_2 degassing and methane production, and the equilibrium fractionation between phrases. Computer code such as "BALANCE"

(Kimball, 1984; Plummer, 1984) and lately NETPATH (Plummer et al., 1994) have been developed in order to account for sources and sinks for ^{14}C. Some of the preceding models assume an open system, in which groundwater ^{14}C interacts and exchanges with the surrounding environment, and some assume a closed system.

Here, in Table 5.1, we have reproduced the result of an extensive correcting exercise on ^{14}C ages in the confined parts of the upper Floridan aquifer using all available models including NETPATH (Plummer and Sprinkle, 2001). This study showed that dissolution of dolomite and anhydrite with calcite precipitation (de-dolomitization), sulfate reduction accompanying microbial degradation of organic carbon, re-crystallization of calcite (isotopic exchange),

TABLE 5.1. Uncorrected and Corrected Ages of Upper Floridan Aquifer Using Six Various Correcting Models

Unadjusted age	Tamers (1975)	Ingerson-Pearson (1964)	Mook (1972)	Eichinger (1983)	Fonts-Garnier (1979)	Netpath	Netpath
9,100	900	2,900	16,600	2,600	4,300		
38,200	32,700	33,600	37,500	33,200	34,300	30,900	30,900
23,800	18,600	20,200	25,600	20,000	21,300	14,200	14,200
9,100	900	2,900	16,600	2,600	4,300	1,400	1,400
22,600	17,100	18,100	22,200	17,700	18,900	16,900	16,900
42,900	37,600	36,900	43,300	35,800	36,800	29,500	30,300
40,000	34,800	31,800	40,300	29,100	31,400	20,500	21,900
47,200	41,700	40,500	46,300	39,300	40,400	28,000	28,000
43,100	37,900	36,200	43,100	34,700	36,000	27,800	27,800
9,100	900	2,900	16,600	2,600	4,300		
25,100	19,700	20,100	7,400	19,600	20,500	18,000	18,200
15,500	10,100	9,900	12,900	9,100	9,900	6,500	7,300
45,300	39,900	36,300	43,300	32,800	35,800	22,800	27,300
50,300	44,600	36,800	49,100	28,800	34,800	−1,600	22,800
39,000	33,600	32,000	36,500	30,200	31,800	20,400	22,000
47,300	41,500	36,000	46,300	31,700	35,100	20,000	25,800
9,100	900	2,900	16,600	2,600	4,300		
12,200	6,800	7,300	7,500	6,800	7,900	6,600	6,800
23,400	18,000	19,500	3,900	19,000	21,000	13,500	13,900
37,700	32,400	33,000	33,100	32,400	33,500	27,200	27,500
32,300	26,900	27,500	26,900	26,900	28,400	20,600	21,700
32,000	26,600	26,700	28,800	26,000	26,900	20,100	20,600
32,400	27,200	27,800	27,600	27,200	28,500	21,200	21,500
8,200	900	2,400	6,500	2,100	3,300	400	400
31,200	25,800	26,200	27,700	24,600	26,600	19,100	20,000
19,000	13,800	15,100	8,900	14,700	16,600	9,800	9,800
24,500	19,500	20,100	20,800	19,500	20,800	18,900	18,900

Source: Plummer and Sprinkle, 2001, with permission from Springer.

and mixing of freshwater with some saline water are the reactions to control the ^{14}C concentration of upper Floridan aquifer. As it is seen from the table, NETPATH ages are generally younger than other models and unadjusted ages are generally older than the rest. For unadjusted ages the initial value of 100 pmC was applied, but for the models a lower initial value of about 40 pmC was used.

At the end of this never fully explained ^{14}C section, we tend to agree with Domenico and Schwartz (1998) when they write, "confident predictions can be made only when the processes affecting the carbon chemistry are absolutely defined—the exception rather than the rule ... taking into account more processes, when analyzing ^{14}C results, will not necessarily lead to a highly accurate age." However, we would like to directly quote this statement as well (Gonfiantini et al., 1998, p. 230): "In any case, the experience accumulated in using ^{14}C for groundwater dating is so vast, the hydrological systems studied so numerous and the data available so abundant, that the experienced isotope hydrologist should be able to make sound judgments on the value and use of ^{14}C data in any possible situation."

5.3.4 Advantages and Disadvantages

Advantages
1. Carbon-14 is an old and well-established method that has been proved and developed by considerable research during the last half-century. It is applicable to a variety of materials and processes.
2. Sampling and analysis for this method are now routine and cheaper than the majority of the dating methods.
3. It is, in one sense, the only method available to date old groundwaters and to fill the dating range between young and very old groundwaters.
4. The deficiencies, the principles, and the positive points of the methods are all well known.

Disadvantages
1. It is an extremely difficult task to determine the correct initial value due to the various processes that modify ^{14}C signature of the percolating rainwater.
2. A large number of geochemical reactions modify the concentration of ^{14}C in the groundwater. It has been tried to alleviate this problem by taking into account a large number of processes, but an appreciable amount of uncertainty surrounds the validity of this approach and the results obtained.
3. Having pointed out the above two major obstacles, it is safe to argue that the ^{14}C method is often a semiquantitative technique.

5.3.5 Case Study

Buckinbah Creek Watershed, New South Wales, Australia A very interesting and convincing ^{14}C dating example has been undertaken by Kazemi (1999) in the Buckinbah Creek Watershed, BCW (see also Section 4.6 for more information). Located within the Macquarie River Basin of eastern Australia and 330 km northwest of Sydney, BCW is 80 km^2 in area (Figure 5.2) and has experienced some degree of dry-land salinization. Geologically, BCW is composed of faulted volcanic, volcanoclastic, and carbonate rocks with a clay-rich soil profile. Groundwater in the watershed occurs in the shallow regolith, which locally is very saline, and in the underlying, deep, regional fractured aquifer.

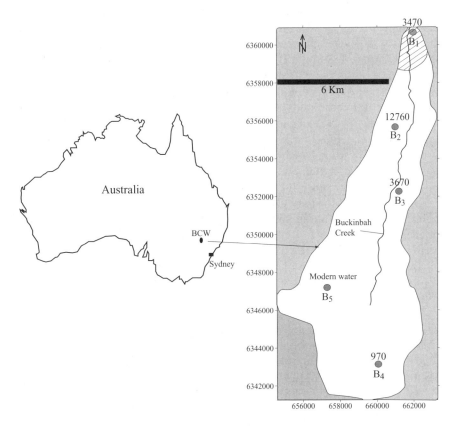

Figure 5.2. Carbon-14 groundwater ages (in year) in the deep aquifer, Backinbah Creek Watershed, eastern Australia. Upstream of B_1 (dashed area) creek disappears in the limestone causing ^{14}C age of the groundwater to be much lower than otherwise should be. Age of B_5 was estimated at modern, which was confirmed by its CFCs content and its low depth (from Kazemi, 1999).

TABLE 5.2. The Results of Groundwater Dating with ^{14}C in the Buckinbah Creek Watershed, Eastern Australia

Sample	B_5	B_4	B_3	B_2	B_1
Depth (m)	16.7	30	50	33	>20
^{14}C (YrBP ± 1σ)	Modern < 200	970 ± 110	3,670 ± 130	12,760 ± 300	3,470 ± 140
δ^{13} C (‰PDB)	−14.8	−5.3	−13.9	−6.4	−10.4
^{14}C (pmC ± 1σ)	98.8 ± 1.2	88.6 ± 1.2	63.3 ± 1	20.4 ± 0.8	64.9 ± 1.2
EC (µS/Cm)	3,770	3,405	1,647	1,269	1,046

Source: Kazemi, 1999.

Hydrogeochemical and isotopic studies have shown that these two aquifers have no or minimal interaction (Kazemi and Milne-Home, 1999). Groundwater samples from the deep (B_1 to B_4) and shallow aquifers (B_5), collected in 1997 and 1998, were analyzed for ^{14}C (Table 5.2) and some for CFCs as discussed in Chapter 4. Sample B_5, analyzed twice for CFCs content, was dated by CFCs method as older than 1975. It has a CFC-12 concentration of less than 100 pg/kg and CFC-11 concentration of 24 pg/kg. As seen from the table, B_5 has correctly been dated as modern by ^{14}C method as well. Shallowness of B_5 is another check to assure the correctness of the ^{14}C dating exercises. Sample B_4 had no CFCs in it, something in perfect agreement with the ^{14}C age of 970 years. BCW study has shown that the two dating methods, ^{14}C and CFCs, fully support and complement each other. Further, it shows that multitracer dating will greatly increase the hydrogeologists and groundwater professionals' confidence. Also, for the ^{14}C ages, minimum correction procedure has been applied because they agree very well with the hydrogeological settings and with each other.

It is seen from Figure 5.2 that the age of the deep groundwater increases from B_4 (located at the top of watershed) to B_2 (located close to the outlet), representing a normal situation. However, the age of B_1 is less than that of B_2 due to the diluting effect of recharge to the groundwater by creek water in this area. Buckinbah Creek disappears midway between B_2 and B_1 into nearby limestone cavities (hatched area in Figure 5.2) and mixes with groundwater around B_1. The interesting point here is that there is a negative correlation between the ages of deep groundwater samples and their salinity levels (albeit excluding B_1 due to the reason mentioned). This is due to the precipitation of some minerals such as kaolinite, alunite, and gypsum as well as dilution by recharging low-salinity water as groundwater travels along the flow path. The dry-land salinity, a secondary manmade process that has affected the upper parts of the BCW to a greater extent, as compared to the lower reaches, may also play a role. It should be added here that groundwater age shows a positive correlation with groundwater salinity (the higher the salinity, the longer the age) in the semi-arid region of Cariri region, Brazil (Mendonca et al., 2005), something that is naturally expected.

5.4 INDIRECT METHODS

As explained in Chapter 4, some approaches for determining the age and residence time of groundwater have, in this book, been classified as indirect methods. This is because these techniques do not directly yield a groundwater age as such. In addition to sampling and study of groundwater, these methods usually require some other measurements and tools such as rainwater sampling, rock mineralogical studies, climatic information, and reference ages obtained by other methods to provide reliable semiquantitative age estimates. For dating old groundwaters, stable isotopes of water (deuterium and oxygen-18) and conservative and reactive ions are discussed here. However, some basic principles may help to identify old groundwaters. These include confining units (confined aquifer are usually old), depth (deep aquifers are probably old), permeability (the lower the permeability, the older the age), chemically oxidizing or reducing (oxidizing groundwaters are young and reducing groundwaters are old), and salinity (high-salinity waters in a similar situation are older than low-salinity waters), etc. If one aquifer has all the indication above, i.e., it is deep, confined, chemically reducing, saline, and not very permeable, then it is likely old.

5.4.1 Deuterium and Oxygen-18

The Quaternary Period, comprising the Pleistocene and Holocene eras, started 1.6 million years ago with Holocene covering the shorter part of last 10,000 years (Table 1.2). Evidence from deep sea sediments, continental deposits of flora, fauna, and loess, and ice cores show that during the late Quaternary period (the past one million years), a series of large glacial-interglacial cycles, each lasting about 100,000 years, shaped the Earth's surface climate (Imbrie et al., 1992). The atmospheric precipitation in each of these cycles is unique as far as its deuterium and oxygen-18 content are concerned (Figure 5.3). Because water isotopes are near conservative, they preserve the signature of atmospheric condition and carry it to the subsurface. Therefore, they can help reveal the origin and age of groundwater. The basis of age determination of old groundwater is this simple reason: "Pleistocene-recharged groundwaters are isotopically lighter than Holocene and modern groundwater". For example, old European groundwaters are depleted in deuterium by ~12% when compared to modern infiltration waters (Rozanski, 1985), and in the Great Hungarian Plain, modern groundwaters possess an oxygen-18 values of more positive than −10% (Deak and Coplen, 1996). Also, old groundwaters (demonstrated by ^{14}C values of less than 20 pmC and tritium values of close to background) in the central plain and littoral plain portions of the North China Plain are isotopically 11% lighter than younger piedmont plain groundwaters (Zongyu, 2005). In addition to the above difference, old groundwater samples from a given aquifer are isotopically less variable as compared to the shallow groundwater samples, which reflect the seasonal variations in the

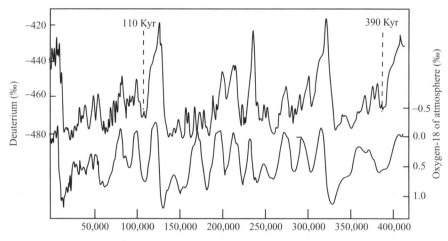

Figure 5.3. Deuterium content of atmospheric precipitation and oxygen-18 content of atmosphere O_2 during glacial-interglacial cycles. The deuterium and oxygen-18 values are based on the measurements from ice cores and entrapped air, respectively (modified after Petit et al., 1999).

isotopic composition of precipitation. Clark and Fritz (1997) also try to distinguish between isotopic composition of old groundwaters in the arid and in the temperate regions. They argue that paleoclimate effect in the arid region is manifested not only in the depleted isotopic composition, but also in the displaced position of meteoric water line (MWL), which has a deuterium excess of 15% to 30% (compared to deuterium excess of 8 for WMWL).

Some Case Studies, Confined Aquifer of Triassic Dockum Group Sandstone, Southern Great Plains, Texas and New Mexico, USA Dutton and Simkins (1989) document a case of a confined aquifer in New Mexico and Texas where deep, old groundwater of Dockum Group is isotopically lighter than modern water of Ogallala and Trinity and Edwards shallow aquifers (Figure 5.4). The Dockum Group, 610 m in thickness, is divided into lower Dockum with sandstone and conglomerate beds as the main lithology and upper Dockum, which is muddy and acts as the confining bed for the confined lower Dockum aquifer. It is overlain by Miocene Ogallala and Blackwater Draw Formations and Edwards and Trinity Group, which together form the High Plains aquifer. Isotopic analysis by Dutton and his colleague showed that mean deuterium (−59‰) and oxygen-18 (−8.3‰) content of Dockum water are significantly lighter than mean deuterium (−41‰) and oxygen-18 (−6.2‰) of groundwater in the Ogallala Formation. Depleted deuterium and oxygen-18 values of Dockum are a reflection of a −3°C cooler recharge water, compared to the Holocene or present-day recharge. After examination of the isotopic data, geomorphic evidence, and limited ^{14}C ages, Dutton and Simkins concluded that high paleorecharge elevations of 1,600–2,200 m (700–1,000 m higher than

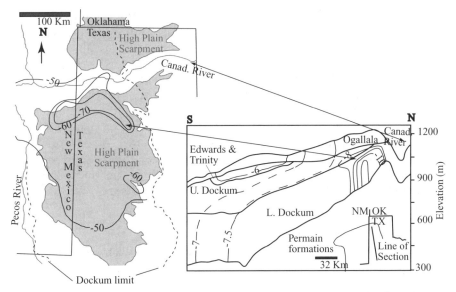

Figure 5.4. Deuterium values (%) in lower Dockum groundwater (left) and cross section of oxygen-18 values (%) of High Plains aquifer and Dockum group (right), Texas and New Mexico, USA (modified after Dutton and Simpkins, 1989).

present-day recharge areas) and Pleistocene climatic variation led to isotopic lightness of Dockum Group aquifer. The age of groundwater in Dockum was then considered to vary widely, between 11,000 to 200,000 years. This case study demonstrates that old groundwaters are isotopically lighter than young groundwaters.

Groundwater in Cone Mine, Yellowknife, Northwest Territories, Canada
Similar to the Dockum Group, groundwater in the deep boreholes in Cone Mine, Yellowknife, Northwest Territories, Canada, are isotopically lighter (oxygen-18 = −28%) than modern groundwater (oxygen-18 = −18.9%). The origin of heavy isotope waters was investigated by Clark and colleagues (2000) to be glacial melt water injected into the subsurface during the ablation (retreat) of the Laurentide Ice Sheet some 10,000 years ago.

Deep-Confined Basalt Aquifers of the Pullman-Moscow Basin of Washington, USA The oxygen-18 values of these basalt aquifers are depleted by 0.4 to 4.9 per mil relative to the overlying unconfined shallow, recently recharged groundwaters (Larson et al., 2000). This suggests that groundwater in the basalt is of Pleistocene age, something that has been also proved by earlier radiocarbon dating studies.

San Joaquin Valley, Central California Davis and Coplen (1987) documented a special case in the San Joaquin Valley (surrounded by Sierra Nevada to the northeast and the Coastal Ranges to the southwest), central California, where old groundwaters ($\delta^{18}O = -9\%$), located deeper, are isotopically heavier than young groundwaters ($\delta^{18}O = -11\%$). This unusual situation has been attributed to the tectonic uplift of the Central Coastal Ranges by about 1,100 m in the past 615,000 years (mid-Pleistocene). The higher altitude resulting from the uplift brought about a cooler climate, compared to early Pleistocene, which in turn led to isotopically lighter precipitation. In addition, before uplift, the atmospheric precipitation from the Sierra Nevada with an oxygen-18 value of about -10% used to mix in intermittent lakes (oxygen-18 = -6%) before being recharged to the aquifer; this changed after uplift. Therefore, waters recharged before 615,000 years ago were found to be isotopically heavier due to the then-lower recharge altitude and mixing with heavy lake waters.

5.4.2 Conservative and Reactive Ions

Conservative ions can be employed to date groundwater where there is a sharp difference between the concentration of these ions in the local rainfall and in groundwater. These ions do not participate in the geochemical reactions, and hence their concentration in the groundwater is largely controlled by the atmospheric input and by the dissolution of in situ minerals. In addition, the concentration of conservative ions in the atmospheric precipitation varies through time because of change in the amount of aerosols in the atmosphere (such as increase due to industrial activities) and because of variations in the source and quantity of rain. If the concentration of a specific conservative ion such as chloride in the groundwater is less than that in the local rainfall, then the possibility that the given groundwater originated from a different type of rain is high. Such rainfall occurred under wetter climatologic condition associated also with less evaporation than the present ones. It is also possible that the rain-bearing, moisture-laden clouds as the source of such rainfall were originated from a different water body with its own specific characteristics. To date groundwater, we need to identify the historic or geologic time interval during which these types of atmospheric, as just discussed, precipitations were prevalent. We then can relate the age of the groundwater to these time intervals. However, it is obvious that we cannot actually calculate an age for the given groundwater sample, but rather we can provide a rough timeframe, i.e., whether the sample is late Pleistocene, early Holocene, etc. Good information about the region's paleoclimate and its effect on the groundwater recharge is needed to have a better estimate of the age range of the given aquifer. The ratio of various conservative ions such as Br/Cl, I/Cl, F/Br are also helpful because these ratios are known and well-defined for different types of waters (rainwater, seawater, etc.). The advantages of using conservative ions as age-tracers are that (1) the sample collection is straightforward and cheaply analyzable and (2) the laboratory facilities are worldwide.

In contrast to the conservative ions, reactive ions are involved in geochemical reactions when present in groundwater and their concentration is, therefore, time-dependent. Those reactive ions that do not precipitate under aquifer conditions (i.e., they have high solubility), their quantity in the groundwater system increases over time because of water–rock interaction. Therefore, the higher the concentration of these ions, the longer the residence time of the ambient groundwater. On this basis, we should be able to find a relationship between the age of the groundwater and the concentration of the reactive ions. Not many elements are eligible to be considered for this purpose; only those with high solubility and known-simple geochemistry such as lithium, rubidium, and strontium are of potential candidates. Of course, few independent checks such as radiocarbon ages are needed to translate the concentrations of reactive ions into ages, i.e., calibrate the data. As compared to the conservative tracers, reactive tracers, which are mostly trace elements, need specialized sampling and filtering techniques and analyzing facilities. A large volume of data set and well-defined hydrogeological conceptual model is also required if one needs to have reliable age estimates. However, this method of age-dating, if widely adapted and fine-tuned, could find appreciable potential in the groundwater dating discipline because of the simpler principles, cheaper costs of analysis, and less interfering-limiting reactions as compared to isotopic techniques. The age range for which this method is applicable is not quite clear as yet, but it could fall within the upper level of old age range and lower level of very old age range. These estimates are based on one case study from the UK, described next.

Case Study in the UK The first attempt to use conservative and reactive ions to quantitatively date groundwater is the study by Edmunds and Smedley (2000) for the East Midlands Triassic (Sherwood sandstone) aquifer in the UK. Sherwood sandstone, a coarse to very fine-grain typical red bed sandstone, is 120–300 m in thickness and contains quartz and K-feldspar as well as post-depositional dolomite and secondary calcites, which exert a strong influence on the chemistry of the ambient groundwater. It is a single hydraulic unit and is confined from above by a thick sequence of mud rocks and from below by a sequence of Permian mudstones, marls, and dolomitic limestones. The formation waters from the confining layers are more saline than the aquifer water and, therefore, easily distinguishable in the case of leakage. The hydraulic conductivity of Sherwood sandstone at outcrop is 3.4 m/day, but historic and present-day flow velocities in the main aquifer are 0.6 and 0.2 m/year, respectively. This suggests a reduction in the flow velocity downgradient, which results in less recharge to the deeper part of the aquifer.

Edmunds and Smedley (2000) measured routine field parameters, major ions, NO_3^-, NH_4^+, F, Br, I, trace elements, ^{18}O, ^2H, ^{13}C, and ^{14}C of 45 samples collected from 31 sites from Sherwood sandstone. They also had accessed and re-evaluated a number of chemical parameters data sets that were compiled

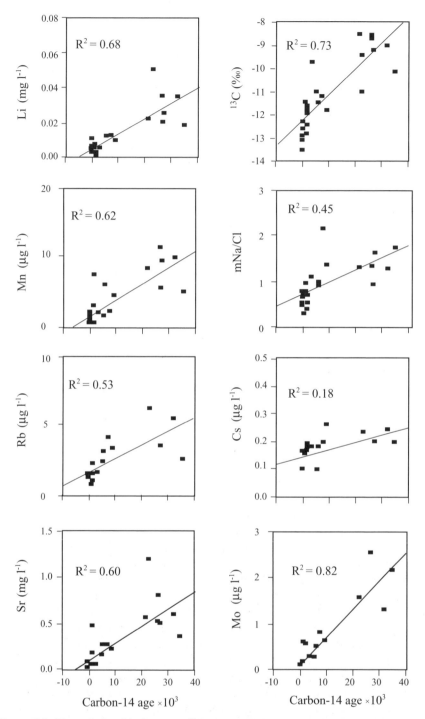

Figure 5.5. The relationship between ^{14}C ages and the concentration of reactive ions in East Midland Triassic sandstone aquifer (Sherwood sandstone), UK. Note that all horizontal axes are ^{14}C ages such as in the bottom diagrams (reproduced from Edmunds and Smedley, 2000).

INDIRECT METHODS

during the earlier studies of the same aquifer. Higher concentration of chloride and other halogens ions in present-day rainfall compared with that in the East Midland aquifer enabled Edmunds and his colleague to define the boundary between old groundwaters and young groundwaters. Old groundwaters of the early Holocene or late Pleistocene age are characterized by low chloride (≤ 10 mg/L) and light deuterium and tritium content. In contrast, young groundwaters have high chloride concentration and show signs of pollution due to industrial activities. The ratio of other halogens such as Cl/Br, F/Cl, etc. has also helped the interpretations. As shown in Figure 5.5, there is a relative correlation between the concentration of five reactive trace elements and groundwater ages obtained by radiocarbon. Equation 5.2 was then developed to deduce the age of the groundwater samples by taking into account the slope and the interception of the regression line (Figure 5.5) as well as the concentration of the element:

$$\text{Chemical age} = [\text{element concentration} - b_0]/b_1 \quad (5.2)$$

where b_0 is the intercept and b_1 is the slope of the regression line. For example, for a sample whose Rb concentration is $5.53 \mu g l^{-1}$, the Rb age would be 38×10^3 years:

$$\text{Rb age} = (5.53 \mu g l^{-1} - 1.69 \mu g l^{-1})/0.101 \mu g \ 10^{-3} \text{ years} = 38 \times 10^3 \text{ years}$$

The intercept and the slope of the regression line for each element were obtained from Figure 5.5. To obtain an age for each sample, the calculated chemical ages from all five elements were averaged (cesium was not used in the calculations because of very low R^2 of 0.11). Finally, it was concluded that the freshwater of the aquifer could have an age of up to 100,000 years. Note that not all samples were plotted on Figure 5.5 because the range of carbon-14 ages (up to 40,000 years) did not allow more saline samples to be plotted.

TABLE 5.3. Advantages and Disadvantages of Various Methods

Method	Advantages	Disadvantages
^{32}Si	Could cover the age gap	Complex geochemistry
	Could be used as a check for other tracers	Limited analytical facilities worldwide
	Well-known atmospheric input	Great loss in the unsaturated zone
		The method is almost discarded
^{39}Ar	Could cover the age gap	Sampling and analysis difficulty
	Relatively well-known C_0	Requires large volume of sample
	Limited underground production except in igneous rocks	Limited laboratories worldwide
		Possibility of introduction of error
	A check for subsurface production of other tracers such as ^{85}Kr and ^{222}Rn	Underground production in igneous rock aquifers
^{14}C	Well established	Initial value problem
	Worldwide analytical facilities	Complex geochemical interpretation
	Wide age range	Semiquantitative nature
	The only well-recognized method to date old groundwaters	Possibility of large differences in results by various models
Indirect methods	Worldwide analytical facilities	Little known
	Wide age range	Rough and semiquantitative
	As a check on other dating method	Applicable in special circumstances
	Data can be used for other purposes	Large volume of data sets required

6

AGE-DATING VERY OLD GROUNDWATERS

This chapter reviews the methods to estimate quantitatively the age of very old groundwaters (beyond ^{14}C dating range to tens of millions of years) including ^{4}He, ^{36}Cl, ^{40}Ar, ^{81}Kr, ^{129}I, and uranium disequilibrium series, especially the ^{234}U/^{238}U method. The principles, limitations, requirements, relevant equations, advantages, case histories, field sampling, and laboratory measurements of each method are described. Included in this chapter are the specifications of the potential aquifers that these methods are applicable to, such as deep sedimentary basins, regional flow systems, and groundwater in the Earth's crust. It also includes reasons why quantitative age measurements of very old groundwaters have not been quite successful as yet. Table 6.7 lists the advantages and disadvantages of each method. Note that ^{4}He is also used for dating young groundwaters (Chapter 4), and some other potential—yet to be proven—very old dating methods such as that based on ^{87}Sr/^{86}Sr are briefly discussed in Chapter 8. With regard to ^{36}Cl and ^{4}He methods, there are substantial reviews by Phillips (2000) and Pinti and Marty (1998), Solomon (2000), Mazor (2004), respectively, which are suggested for consultation if in-depth understanding of these methods is sought. For all aspects of noble gases including groundwater dating, Porcelli et al. (2002) is recommended for study and research purposes.

Groundwater Age, by Gholam A. Kazemi, Jay H. Lehr, and Pierre Perrochet
Copyright © 2006 John Wiley & Sons, Inc.

6.1 KRYPTON-81

General information about krypton and its isotopes is found in Section 4.5, where ^{85}Kr is described as a tracer to date young groundwaters. Krypton-81 has since long been proposed as a groundwater dating tool (Loosli and Oeschger, 1969). It has, however, not been used as a tracer in other hydrological fields.

6.1.1 Production of ^{81}Kr

Krypton-81 is produced in the upper atmosphere by cosmic-ray-induced spallation of five heavier Kr isotopes (from ^{82}Kr to ^{86}Kr), e.g.,

$$^{83}_{36}Kr + P \rightarrow {}^{81}_{36}Br + P + \alpha$$

and through neutron capture by ^{80}Kr:

$$^{80}_{36}Kr + n \rightarrow {}^{81}_{36}Kr + \gamma$$

So far, all research suggests that there is neither significant subsurface production nor an appreciable anthropogenic source for ^{81}Kr. It can, therefore, be viewed as a solely atmospheric tracer and an ideal one because of its long half-life, largeness of the atmospheric reservoir, low direct yield from spontaneous fission of ^{238}U (i.e., insignificant subsurface production), and shielding from other decay products by ^{81}Br (various references). Studies cited throughout this section suggest some possible diffusion of ^{81}Kr from the surrounding shale formations into the groundwater reservoirs.

The inventory and/or concentration of ^{81}Kr in different reservoirs are (sources: Collon et al., 2000; Du et al., 2003)

In the atmosphere: 6.4×10^{25} atoms
[^{81}Kr/Kr = $(5.2 \pm 0.4) \times 10^{-13} - (1.1 \pm 0.05) \times 10^{-12}$]
In modern groundwater: 1,100 atoms/L (= initial value for calculations of groundwater ages)
In the world's oceans: 1.2×10^{24} atoms   880 atoms/L

6.1.2 Sampling, Analysis, and Reporting the Results

Analyzing groundwater for its ^{81}Kr content is a huge task because there are fewer than 1,100 ^{81}Kr atoms in every liter of water. This low concentration represents an activity of only 1.1×10^{-10} becquerel, about one disintegration in every 300 years. Therefore, ^{81}Kr atoms in a water sample should be counted before they decay. Until now, groundwater samples have been analyzed for their krypton content through four methods. (Note that low-level counting, LLC, was initially used by Loosli and Oeschger in 1969 to detect ^{81}Kr and measure its abundance in the atmosphere; it is no longer considered as a measurement technique because of groundwater samples' low ^{81}Kr concentration.) All these methods have been used to a limited extent only. We therefore refrain from describing them in detail because their reliability has not been fully tested as yet and new techniques may come into use in the near future. The latest technique, Atom Trap Trace Analysis, or ATTA, is only briefly described. Similarly, sample requirements for ^{81}Kr analysis are not dealt with here because they vary depending on the analytical method chosen. Some methods require up to 20 m^3 of a water sample, while some others require only 2 m^3. Groundwater samples are analyzed for ^{81}Kr by

1. Laser resonance ionization mass spectrometry (RIMS) (Thonnard et al., 1987; Lehmann et al., 1991).
2. Photon burst mass spectrometry (PBMS) (Fairbank, 1987; Lu and Wendt, 2003).
3. AMS with a cyclotron (Collon et al., 2000).
4. Atom Trap Trace Analysis (ATTA). This method is explained in detail by Chen (1999) and Du et al. (2003). The basis of this latest technique, which may find widespread application in the future, is on selectively capturing the atom of the isotope of interest by a magneto-optical trap and detecting it by observing its fluorescence. The trapped atoms are then counted. A control isotope such as ^{83}Kr or ^{85}Kr is simultaneously counted by the system to subtract the noise and to act as a reference to increase the reliability and accuracy of the analysis.

The results of ^{81}Kr analyses are reported in a fashion similar to reporting deuterium and oxygen-18 isotopes. The ratio of water isotopes (^{2}H, ^{18}O) in a sample is compared to that of Standard Mean Ocean Water (SMOW). In the case of ^{81}Kr, the ^{81}K/Kr ratio in the water sample is divided by the same ratio (i.e., ^{81}K/Kr) in the air, so the result would be as follows:

$$\frac{R}{R_{air}} = \frac{\left[^{81}Kr/Kr\right]_{sample}}{\left[^{81}Kr/Kr\right]_{air}}$$

This ratio can be directly inserted into the radioactive decay equation as C/C_0 to yield ages. It is similar to percent modern carbon unit. Krypton-81 analysis can also be reported in atoms/L and ^{81}K/Kr. These units are convertible to each other if we know the concentration of krypton in the sample.

6.1.3 Age-Dating Groundwater by ^{81}Kr

Once atmospheric precipitation enters the aquifer through rainfall or snowfall recharge, krypton-81 starts decaying to bromide-81, a stable isotope of bromine.

$$^{81}_{36}Kr + {}^{0}_{-1}e \rightarrow {}^{81}_{35}Br + {}^{0}_{0}\nu$$

Therefore, as time goes by, the concentration of ^{81}Kr in the groundwater decreases in accordance with the radioactive decay principle. To age-date a groundwater sample, one needs the initial concentration of ^{81}Kr (Kr_0) in the recharge water and its present concentration in the groundwater (Kr), i.e., $^{81}Kr = {}^{81}Kr_0\, e^{-\lambda t}$.

Example: Concentration of ^{81}Kr in a groundwater sample is 900 atoms/L. What is the age of this sample?

$$900 \text{ atoms/L} = 1{,}100 \text{ atoms/L} \times e^{-\lambda t} \rightarrow$$

$$t = -\ln\frac{900}{1{,}100} \Big/ \lambda = -\ln\frac{900}{1{,}100} \Big/ 3.03 \times 10^{-6} = 66{,}297 \text{ years}$$

Example: If krypton-81's R/R_{air} of a groundwater sample is 40%, what is the age of this sample?

$$t = -\frac{\ln 0.4}{\lambda} = -\frac{\ln 0.4}{3.03 \times 10^{-6}} = 302{,}722 \text{ years}$$

Krypton-81 technique suits dating groundwaters whose age ranges from 35,000 to 761,000 years, assuming 10% and 90%, respectively, of the initial atoms are decayed to ^{81}Br. However, these ranges need to be verified in practice.

6.1.4 Advantages and Disadvantages

Advantages
1. Subsurface and anthropogenic sources of ^{81}Kr are minimal and trivial.
2. Chemical reactions involving ^{81}Kr are limited because of its inertness.
3. It could be the only reliable method to quantitatively date very old groundwaters.

Disadvantages
1. The method, although proposed long ago, has not received widespread application because of the technical difficulty in analyzing groundwater samples.
2. At present, worldwide laboratory facilities for analysis are limited to one to two institutions.
3. Worldwide, there are not many extensive deep regional freshwater aquifers, similar to the Great Artesian Basin of Australia, to which this technique can be applied.

6.1.5 Case Studies

Three aquifers have so far been investigated by ^{81}Kr dating technique:

Milk River Aquifer, Canada Milk River aquifer is an extensively studied (*Applied Geochemistry*, Volume 6:4), 30–60-m thick Cretaceous sandstone located within the Milk River formation, southern Alberta, Canada. It is a deep regional aquifer confined from above and below by thick shale formations. The dominant recharge area for the aquifer is Sweetgrass Hills, Montana, where described Cretaceous sandstone crops out. In the dating study by Lehmann et al. (1991), only one sample from well no. 9 of this aquifer was successfully analyzed for ^{81}Kr content though four samples were collected and processed for this purpose. This sample had ^{81}Kr/Kr (R/R_{air})% of 82 ± 18%. The age of this sample calculated on the basis of the presently known half-life of 229,000 years would be about 66,000 years:

$$t = -\ln 0.82 / 3.03 \times 10^{-6} = 65,564 \text{ years}$$

If the lower limit of the ^{81}Kr/Kr ratio is taken (i.e., 82% − 18% = 64%), then the age would be 147,000 years and the upper limit of the ratio (i.e., 82% + 18% = 100%) would give a zero age.

$$t = -\ln 0.64 / 3.03 \times 10^{-6} = 147,443 \text{ years}$$

However, if the then-accepted ^{81}Kr half-life of 210,000 years is used in the calculations, the upper limit of the age of the sample would reduce to 135,000. This is slightly lower than Lehmann and his colleagues's estimate of 140,000 years, showing a small inaccuracy in their calculations. ^{36}Cl and hydrodynamic-derived ages for the same sample are about 100,000 years.

Great Artesian Basin, Australia Great Artesian Basin (GAB) is one of the world's few aquifers that has been studied by all dating techniques. Collon et al. (2000) measured the ^{81}Kr/Kr ratio of four samples from the GAB at [1.54, 1.7, 2.19, and 2.63] × 10^{-13}. By taking into account an initial value of 5.2×10^{-13}, ages of 402,000, 354,000, 287,000 and 225,000 years, respectively, were obtained (with error bar of 15%). These ages are shorter than previously measured ^{36}Cl ages. Lehmann et al. (2003) used ^{81}Kr ages of this study to calibrate ^{36}Cl and ^{4}He dating methods and to gain an insight into the rate of the underground production of these tracers. They found that for two samples, whose initial ^{36}Cl values are high, ^{4}He accumulation rate is low (0.2×10^{-10} cm^3 STP/cm^3 water year). Therefore, it is possible to date these samples using the ^{36}Cl dating technique. Nevertheless, for the other two samples with low initial ^{36}Cl, ^{4}He production rate is 10 times higher, and ^{36}Cl groundwater dating is problematic. Therefore, it was concluded that the success of ^{4}He and ^{36}Cl dating techniques depends on the dominant process by which the concentration of these isotopes in the aquifer is controlled.

Nubian Sandstone Aquifer, Sahara, Egypt (Sturchio et al., 2004) The Nubian Aquifer in the Dakhla Basin of the western desert of Egypt underlies the very arid northeast corner of the Sahara Desert. It contains ~50,000 km^3 of groundwater in a thick (up to 3,000 m) sandstone formation with thin interbedded shales. Six samples from this aquifer, analyzed for ^{81}Kr content by Sturchio et al. (2004), yielded ^{81}Kr/Kr (R/R_{air})% ratios of 52.6, 36.5, 4.8, 30.6, 22.8 and 12.8. With a simple radioactive decay model (no subsurface addition, no anthropogenic sources), these ratios results in ages of [2.1, 3.3, 10, 3.9, 4.9, and 6.8] × 100,000 years, respectively. Four of these ages showed very good agreement with ^{36}Cl/Cl ratios of the samples, i.e., the higher the ^{36}Cl/Cl ratio of the samples, the lower the ^{81}Kr ages. ^{36}Cl/Cl values of the other two samples were either affected by the initial value problems or by subsurface production of ^{36}Cl. Krypton-81 ages were also supported by the hydrogeological setting of the aquifer and by hydrodynamically estimated ages. This study, which was highly promoted as the first study to demonstrate the existence of a one-million-year-old groundwater, is also the first of its kind to use the ATTA technique to measure ^{81}Kr content of the groundwater samples.

6.2 CHLORIDE-36

A halogen and a World War I war gas, chlorine was discovered by Carl Wilhelm Scheele in 1774. (Note that chlorine is the element and chloride is the ion. In nature, chlorine is found only as the chloride ion. Hence, chloride and chlorine are often used interchangeably.) It has an atomic weight of 35.45 and an atomic number of 17. Chlorine is abundant in nature, is necessary to most forms of life, and is a powerful oxidizing, bleaching, and disinfecting agent. About 1.9% of the mass of seawater is chloride ion. Chlorine can be manufactured via the electrolysis of a sodium chloride solution such as a brine. Compounds of chlorine include chlorides, chlorites, chlorates, perchlorates, and chloramines. In natural waters, the chloride ion behaves conservatively, does not participate in many geochemical reactions, is highly soluble and mobile, and often constitutes the most dominant ionic species.

Chlorine has 16 isotopes whose mass numbers range from 31 to 46, but only three of these occur naturally. The remaining 13 have half-lives shorter than 1 hour and this is why they do not occur in nature. The three abundant isotopes of chlorine include two stable isotopes, chlorine-35 and chlorine-37, with 75.53% and 24.47% abundances, respectively (Marques et al., 2001), and one radioactive isotope, ^{36}Cl, with a half-life of 301,000 ± 2,000 years (some references give ± 4,000 years; e.g., Attendorn and Bowen, 1997). The natural mixture of these three isotopes makes up environmental chlorine. Chlorine-36, in this book, is treated as a tracer to age-date young (Chapter 4) and very old groundwaters only. However, it has a number of other scientific applications such as geochronological study of young volcanic rocks (Leavy et al., 1987), estimation of groundwater recharge (Cook and Robinson, 2002; Cecil et al., 1992), calculation of dispersivity (Cecil et al., 1999), mixing study of water bodies (Calf et al., 1988), dating of ice and sediments, vadose zone tracing, salinity balance studies, and geothermal system research (Cecil, 2000; Phillips, 2000). A complete coverage of the ^{36}Cl subject is given in the Ph.D. thesis by L. D. Cecil (2000), which can be accessed by contacting the University of Waterloo or the author at the USGS. Cecil's thesis shows that up to 1997, the list of global research using ^{36}Cl as a tracer in various geologic and hydrologic environments stood at 60 references.

6.2.1 Production of ^{36}Cl

Chlorine-36 is produced through the following processes. Some of these processes are illustrated in Figure 6.1, which also shows other sources of ^{36}Cl in groundwater.

1. In the atmosphere, by cosmic-ray splitting of ^{40}Ar and neutron activation of ^{36}Ar,

$$^{40}_{18}Ar + P \rightarrow {}^{36}_{17}Cl + n + \alpha$$
$$^{36}_{18}Ar + n \rightarrow {}^{36}_{17}Cl + P$$

The amount of ^{36}Cl produced through these processes depends mainly on the geographic latitude and is greater in middle latitudes. For example, the fallout rate of ^{36}Cl at Stripa in Sweden at 57°N latitude is 14 ± 3 atoms m^{-2}s^{-1} (Bowen, 1988); the mean atmospheric flux of ^{36}Cl over the United States is 19.6 ± 4.5 atoms m^{-2}s^{-1}; and the mean global ^{36}Cl flux is 30.5 ± 7 atoms m^{-2}s^{-1} (Moysey et al., 2003). It is estimated that splitting of ^{40}Ar and activation of ^{36}Ar are responsible for 67% and 33% of total natural atmospheric production of ^{36}Cl, respectively (Cecil, 2000). Some ^{36}Cl are also injected into the atmosphere by transportation of marine aerosols. The residence time of ^{36}Cl in the atmosphere is about one week. In addition to spatial variations, atmospheric ^{36}Cl production may have also varied temporally over long time scales (Love et al., 2000).

2. Weapon testing of fusion devices starting in late 1952 and extending through mid-1958 has led to the production of up to 70,000 atoms m^{-2}s^{-1} (Bentley, 1986). The reaction occurring through this activity is

$$^{35}_{17}Cl + n \rightarrow {}^{36}_{17}Cl + \gamma$$

Chlorine-36 produced by this phenomenon is used as a marker to age-date young groundwaters that are less than 50 years old (see Chapter 4).

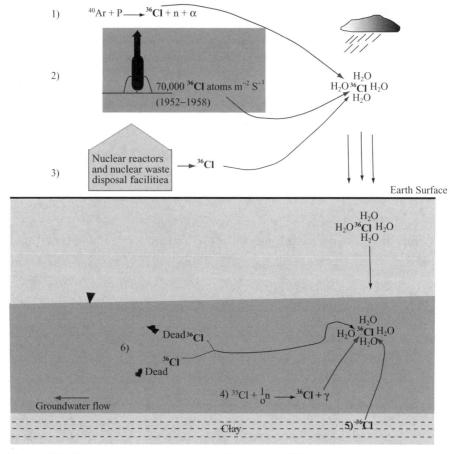

Figure 6.1. Aboveground and underground sources of ^{36}Cl and how rainwater obtains its ^{36}Cl content as it enters and moves in the aquifer.

3. From nuclear power and nuclear fuel re-processing facilities. Chlorine-36 together with iodine-129 are the only two elements that can escape from the nuclear fuel waste disposal repositories and enter the surrounding environment. Almost all other radionuclides will remain trapped, either in the fuel itself or within a very short distance of it (Wiles, 2002). It has also been recently suggested that the operation of two nuclear reactors in the United States has led to the production of ^{36}Cl (Davis et al., 2003).

4. By neutron activation of stable ^{35}Cl at the Earth's surface or in the shallow subsurface (epigenic = insignificant) and in the deep subsurface (nucleogenic = significant):

$$^{35}_{17}Cl + n \rightarrow {}^{36}_{17}Cl + \gamma$$

The rate of production of ^{36}Cl through this process varies depending on the type of rocks, minerals, and solutions, and the availability of neutron sources (neutrons are produced as a result of bombardment of rocks with alpha particles as well as through spontaneous fission of ^{238}U). In general, it is higher in rocks with higher uranium and thorium contents such as uranium ore deposits and is lower in lithologies like basalt and sandstone. For instance, the ^{36}Cl production rate at the top of Murray Group Limestone was calculated by Davie et al. (1989) (using Equation 6.1) to be 0.09 atoms/cm^3/year.

$$P_{36} = \frac{^{36}Cl}{^{40}Ca} \times \frac{\rho}{MW} \times N_A \times \frac{\log_e 2}{t_{0.5}} \tag{6.1}$$

where P_{36} is the production rate of ^{36}Cl, $^{36}Cl/^{40}Ca$ is the equilibrium ratio of ^{36}Cl to ^{40}Ca, ρ is the density of the limestone, MW is the molecular weight of limestone, N_A is Avogadro's number, and $t_{0.5}$ is the half-life of ^{36}Cl. The equations to calculate the in situ production rate of ^{36}Cl in the aquifer are well documented; for a detailed description, the reader is referred to Bentley et al. (1986b) and Phillips (2000).

5. In basalt, rhyolite, sandstone, and carbonate rocks, the following reactions can contribute to in situ production of ^{36}Cl (Cecil, 2000):

$$^{39}_{19}K + n \rightarrow {}^{36}_{17}Cl + \alpha \quad \text{(Neutron capture reaction)}$$
$$^{40}_{20}Ca + \mu^- \rightarrow {}^{36}_{17}Cl + \alpha \quad \text{(Muon capture reaction)}$$

The latter reaction is of minor importance, especially in the higher altitudes (Zreda et al., 1991).

6.2.2 Sampling, Analysis, and Reporting the Results

To analyze water samples for ^{36}Cl content, the following steps are undertaken (Davie et al., 1989; Cecil, 2000; Davis et al., 2003): A specified volume of water sample, depending on its chloride concentration, is collected. The selected volume should contain a minimum of 1–2 mg of chloride, but up to 25 mg may be required. In some studies, sample volumes of up to 4 liters have been collected. If samples are too low in chloride, preconcentration must be undertaken. Collected samples are then prepared for target loading. This means preparation of about 8–10 mg of pure AgCl and pressing it into the copper sample holders. The samples are prepared for target loading through addition of $AgNO_3$ solution to the samples to precipitate AgCl. The precipitated (AgCl) is then purified by dissolving it with NH_4OH. Next, $Ba(NO_3)_2$ is added to precipitate unwanted sulfur as $BaSO_4$. HNO_3 is added next to neutralize the solution and in the last stage, the AgCl is reprecipitated with $AgNO_3$ for target loading. If the Cl$^-$ concentration of the sample is very low, a ^{36}Cl free carrier must be added to it. The carrier may be AgBr of very low sulfur and chlorine content (because the most common interference for ^{36}Cl measurements is from sulfur-36). The prepared water samples are then analyzed for ^{36}Cl content using an accelerator mass spectrometer (AMS). Usually standard and blank samples are tested to identify the reproducibility and accuracy of the measurements. Davie et al. (1989) report that ^{36}Cl measurements by AMS have good reproducibility of about 6% and a sensitivity for ^{36}Cl/Cl of a few parts in 10^{15}. Before 1979, ^{36}Cl in environmental samples was measured by counting beta-particle emissions during radioactive decay, a difficult practice due to the ^{36}Cl long half-life. This method required tens of grams of chloride and counting times as long as a week. With AMS the counting times is about 30 min with 10% precision. Sensitivity has also improved with AMS. β-counting methods have a sensitivity of about one ^{36}Cl atom in 10^{12} Cl atoms, and AMS methods have a sensitivity of about five atoms of ^{36}Cl in 10^{15} Cl atoms (some references claim 2 atoms in 10^{16} atoms; e.g., Phillips, 2000). Presently, cost of analysis for ^{36}Cl at PRIME Lab, Purdue University, is about 350 USD, not including sample preparation.

The measured chloride-36 is reported in two ways:

1. As a ratio of ^{36}Cl/Cl × 10^{15} (the number of ^{36}Cl atoms to the total number of chlorine atoms). The multiplication factor (10^{15}) is used because the actual concentration of ^{36}Cl is very small.
2. As atoms of ^{36}Cl per liter of water using the following equation (Davis et al., 2003):

$$^{36}\text{Cl (in atoms per liter)} = 1.699 \times 10^4 \times B \times R \quad (6.2)$$

where B is the concentration of chlorine in mg/l and R is the (^{36}Cl/total Cl) × 10^{15} ratio. Therefore, to determine the concentration of ^{36}Cl, the concentration of Cl⁻ in water samples must be measured with conventional methods such as silver nitrate titration, ion selective electrode, and ion chromatography.

Example: The ^{36}Cl/Cl × 10^{15} of a water sample is 150 and its chlorine concentration is 80 mg/L. What is the ^{36}Cl concentration in this sample?

^{36}Cl (in atoms per liter) = $1.699 \times 10^4 \times B \times R$

^{36}Cl (in atoms per liter) = $1.699 \times 10^4 \times 80 \text{ mg/L} \times 150 = 2.04 \times 10^8$ atoms/liter

6.2.3 Groundwater Dating by ^{36}Cl

The principle of this method is simple. It is based on the radioactive decay of ^{36}Cl in the subsurface groundwater system. Above the Earth's surface, ^{36}Cl (atmospheric ^{36}Cl) with an initial value, ^{36}Cl$_0$, enters groundwater by rainwater infiltration. After time t, it decays to ^{36}S and ^{36}Ar

$$^{36}_{17}\text{Cl} \rightarrow {}^{36}_{18}\text{Ar} + \beta^- + \overline{\nu}$$
$$^{36}_{17}\text{Cl} \rightarrow {}^{36}_{18}\text{S} + \beta^- + \overline{\nu}$$

and reaches a new concentration, ^{36}Cl, according to the decay equation

$$C = C_0 e^{-\lambda t}$$

If one knows the initial concentration, C_0, and the present concentration, C, then one can calculate the length of time that the ^{36}Cl has resided in the subsurface groundwater system. For example, if the concentration of ^{36}Cl in a groundwater sample is 6,000 atoms/liter and its initial concentration (in the infiltrating rainwater) was 8,000 atoms/liter, then the time "t" or age of the groundwater would be 124,926 years. This means that the water molecule that contains the ^{36}Cl in question entered the groundwater system 124,926 years ago. Chlorine-36 dating method is capable of dating groundwaters with an age range of 46,000 to 1,000,000 years if we assume that 10% and 90%, respectively, of the original ^{36}Cl atoms are disintegrated.

However, many inaccuracies and problems surround C_0 and C values, the initial and the present concentration of ^{36}Cl, respectively. We do not accurately know what the concentration of ^{36}Cl in rainwater was when it entered the subsurface system. This is the biggest obstacle for the ^{36}Cl dating method (Love et al. 2000). Davis et al. (1998) explain 6 ways to estimate the C_0 value and argue that all these ways contain weaknesses and inaccuracies. These approaches include

1. Calculation of theoretical cosmogenic production and fallout,
2. Measurement of ^{36}Cl in the present-day atmospheric precipitation and use it as C_0,
3. Assuming that shallow groundwater contains a record of C_0,
4. Extraction of ^{36}Cl from vertical depth profiles in desert soils,
5. Recovering ^{36}Cl from cores of glacial age, and
6. Calculation of subsurface production of ^{36}Cl for water that has been isolated from the atmosphere for more than one million years.

As with regard to the measured C, the situation is not satisfying, too. The assumption in using the decay equation is that the ^{36}Cl atoms that enter the subsurface groundwater system behave as an isolated packet when they migrate through the flow system (piston flow theory). As explained above, ^{36}Cl is not only produced in the atmosphere, but it is also produced in the subsurface. There are other subsurface sources of ^{36}Cl such as what is called dead chlorine: "chlorine present in the fluid inclusions, ancient formation waters, saline water from the compacting clays, and chlorine from salt beds and evaporites." In addition, subsurface mixing (mixing of low and high ^{36}Cl waters), cross-formational flow, diffusion between aquitard and aquifer, and dilution and evaporation processes complicate the task of finding which C should be used in the decay equation: the C measured in the laboratory or the C obtained when the contributions/effects of all the above factors have been eliminated.

It should be pointed out that apart from ^{81}Kr, ^{36}Cl is the only method to quantitatively age-date very old groundwaters using the radioactive decay principle. Also, the expectation was/is that ^{36}Cl plays an increasing role in dating very old groundwaters, especially for studying water dynamics in regional aquifers and in impervious rock formations (Gonfiantini et al., 1998). However, it is safe to report that only in a very few cases, groundwater ages were satisfactorily determined by this method. In many instances, the ^{36}Cl dating study was not successful due to many factors such as the uncertainty in the estimation of $^{36}Cl_0$, the effect of dilution/evaporation processes on the value of C, and the much longer half-life of ^{36}Cl as compared with the age of groundwater (Turner et al., 1995; Davie et al. 1989; Yechieli et al., 1996). Other counterproductive factors include relatively recent full introduction of the method (virtually after 1980), limited number of well-defined regional aquifers worldwide for which this method is appropriate, many analytical considerations and unavailability of equipment in a large number of countries, difficulty in interpreting the results (initial value problem, local interferences, many sources of ^{36}Cl, etc.), and inherent inadequacies (for example, Park et al., 2002, show that if chlorinity of the water sample exceeds ~75–150 mg/kg, the ^{36}Cl method cannot be used for groundwater dating). These may be reasons why an important textbook like that by Mazor (1991) does not include a section on ^{36}Cl to describe it as an age-dating tool, though the method was introduced before 1991. A valuable paper to describe the principles and the limitations of age-dating very old groundwater by ^{36}Cl is that by Park et al. (2002).

An Exceptional Case Study The age of groundwater in the Cigar Lake uranium ore deposit, Saskatchewan, Canada, was calculated using a principle different from radioactive decay (Cornett et al., 1996). Here, the concentration of ^{36}Cl in the matrix of ore deposit and in the groundwater within the ore deposit were measured. Equation 6.3 was then used to calculate the age of groundwater:

$$t = \frac{1}{\lambda}\left[1 - \frac{C_{gw}}{C_{ore}}\right]^{-1} \quad \text{(Cornett et al., 1996)} \quad (6.3)$$

where t is the age, λ is the ^{36}Cl decay constant, C_{gw} and C_{ore} are the $^{36}Cl/Cl$ ratio in groundwater and in deposit's rock matrix. The basis of Equation 6.3 is that ^{36}Cl content of groundwater increases (called ^{36}Cl ingrowth) when it comes in contact with the uranium ore body (uranium ore body produces a high amount of ^{36}Cl through neutron capture of ^{35}Cl and ^{39}K; see above). Therefore, the longer the contact time, the higher the concentration of ^{36}Cl in groundwater. In order for Equation 6.3 to be applicable satisfactorily, the following conditions must be met:

1. The background ^{36}Cl content of the groundwater before it comes in contact with the ore deposit must be subtracted from C_{gw}, so that only ^{36}Cl ingrowth is included in the calculations.
2. The ^{36}Cl production and groundwater flow occur continuously, i.e., they happen nonstop.
3. The ^{35}Cl is subject to the average neutron flux, i.e., heterogeneity in the ore and in the neutron flux are insignificant.
4. The ^{36}Cl input, from other sources into the ore or into the groundwater, is small.
5. Because this method is based on the transfer of chloride from the ore body into the groundwater, the movement rate of chloride must take place at the same rate of groundwater.

This way of dating groundwater by ^{36}Cl is similar to dating by ^{4}He and ^{40}Ar methods, which are described in the following sections.

6.2.4 Advantages and Disadvantages

Advantages
1. The laboratory facilities, though still expensive and limited, are, comparatively speaking, more widespread than ^{81}Kr or some other very old dating methods.
2. The long half-life of ^{36}Cl makes this method suitable for dating very old groundwaters.
3. Measured ^{36}Cl values may be used for other hydrologic applications, in addition to their use in dating.

Disadvantages
1. The too much, above-mentioned ground sources for ^{36}Cl as well as the initial value problem.
2. Underground production and sources and sinks for ^{36}Cl. This has prompted Phillips (2000) to suggest limiting the dating range of ^{36}Cl to 500,000- to 1,000,000-year-old groundwaters only.
3. There are not many aquifers for which this method can be applied, because it is suited for deep regional aquifers only.
4. This method is not applicable to saline waters with chloride concentration of more than 150 mg/L.
5. High cost of analysis and sample preparation, and inadequate laboratories worldwide.

6.2.5 Case Studies

Although the ^{36}Cl method was proposed in the 1960s, it has not seen considerable development and expansion. Before 1980, there was difficulty in meas-

TABLE 6.1. Case Studies Involving ^{36}Cl Dating Method

Nubian Sandstone aquifer, Sahara, Egypt	Sturchio et al., 2004
Great Artesian Basin, Australia	Bentley et al., 1986b; Torgersen et al., 1991; Love et al., 2000
Paleochannel groundwater, Kalgoorlie, Australia	Turner et al., 1995
Clare Valley, South Australia	Cook, 2003; Cook et al., 2005
Ngalia basin aquifers of central Australia	Cresswell et al., 1998 and 1999
Murray basin, Australia	Davie et al., 1989; Kellet et al., 1993
Milk River aquifer, Alberta, Canada	Phillips et al., 1986, Notle et al., 1991
USA nationwide	Davis et al. and Moysey et al., 2003
Nevada test site	Rose et al., 1997
Columbia Plateau flood basalts	Gifford et al., 1985
Aquia aquifer, Maryland Coastal Plain, USA	Purdy, 1991; Purdy et al., 1996
Carrizo Aquifer, southern Texas, USA	Bentley et al., 1986a
Hydrothermal waters—Eastern Clear Lake area—USA	Fehn et al., 1992
Northern Switzerland	Pearson et al., 1991
Midlands sandstone aquifer, UK	Andrews et al., 1994
Dead Sea	Yechieli et al., 1996
Mazowsze Basin, Poland	Dowgiallo et al., 1990
East Africa Rift Zone	Kaufman et al., 1990
Stripa site, Sweden	Andrews et al., 1986
Germany	Lodemann et al., 1997
Australia and Canada	Fabryka-Martin et al., 1987, 1988
Canada	Cornett et al., 1996

uring ^{36}Cl concentration. After that, there were problems of subsurface production, ^{36}Cl sinks, and initial value. Table 6.1 lists worldwide aquifers that have been subjected to dating by the ^{36}Cl method. From these aquifers, GAB and Milk River are the most studied aquifers as far as the ^{36}Cl method is concerned. It is interesting to note that nearly all publications dealing with ^{36}Cl dating have been produced by a group of authors, which may be an indication of the complexity of the topic and the need of teamwork to obtain compelling results. A brief introduction to some of the case studies follows Table 6.1.

Great Artesian Basin, Australia: Southwest Margin (Love et al., 2000) and Deep Jurassic Aquifer of Eastern Margin (Bentley et al., 1986b) Love et al. (2000) analyzed ^{36}Cl in groundwater samples from 21 flowing and pumped boreholes from the southwest margin of the GAB. The data and the calculated ages based on three different equations (6.4, 6.5, 6.6) are given in Table 6.2. After applying the three different correcting schemes and reaching large discrepancies, Love and his colleagues (p. 1572) support what we stated in the previous section: "estimates of absolute groundwater ages using ^{36}Cl data are subject to large uncertainties because of problems of estimating the initial con-

TABLE 6.2. Results of Groundwater Dating by ^{36}Cl in the Southwestern GAB, Australia

Sample	^{36}Cl/Cl × 10^{-15}	^{36}Cl × 10^6 atoms/L	Age × 1,000 years[1]	Age × 1,000 years[2]	Age × 1,000 years[3]
Northern Transect (NW to SE)					
Lambina Homestead	129	1,888			
Warrungadinna	132	2,520			
Lambina Soak	108	2,026			
Marys Well 3	102	1,162	94	58	
Murdarinna 2	109	1,186	63	47	190
Midway Bore	89	934	160	160	290
Oodnadatta Town Bore 1	52	594	410	380	490
Watson Creek 2	25	439	790	550	620
Duckhole 2 (24701)	31	444	670	540	620
Appatinna Bore*	134	780			
Southern Transect (W to E)					
C. B. Bore	115	1,030			
Ross Bore	54	1,478			
Evelyn Downs Homestead	54	1,296			
Woodys Bore Windmill	47	1,914	460		
Robyns Bore 2	41	1,691	530	46	30
Rick Bore 2	37	1,295	580	110	150
Paulines Bore	40	1,149	540	230	200
Nicks Bore	46	1,135	470	280	200
Leos Bore	53	1,012	400	320	250
Fergys Bore	61	969	340	330	270
Lagoon Hill Drill Hole 15	29	461	710	710	600

*: Data from this shallow groundwater sample were used to determine the initial value.
Source: Love et al., 2000.

centration of ^{36}Cl prior to nuclear weapons testing and correcting for source or sinks of chloride." The three equations, from Bentley et al. (1986b), used by Love and his colleagues to correct for these two major problems (initial value and real present value) are given below.

1. Age calculated by

$$t = -\frac{1}{\lambda} \ln \frac{R - R_{se}}{R_0 - R_{se}} \quad (6.4)$$

where R_{se} is the secular equilibrium ^{36}Cl/Cl ratio and $R = {}^{36}$Cl/Cl.

2. Age calculated by

$$t = -\frac{1}{\lambda} \ln \frac{Cl_m}{Cl_{ET}} \frac{R - R_{se}}{R_0 - R_{se}} \quad (6.5)$$

where Cl_m is the chloride concentration measured in groundwater, Cl_{ET} is the estimated chloride concentration governed by evapotranspiration only and free of subsurface Cl⁻ addition (i.e., chloride concentration in recharge water).

3. Age calculated by

$$t = -\frac{1}{\lambda} \ln \frac{C - C_{se}}{C_0 - C_{se}} \tag{6.6}$$

Equations 6.4 and 6.6 are similar with the exception that in Equation 6.6, ^{36}Cl values in atoms/L are used instead of $^{36}Cl/Cl$ in Equation 6.4.

The age of deep Jurassic confined aquifer in the GAB was estimated by Bentley et al. (1986b) to be $(1,200 \pm 500) \times 10^3$ years (much older than the hydraulic age). Through this study, which is the first major ^{36}Cl dating research, ^{36}Cl content of 26 groundwater samples were shown to range from $1.4-55.7 \times 10^7$ atoms/L.

Murray Basin, Australia One of the earlier studies dealing with ^{36}Cl in groundwater is that by Davie et al. (1989), through which ^{36}Cl of limestone groundwater in the Murray Basin of Australia was measured. Eighteen samples were analyzed, showing ^{36}Cl values of 85 to $7,200 \times 10^6$ atoms/L. Murray's samples $^{36}Cl/Cl$ ratio are low ($13-43 \times 10^{-15}$), generally lower than those reported by Davis et al. (2003) for shallow groundwaters of the United States, which might reflect either the inevitable additive impact of nuclear detonations in North America or geologic-geographic differences. Davie et al.'s study showed that the ratio of $^{36}Cl/Cl$ in the aquifer increases along the flow path most probably because of the significant localized recharge, which contains high $^{36}Cl/Cl$ concentration.

Paleochannel Groundwater, Kalgoorlie, Western Australia Paleochannels, river channels covered by overburden locally known as regolith, in the Kalgoorlie region of Western Australia are a significant source of brackish groundwater for the region's extensive mining activities. Sometimes the geochemistry of paleochannel's groundwater aids mining exploration. By analyzing 17 samples from water wells and seeps, Turner et al. (1995) showed that the $^{36}Cl/Cl$ ratios and ^{36}Cl content of the paleochannel groundwater range from 32 to 129×10^{-15}, and 1,460 to 79,000 atoms/L, respectively. They were not able to date the paleochannel, because the half-life of ^{36}Cl, compared to the age of groundwater, is too long.

A Significant Case Study in the USA Davis et al. (2003) measured the concentration of ^{36}Cl in many groundwater samples throughout the United States to determine what they call the "pre-anthropogenic levels" (natural = Pristine = background level) ^{36}Cl concentration of groundwater. This study, which was subsequently completed by Moysey et al. (2003), is perhaps the most comprehensive study worldwide of ^{36}Cl in groundwater so far, through which a total of 183 water samples from unpolluted wells and springs were analyzed. Davis at al. (2003) showed that natural ratios of ^{36}Cl/Cl are lowest near the coast and increase to a maximum in the central Rocky Mountains of the US (less than 50×10^{-15} in Florida to more than $1,200 \times 10^{-15}$ in the Rocky Mountains). They attribute this pattern to an inland decrease in the stable chlorine content of the atmospheric precipitation, i.e., the increase in the ratio of ^{36}Cl/Cl inland is due to the inland decrease of stable chlorine, not an increase of ^{36}Cl.

6.3 HELIUM-4

Helium was discovered in 1868 by J. Norman Lockyear. It is an inert and non-toxic gas with an atomic number of 2 and an atomic weight of 4. Its boiling and melting points are the lowest among the elements. It is the second most abundant element in the known universe after hydrogen and constitutes 23% of all elemental matter measured by mass. In the Earth's atmosphere, however, the concentration of helium by volume is 5.24 parts per million only (Verniani, 1966). Helium is the 71st most abundant element in the Earth's crust. There are tiny amounts of helium in mineral springs, volcanic gas, and iron meteorites. The greatest concentrations (trace amounts up to 7% by volume) of helium on the Earth are in natural gas fields, from which most commercial helium is derived (Rogers, 1921).

There are eight known isotopes of helium, ranging in mass from 3 to 10, but only ^{3}He and ^{4}He are stable. The unstable ones have half-lives of less than milliseconds. ^{4}He is an unusually stable nucleus because its nucleons are arranged into complete shells. In the Earth's atmosphere, the concentration of ^{4}He is 5.24 ppmv (very close to total helium) and there is only one ^{3}He atom for every 730,000 ^{4}He. The ^{3}He/^{4}He ratio in the atmosphere (R_a) is thus 1.384×10^{-6} (this value is slightly different in various references). Rocks from the continental crust have typical ^{3}He/^{4}He ratios of 0.02–0.03 R_a, while the depleted mantle has an average value of 8 R_a (Graham, 2002) and subcontinental mantle 6.5 R_a (Dunai and Porcelli, 2002). Helium is unusual in that its isotopic abundance varies greatly depending on its origin. Rocks from the Earth's crust have isotope ratios varying by as much as a factor of ten; this is used in geology to study the origin of such rocks. Heliums-3 and -4 are both used in age-dating groundwater. In addition, the concentration of helium-4 in the soil gas and in the pore fluids has been used to locate areas of deep groundwater discharge into watersheds (e.g., Gascoyne and Sheppard, 1993). Helium-3 is also used to help quantify the renewal rate and mixing dynamics and environmental states of the lakes through experiments on sediment pore waters (Strassmann, 2005).

6.3.1 Production and Sources of ^{4}He

As shown in Figure 6.2, there are four different sources of ^{4}He in groundwater.

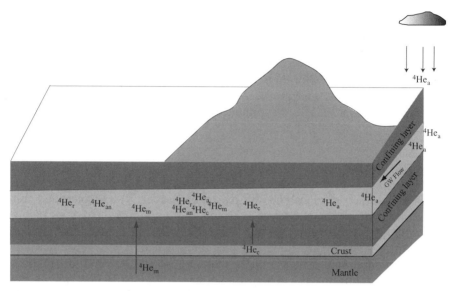

Figure 6.2. Sources of ^{4}He in a confined aquifer. ^{4}He$_a$ = atmospheric helium, ^{4}He$_{an}$ = ancient helium, ^{4}He$_r$ = radiogenic helium, ^{4}He$_c$ = crustal helium, ^{4}He$_m$ = mantle helium. Note that radiogenic helium and crustal helium are produced by the same mechanism.

1. Atmospheric helium: Like all other noble gases, a chief reservoir for ^{4}He is the atmosphere. Atmospheric helium is dissolved in rainwater and is carried into the groundwater. As stated, the ratio ^{3}He/^{4}He in the air is 1.384×10^{-6}, but due to small fractionation, it reduces to 1.36×10^{-6} when helium dissolves in precipitation (Benson and Krause, 1980). Hence, the concentration of ^{4}He in recharging groundwater is about $48\,\mu\text{cm}^3$ (STP) kg^{-1} (Solomon, 2000). A portion of atmospheric ^{4}He in groundwater is the result of excess air entrainment during recharge. This leads to higher ^{4}He concentration in recharging water. To date groundwaters, we must separate ^{4}He resulting from solubility and ^{4}He from excess air. This is done by measuring neon concentration in groundwater and comparing helium to neon ratios in air (0.2882) with that of those in the groundwater sample.
2. Radiogenic helium or crustal helium. Helium-4 is produced from aquifer matrix and from the sediments grains by alpha decay of uranium and thorium; the alpha particles that emerge are fully ionized ^{4}He nuclei:

$$^{238}_{92}\text{U} \rightarrow {}^{206}_{82}\text{Pb} + 8\,{}^{4}_{2}\text{He} \qquad {}^{238}_{92}\text{U} \rightarrow {}^{234}_{90}\text{Th} + {}^{4}_{2}\text{He}$$
$$^{235}_{92}\text{U} \rightarrow {}^{207}_{82}\text{Pb} + 7\,{}^{4}_{2}\text{He} \qquad {}^{232}_{92}\text{Th} \rightarrow {}^{208}_{82}\text{Pb} + 6\,{}^{4}_{2}\text{He}$$

Helium produced within the Earth's crust through the above reactions enters groundwater system, too, the so-called crustal flux of helium

(Torgersen and Clarke, 1985). Some references prefer to refer to this type of helium as terrigenic helium similar to mantle's ^{4}He (Kipfer et al., 2002). Typical ^{3}He/^{4}He ratios for the continental crust are 2.77×10^{-9} – 4.1×10^{-9}, or 0.02–0.03 R_a.

3. Mantle or terrigenic helium. There has recently been strong evidence to suggest that ^{4}He produced deep in the Earth's mantle can find its way to enter relatively shallow groundwaters. For example, Kulongoski et al. (2005) have shown that seismicity and fracturing are important in transport of mantle helium through fault zones to the upper crust and shallow-level groundwaters. As mentioned above, the ratio of ^{3}He/^{4}He in the mantle helium is on average 6.5–8 R_a, but it can vary from 4 to 37 R_a depending on the location (Sapienza et al., 2005). However, Porcelli and Ballentine (2002) propose the value of 6.5–6.7 R_a for calculating the mantle contribution of helium in crustal fluid.

4. Helium-4 atoms that were entrapped in the crystal lattices of sediments or rock strata in the course of deposition may enter groundwater through solid-state diffusion process (Solomon et al., 1996; see Chapter 4). Ancient ^{4}He is the name for this type of helium.

The ratio of ^{3}He to ^{4}He (^{3}He/^{4}He) is a key indicator to differentiate among ^{4}He from any of the above-mentioned sources and in groundwater dating. Similarly, the ratio of ^{4}He to ^{40}Ar (^{4}He/^{40}Ar) varies depending on the source of helium; the higher this ratio, the more significant the role of radiogenic sources in the concentration of ^{4}He.

6.3.2 Sampling, Analysis, and Reporting the Results

Like all noble gases, the main point to observe when sampling groundwater for helium is to avoid mixing it with air. Air helium could diffuse into the sample. Also, correct sample pressure has to be maintained to avoid helium degassing from the water samples (Solomon, 2000). The main method used for sampling for ^{4}He is that using the copper tube welded at both ends with clamps. A copper tube is impermeable to helium, and samples can be stored, without contamination and degassing, for a year. Further, the malleability of copper allows convenient sample opening for degassing it in an extraction line (D. L. Pinti, University of Quebec, personal communication, 2005). Use of the copper tube is an almost standard method for sampling those substances that are present in the atmosphere in high quantity and in the groundwater in small quantity.

Sampled groundwater is degassed in a vessel under high vacuum and the helium is purified from reactive gases (N_2, CO_2, water vapor, hydrocarbons) in ultrahigh vacuum extraction and purification lines (Beyerle et al., 2000; Pinti and Marty, 1998b). At the end, separated helium is analyzed using quadrupole mass spectrometry or noble gas mass spectrometry (Marty et al., 1993). The unit to report ^{4}He concentration is the cm^3 STP/g of water or the cm^3 STP/L of water. Less appropriate units of mole/g or mole/L are still in use in the literature.

6.3.3 Age-Dating Groundwater by ^{4}He

Helium accumulates in the groundwater from the in situ radioactive decay of the uranium and thorium in the aquifer matrix, as well as from any flux into the groundwater from the underlying crust and mantle (Torgersen and Clarke, 1985; Marty et al., 1993; Zhao et al., 1998). Therefore, the concentration of helium in the aquifer increases with time. If the rate at which the in situ production and crustal flux have supplied helium to the flowing groundwater is known, then it is possible to calculate the length of time that the groundwater of known helium content has resided in the aquifer. The equation for such a purpose is

$$t = \frac{^4\text{He}_{rad}}{J_{He}} \quad \text{(Kipfer et al., 2002)} \quad (6.7)$$

where $^4\text{He}_{rad}$ and J_{He} are the measured concentration and the production rate (accumulation rate) of ^{4}He in groundwater, respectively. It is relatively simple to measure the concentration of ^{4}He in groundwater. But it takes a considerable amount of efforts to calculate the accumulation rate of helium, i.e., J_{He}. To calculate the accumulation rate, the following procedure must be undertaken:

1. The sources of helium must be identified, whether there is only one in situ, crust, or mantle source or two or all of these sources.
2. In situ production rate must be calculated by

$$J_{He} = \frac{\rho_r}{\rho_w}(C_U \times P_U + C_{Th} \times P_{Th}) \times \left(\frac{1-\theta}{\theta}\right) \quad \text{(Modified from Kipfer et al., 2002)}$$
$$(6.8)$$

where ρ_r and ρ_w are the densities of the aquifer material and the water, C_U and C_{Th} are uranium and thorium concentration in rocks (μg/g), and θ is the porosity of aquifer. Production rates from uranium and thorium decay are $P_U = 1.19 \times 10^{-13}$ cm^3 STP per μg_U^{-1} yr^{-1} and $P_{Th} = 2.88 \times 10^{-14}$ cm^3 STP per μg_U^{-1} yr^{-1}. Equation 6.8 is based on 100% transfer of produced ^{4}He into groundwater, which comes true after an equilibrium time between rock and fluid in the order of thousands of years. A number of other equations have been developed to calculate the in situ production rate of ^{4}He. Ballentine et al. (2002) report that the accumulation rate of radiogenic ^{4}He in groundwater from the decay of uranium and thorium in aquifer rocks is typically 10^{-16}–10^{-17} mole g^{-1} year^{-1}.

3. Crustal and mantle sources of helium need to be calculated as well, and this is the main obstacle for quantitatively dating groundwater by ^{4}He. Some researchers doubt the influx of crustal helium in groundwater and argue that upward moving crustal helium escapes into the Earth's surface (Torgersen and Clark, 1985). Some others (Mazor, 2004) argue that all crustal- and mantle-produced helium is dissolved in the aquifer immediately above the crust, and the chance of helium escaping to the upper aquifers is zero. If this assumption is accepted, then the ^{4}He content of the upper aquifer can be attributed to atmospheric and in situ sources only. There are some intermediate views as well. For example, Pinti and Marti (1998) identified at least three sources of ^{4}He in the Paris basin.

Note The age range of the ^{4}He method is 10^4 to 10^8 years, and due to the semiquantitative nature of the method, it is referred to by some authors as age-indicator rather than as a dating method. In addition to the problem of calculating the accumulation rate of ^{4}He, other issues like mixing of groundwater with different ^{4}He concentration (and hence different ages), excessive ^{4}He content due to excess air phenomenon, and sampling related errors surround dating by ^{4}He. A recommended exercise is to check the ^{4}He ages against (1) ages obtained by other methods, (2) depth of the confinement, (3) temperature of the aquifer, (4) concentration of radiogenic helium and ^{40}Ar and geological setting (Mazor, 2004).

6.3.4 Advantages and Disadvantages

The helium-4 dating method has the potential to be applied to a wide range of ages, i.e., from thousand-year-old groundwaters to millions of years old, and also to young groundwaters. There is plenty of information about the method in the literature, and the method has been applied to a large number of aquifers worldwide. However, the many sources of ^{4}He make the method prone to considerable error, and vulnerable, especially if no other calibrating ages are available.

6.3.5 Case Studies

Helium-4's dating research until 1998 is presented in Table 6.3. As seen from the table, helium ages are either older or much older than ages obtained by other dating techniques. This discrepancy is due to the heterogeneity of the aquifers, possible role of mantle-derived ^{4}He, and mixing of water with different ages (Pinti and Marty, 1998). However, Mazor and Bosch (1992) disagree with such reasoning and argue that the hydraulic ages are younger than the real ages, not the other way around, i.e., the ^{4}He ages are closer to the real ages. The youth of the hydraulic ages is, then, attributed to the hydraulic discontinuities and immobilized sections in the deep aquifers. Let us now briefly

TABLE 6.3. Helium-4 Groundwater Dating Case Studies Until 1998

Aquifer	Formation age and lithology	Hydrologic or isotopic age		^{4}He age ($\times 10^6$ yr)	Ref.
		Age ($\times 10^3$ yr)	Method		
Precambrian Shield, Canada	Precambrian, leucogranite	11	^{14}C	4–200	1
East Bull Lake, Ontario, Canada	Precambrian, Gabbro-anorthosite	10	^{14}C	120	2
Stripa, Sweden	Precambrian, granite	25	^{14}C, ^{3}H	0.38	3
Stampriet, South Africa	Ordovician, sandstone	2.7–38	^{3}H	0.027–0.38	4
Uitenhage, South Africa	Ordovician, quartzite	0.4–35	^{3}H	0.35–35	4
Retford, UK	Triassic, sandstone	2	^{14}C	0.004	5
Gainsborough, UK	Triassic, sandstone	36	^{14}C	0.06	5
Wallal, GAB, Australia	Jurassic, sandstone	35.6	^{14}C, ^{3}H	6	6
Wiringa, GAB, Australia	Jurassic, sandstone	20.6	^{14}C, ^{3}H	1.4	6
Paris Basin, France	Mid-Jurassic, limestone	2–1,000	^{14}C, ^{3}H	10–140	7
Innviertel, Molasse Basin, Austria	Miocene, marl & shale	30–40	^{14}C, ^{3}H	0.46	8
Strugeon Falls, Ontario, Canada	Pleistocene, sandstone	0.250	CFCs, ^{3}H	0.07	9

1: Bottomley et al. (1984) 2: Bottomley et al. (1990) 3: Andrews et al. (1982) 4: Heaton (1981) 5: Andrews and Lee (1979) 6: Torgersen and Clarke (1985) 7: Marty et al. (1988) 8: Andrews et al. (1982) 9: Solomon et al. (1996).
Source: Modified from Pinti and Marty, 1998, with the permission of the Geological Society, London.

discuss the latest case studies from the Kobe area of Japan and Paris Basin of France.

Kobe Area, Southwest Japan As shown in Figure 6.3, the age of deep groundwaters (brine) mixed with percolating meteoric water in the Kobe area, Japan, has been determined by a revised version of the ^{4}He dating method that takes into account the ratio of ^{3}He/^{4}He, the production rate, and the concentration of ^{3}He and ^{4}He. The calculated ages of the samples are 25, 33, 48, 62, 78, 110, and 230 [×1,000] years. The following equation, which accommodates all these parameters, has been used for this purpose:

$$t = C(^4\text{He})_0 \left(1 - \frac{R_0}{R_{\text{ext}}}\right) \frac{p\rho_w}{\left(P(^4\text{He}) + \frac{F(^4\text{He})}{h}\right)} \quad \text{Morikawa et al. (2005)} \quad (6.9)$$

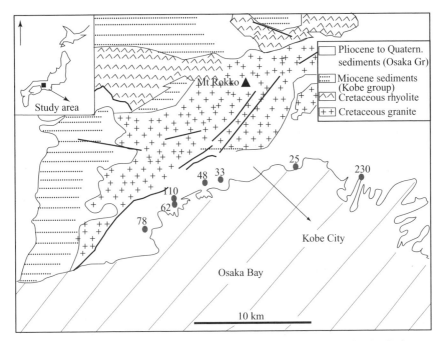

Figure 6.3. Helium-4 ages (in million years) of deep groundwater in the Kobe area, Japan (reproduced with modifications from Morikawa et al., 2005).

where $C(^4He)_0$ is the concentration of ^4He in flux out of the reservoir per units of weight ($cm^3\, STP\, g^{-1}\, H_2O$), R_0 is the helium isotopic ratio ($^3He/^4He$) in flux out of the reservoir, R_{ext} is the helium isotopic ratio ($^3He/^4He$) in water with extra component, p is porosity of the reservoir, ρ_w is the density of water ($g\, cm^{-3}$), $P(^4He)$ is the in situ production rate of ^4He ($cm^3\, STP\, cm^{-3}\, yr^{-1}$), $F(^4He)$ is the crustal ^4He flux from the bottom of the reservoir ($cm^3\, STP\, cm^{-2}\, yr^{-1}$), and h is the thickness of the reservoir (cm).

The in situ production rate of ^4He, $P(^4He)$, was calculated by Equation 6.10:

$$P(^4He) = (1-p) \cdot \rho_R \cdot \{1.2 \times 10^{-13}[U] + 2.9 \times 10^{-14}[Th]\} \quad \text{Morikawa et al. (2005)} \tag{6.10}$$

where [U] and [Th] are the concentration of the the U and Th of aquifer rock in $\mu g\, g^{-1}$, respectively.

Paris Basin, France The Paris Basin, a multilayered, well-studied basin, has been subjected to a number of ^4He dating research (Marty et al., 1998, 2003; Pinti and Marty, 1998). In the eastern recharge area of the basin, the deeper Trias sandstone aquifer, which lies above the crystalline basement, is separated from the overlying Dogger aquifer by an aquitard layer consisting of about

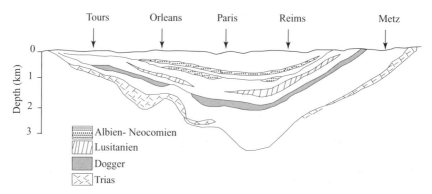

Figure 6.4. Startigraphic setting of Paris Basin, France (reproduced from Marty et al., 1993).

600 m succession of shales and clays (Figure 6.4). Marty and others (2003) used the helium isotopes dating technique to show that in the eastern recharge area, Dogger aquifer is stagnant and has been extremely well isolated from the Trias over the past several million years. In contrast, Trias water has been flowing, since the Miocene age, at a present-day velocity of $4\,m\,yr^{-1}$. They have shown that ^{4}He content of the Trias aquifer is higher than what could be derived from atmospheric, crust, and in situ sources. Another major source, such as the mantle, is needed to account for such high concentration of helium and ^{3}He/^{4}He ratios. The mantle-derived flux of ^{4}He and ^{3}He were calculated at 1.5 × $10^{-12}\,mol\,m^{-2}\,yr^{-1}$, respectively. The age data from this study proved that mass transfer across aquitards (say, for example, transfer of ^{4}He from Trias to Dogger) is extremely small and is limited to the fractures only.

Pore Waters in the Opalinus Clay, Mont Terri, Switzerland Helium-4 concentrations of up to $1 \times 10^{-4}\,cm^3\,STP/g$ pore water were measured across the Opalinus Clay and adjacent formations at Mont Terri, Switzerland, which were thought to result from in situ production (Rübel et al., 2000). The ^{4}He ages of pore water samples were determined to range from 200 to 277,000 years, which showed a very good correlation with the flow distance, i.e., the longer the flow distance, the higher the pore water ages.

6.4 ARGON-40

The description about argon is given in Chapter 5, the section on argon-39. Argon-40, a stable isotope of argon with an abundance of 99.6%, is 295.5 times that of argon-36 in the atmosphere. It is dissolved in the atmospheric precipitation and is subsequently carried to the subsurface by recharge processes. Overall, groundwater contains about (4 to 5) × $10^{-4}\,cm^3\,STP$ Ar per cm^3 of water from the solution of atmospheric argon (Pearson et al., 1991). The atmospheric ratio of $^{40}Ar/^{36}Ar$ (295.5) is a reference to find if there is a source, other than the atmosphere, for ^{40}Ar in the aquifer. In addition to atmospheric reservoir, the main source for ^{40}Ar is the decay product of potassium-40. About 11.2% of naturally occurring ^{40}K with a half-life of 1.25×10^9 years decays to stable $^{40}Ar*$ ("*" is indicative of radiogenic argon) by electron capture and positron emission reactions.

$$^{39}_{20}K + ^{0}_{-1}e \rightarrow ^{40}_{18}Ar + h\nu \quad \text{(electron capture)}$$

$$^{39}_{20}K \rightarrow ^{40}_{18}Ar + ^{0}_{+1}e \quad \text{(positron emission)}$$

Therefore, the concentration of ^{40}Ar increases with time. The production rate of ^{40}Ar is calculated through the following equation:

$$\text{Ar atoms g}^{-1}\text{yr}^{-1} = 102.2[K] \quad \text{(Ballentine and Burnard, 2002)} \quad (6.11)$$

where [K] is the concentration of potassium in rocks in ppm. The continental crust has been estimated to contian 2.59% potassium, of which about 0.01167% is ^{40}K, or approximately a total of 5×10^{21} grams of ^{40}K.

6.4.1 Sampling, Analysis, and Reporting the Results

The procedure to sample groundwater for ^{40}Ar analysis is similar to sampling for helium and other noble gases. The most common problem that has to be avoided during sampling is exclusion or inclusion of air. Therefore, leak-tight equipment must be used. The websites of the Noble gas laboratory of the University of Utah (http://www.inscc.utah.edu/~ksolomon/) and that of the EAWAG (http://www.eawag.ch/research_e/w+t/UI/methods_e.html) contain a good description of the sampling procedure and precautionary principles that must be observed. Sample analysis include extracting gases from the water sample, separating and purifying extracted noble gases, and quantitative determination of each gas by mass spectrometers. The results of ^{40}Ar analysis are reported as the ratio of $^{40}Ar/^{36}Ar$, which is most often higher than the atmospheric ratio of 295.5. They are also reported in cm^3 STP ^{40}Ar per g of water or cm^3 STP ^{40}Ar per cm^3 water.

6.4.2 Age-Dating Groundwater by ^{40}Ar and the Obstacles

There are some suggestions in the literature that ^{40}Ar is a potential tracer to help date very old groundwaters (Pearson et al., 1991; Clark and Fritz, 1997; Osenbrück et al., 1998; Mazor, 2004). The basis of such arguments is that the concentration of ^{40}Ar in groundwater increases in accordance with the age of groundwater. The higher the concentration of ^{40}Ar in the groundwater, the longer the groundwater age. Therefore, we would be able to date groundwater if we know the concentration of ^{40}Ar and its annual production rate. The $^{40}Ar/^{36}Ar$ should reach at least 303 in groundwater to conclusively indicate underground production of $^{40}Ar^*$. It may take a minimum of about 10^6 years to reach such a level (Metcalfe et al., 1998). The main obstacle in this regard is the retention of produced ^{40}Ar in the parent rocks, i.e., a large fraction of the produced $^{40}Ar^*$ atoms are retained in the rock matrix. In other words, the amount of $^{40}Ar^*$ produced in the rocks that can reach the groundwater is very low. This is because minerals retain argon rather than helium; very deep waters in a high thermal gradient basin may be warm enough to exceed the closure temperature of minerals and release the produced $^{40}Ar^*$. Another obstacle is the slow and insignificant rate of underground production of ^{40}Ar as compared to the atmospheric input. The third problem facing this method is that in some measurements, it has been revealed that the $^{40}Ar/^{36}Ar$ ratio in some groundwaters is higher than the potential result of underground production (Tolstikhin et al., 1996). Careful hydrogeologic analysis is therefore needed to find out the source of this extra component. In the case of aquifers in Northern Switzerland (Tolstikhin et al., 1996), intrabasin sources of helium and argon, i.e., diffusion of helium and argon from older waters into younger waters, was found to be the cause of excessive $^{40}Ar/^{36}Ar$ ratio. In spite of these difficulties, it should be, however, pointed out that $^{40}Ar\text{-}^{40}K$ technique is a well-established dating technique for rocks in geological contexts.

6.4.3 Case Studies

The ratio of $^{40}Ar/^{36}Ar$ in groundwater has been measured for deep brines of the Palo Duro Basin, Texas (Zaikowski et al., 1987; $^{40}Ar/^{36}Ar$ of up to 2,090 has been reported in this study, which is the worldwide maximum measured so far); Stripa granite, Sweden (Andrews et al., 1989); Great Artesian Basin, Australia (Torgersen et al., 1989; Beyerle et al., 2000); Permian sediments and underlying crystalline basement, northern Switzerland (Tolstikhin et al., 1996); Milk River aquifer, Canada (Andrews et al., 1991); pore waters in the Opalinus Clay, Mont Terri, Switzerland (Rübel et al., 2002, $^{40}Ar/^{36}Ar$ of up to 334; Ballentine and Sherwood-Lollar, 2002). Note that not many of these case studies led to conclusive ages. However, the main scientific contribution made by ^{40}Ar measurement is in the form of a $^4He/^{40}Ar$ ratio, i.e., the amount of the $^4He/^{40}Ar$ ratio is a key parameter to estimate helium groundwater ages. Here, we describe the case of Morsleben in northern Germany (Osenbrück et al., 1998) where pore waters from a 735-m core (from a cap rock over a salt dome) of Triassic sedimentary rocks were analyzed for ^4He, ^{40}Ar, and ^{136}Xe. The core from top to bottom consists of

1. Keuper claystones and dolomite with a thickness of 300 m,
2. Claystones with remnants of gypsum and anhydrite divided by a 40-m-thick compacted sandstone (290 m in thickness), and
3. Limestone and dolomite from the Muschelkalk + a salt layer of 10 m (145 m in thickness).

Osenbrück and his colleagues measured the $^{40}Ar/^{36}Ar$ ratio of 29 pore water samples from the described core, which range from 295 at the top of the core (70 m below surface) to 657 close to the core bottom. The high content of ^{40}Ar in the pore waters from the depth below 450 m were considered to have been produced inside K-rich salts in the salt deposits and entered the cap rock by diffusion. Therefore, this study did not conclusively show the applicability of ^{40}Ar in dating, but rather ^{40}Ar results helped to interpret and analyze ^{4}He analysis, together yielding an age of 6 million years for saline pore fluids. The $^{4}He/^{40}Ar$ is useful because it can show whether both ^{4}He and ^{40}Ar are released in the fluids (the $^{4}He/^{40}Ar$ production ratio in the rocks equals that in fluids) or Ar is retained in minerals (in this case, the resulting $^{4}He/^{40}Ar$ is much higher than the production ratio in the crust or the aquifer).

6.5 IODINE-129

Iodine, a halogen with the atomic number of 53 and atomic weight of 126.9, was discovered by Barnard Courtois in 1811. It was the first element recognized as being essential to humans; and the disease of endemic goiter, which results from iodine deficiency, was the first to be related to environmental geochemistry (Fuge, 2005). It has been estimated that nearly 30% of the world's population is at risk for some from of iodine deficiency disorder (Dissanayake, 2005). In the water, iodine exists mainly in two states, I^- and IO_3^-, both of which behave conservatively. Seawater is the biggest reservoir of iodine (average concentration of 60 µg/L), sedimentary rocks have much higher iodine content as compared to igneous rocks, and the transfer of iodine from marine to the terrestrial environment is via the atmosphere (Fuge, 2005). The residence time of iodine in the ocean and in the atmosphere is about 350,000 years and 18 days, respectively. Iodine has 37 isotopes, whose atomic weight ranges from 108 to 144, but only iodine-127 is stable, forming almost 100% of the mass. Short-lived isotopes of ^{125}I, ^{131}I, and ^{133}I, released from human nuclear-related activities, represent some threats to human health, with ^{133}I posing the greatest risk. It has been estimated that in the Chernobyl accident, 35 MCi of ^{131}I were injected into the atmosphere (Fuge, 2005). The most useful radioactive isotope of iodine is ^{129}I. It has a half-life of 15.7 ± 0.04 million years and it is one of 34 known radioisotopes with half-lives of greater than one million years. In radioactive waste disposal business, ^{129}I is recognized as a bad radionuclide because its very long half-life makes it persist and because it is anionic and is therefore not delayed by any of the mineral barriers (Wiles, 2002). In many ways, ^{129}I is similar to ^{36}Cl. It is a soluble halogen, is fairly nonreactive, exists mainly as a nonsorbing anion, and is produced by cosmogenic, thermonuclear, and in situ reactions. As is the case with $^{36}Cl/Cl$, $^{129}I/I$ ratios in nature are quite small, 10^{-14} to 10^{-10}. The possibility of using ^{129}I as an environmental tracer was suggested a long time ago (Edwards, 1962), but our arena of discussion is limited to the very old groundwater-dating application of ^{129}I, which was realized much later. Iodine-129 was also used in rainwater studies following the Chernobyl accident. Anthropogenic ^{129}I such as those produced by radioactive wastes may also serve as a dating tool for young groundwater and may help to estimate groundwater recharge. Recently, Schwehr et al. (2005) have demonstrated the possibility of using ^{129}I (together with ^{127}I) as a hydrological tracer for past river flow condition. Xenon-129, the daughter product of ^{129}I, is fundamental to give an estimate in the closure of main geochemical reservoirs on Earth (crust-mantle-atmosphere) and on the differentiation of our planet (D. Pinti, personal communication).

6.5.1 Production of ^{129}I

Iodine-129 is produced in seven different ways:

1. By spallation of stable Xe isotopes in the stratosphere, e.g.,

$$^{132}_{54}Xe + P \rightarrow {}^{129}_{53}I + P + \alpha$$

2. By cosmic-ray spallation of Ce, Ba, and Te isotopes and importing the produce through cosmic dust and meteorites (Fabryka-Martin et al., 1989),
3. By volcanic emission of ^{129}I which was produced as a result of the fission of uranium isotopes in the Earth's interior,
4. In the lithosphere, from spontaneous fission of uranium-238 and neutron-induced fission of uranium-235, the former reaction being the main contributor,
5. From aboveground nuclear testing during 1960s and early 1970s; peak ^{129}I/I during this period reached about 10^{-7},
6. From the Chernobyl accident in 1986 (about 1.3 kg), and
7. From nuclear power and nuclear fuel re-processing facilities. Iodine-129, together with ^{36}Cl, are the only elements that can escape from the nuclear fuel waste disposal repositories and enter the surrounding environment. Almost all radionuclides will remain trapped, either in the fuel itself or within a very short distance of it (Wiles, 2002). It is estimated that about 8% of ^{129}I produced in the radioactive waste may be released to the groundwater (Wiles, 2002).

Notes

a. The natural atmospheric production of ^{129}I (i.e., items 1–3 above) leads to a steady-state isotopic ^{129}I/I ratio of $1.1 \pm 0.4 \times 10^{-12}$ or 2×10^4 atoms/L if iodine concentration in rainfall is assumed to be 3 ppb (Fabryka-Martin et al., 1987). This value is considered as pre-bomb recharge water concentration (initial value),

b. The lithospheric production of ^{129}I depends on the uranium content of the given rocks. Sedimentary rocks produce insignificant amounts, while uranium-bearing rocks such as igneous rocks produce ^{129}I, some of the produced ^{129}I decays. Therefore, the amount of ^{129}I available for migration into groundwater depends on the balance between production and decay. Also, not all ^{129}I present in the rock can migrate into groundwater. Having said these, it is obvious that calculating the amount of lithospheric production, and subsequent migration, is a huge task. A number of equations and a thorough, albeit incomplete, discussion can be found in Fabryka-Martin et al. (1985, 1989). One of these equations is reproduced here (6.12) to alert the reader of the complexity of the parameters needed to calculate the rate of such production. The production rate P of ^{129}I atoms per unit volume of rock is calculated by

$$P = \frac{f_u N_A \lambda_{sp} Y_{129} \rho g (1-\theta)}{\alpha u} - \lambda N_{129} \quad (6.12)$$

where

N_{129} = concentration of ^{129}I in unit bulk volume of rock (atoms/cm^3),
f_u = fractional concentration of uranium in rock (g/g),
λ_{sp} = spontaneous fission decay constant (yr^{-1}),
Y_{129} = spontaneous fission yield at mass 129,
N_A = Avogadro's number (6.02 × 10^{23} atoms/mole),
α_u = molecular weight of ^{238}U (g/mole).

Moran et al. (2002b) summarize the inventory and flux of ^{129}I in the environment as in Table 6.4.

TABLE 6.4. Inventory and/or Flux of ^{129}I

Resevoir/Source	Inventory (kg)	Flux
Natural inventory in the surface environment	100 (of this 0.5 g is in the atmosphere)	
Atmospheric bomb tests	100	
Chernobyl reactor accident	1.3	
Savannah River Nuclear fuel reprocessing plant, USA		2.8 kg/year (1964–65) decreased to 0.7 in the late 1970s
Three other major and a few minor fuel reprocessing plants in the USA		Unknown amount
Fuel reprocessing plants at Sellafield, England, and Cap de La Hague, Europe	***2,360** between 1966 and 1997 (= 420 Curie)*	*3–9% of this was escaped to the atmosphere*

6.5.2 Sampling, Analysis, and Reporting the Results

Normal filtering of the samples is necessary to avoid particulate matters. Sample volume required depends on the iodine concentration; about 1–2 mg of iodine is needed, but higher amounts are preferred to handle them confidently and to be able to perform replicate analysis. However, because iodine concentration in groundwater is usually low, sample volumes are generally not small. Water samples to be analyzed for ^{129}I have to be prepared for target loading by AMS (earlier, measurement of ^{129}I was undertaken by neutron activation analysis). A good reference regarding analytical procedure for measurement of ^{129}I of groundwater sample is that by Schwehr et al. (2005). The following steps are to be followed to extract the iodine from the water sample for AMS loading (from Moran et al., 2002b, and Fehn et al., 1992):

1. Pre-concentration of the sample by rotary vacuum distillation,
2. Addition of a 2–4-mg low ^{129}I carrier (^{129}I/I of about 10^{-15}),
3. Acidification of samples by adding HNO_3,
4. Addition of 5–15 mL of CCl_4 (carbon tetrachloride),
5. Oxidization of I to I_2 as well as the dissolved organic matter with about 5-mL H_2O_2 (hydrogen peroxide),
6. Addition of 20–30 mL of 1 mole NH_4OH-HCl (hydroxylamine hydrochloride) to reduce any IO_3^-,
7. Back-extracting sample iodine by using a 0.1 M $NaHSO_3/H_2SO_4$ (sodium bisulfite/sulfuric acid) acid solution,
8. Addition of 2 mg of Cl^- and co-precipitation of AgI and AgCl using $AgNO_3$,
9. Dissolution of AgCl using NH_4OH (ammonium hydroxide), and
10. Centrifuging the remaining AgI, rinsing it with deionized water and drying it.

About 80% of the iodine is extracted from the water sample using the above procedure. The final prepared sample is an AgI pellet weighing 2–5 mg, which can be loaded to AMS for ^{129}I analysis. Only a limited number of laboratories worldwide, as cited in Table 2.2, have the capacity to analyze water samples for ^{129}I. The cost of ^{129}I analysis by AMS at PRIME Lab, Purdue University, USA, is USD 350 if samples are prepared by the customer. Sample preparation, however, costs almost USD 500 if only one sample is submitted, but it is reduced with an increase in the number of samples.

In the hydrological studies, ^{129}I concentrations are usually reported as the ratio of ^{129}I to the total I, i.e., ^{129}I/I (note that I is virtually all iodine-127). Due to the smallness of this ratio, it is multiplied by a coefficient, i.e., ^{129}I/I $\times 10^x$. However, the coefficient is not well established as yet, whether it is 10^{12} or 10^{14} (unlike chlorine-36, which the coefficient is usually 10^{15}). Here, in the calculations, we have chosen 10^{12}. Iodine-129 analysis results can be reported in atoms/liter as well. ^{129}I/I value is converted to atoms/L using Equation 6.13:

$$^{129}\text{I}(\text{in atoms per liter}) = 4{,}700 \times B \times R \qquad (6.13)$$

where B is the concentration of iodine in the sample in µg/L and R is $^{129}\text{I}/\text{I} \times 10^{12}$.

Example: If the $^{129}\text{I}/\text{I} \times 10^{12}$ in a water sample is 41 and the concentration of I$^-$ is 14 µg/L, what would be the concentration of ^{129}I in the sample?

$$^{129}\text{I}(\text{in atoms per liter}) = 4{,}700 \times B \times R = 4{,}700 \times 14 \times 41 = 2{,}697{,}800\,\text{atoms/L}$$

Note that ^{129}I concentration in groundwater can also be reported as radioactive activities, i.e., mBq/L or dpm/L, but this has not seen serious adaptation in the literature.

6.5.3 Age-Dating Groundwater by ^{129}I

Iodine-129 decays by beta emission to ^{129}Xe,

$$^{129}_{53}\text{I} \rightarrow\, ^{129}_{54}\text{Xe} + \beta^- + \bar{\nu}$$

After entering the groundwater system, ^{129}I starts disintegrating into ^{129}Xe. If there is neither subsurface production nor leaching of ^{129}I from the surrounding formations, as time goes by, the concentration of ^{129}I in the groundwater decreases following the decay principles. To calculate the age of groundwater, we need to have the concentration of ^{129}I in the recharging water (initial value) and in the sample and substitute these two in the simple decay equation:

$$^{129}\text{I} =\, ^{129}\text{I}_0 e^{-\lambda t}$$

For dating groundwaters, the initial $^{129}\text{I}/\text{I}$ value of $1.1 \pm 0.4 \times 10^{-12}$, corresponding to 2×10^4 atoms/L, is adapted. For dating pore waters, the initial $^{129}\text{I}/\text{I}$ value of 1.5×10^{-12}, which is that of the recent sediments below the zone of bioturbation (Moran et al., 1998), is used.

Example: If the concentration of ^{129}I in a groundwater sample is 14,000 atoms/liter, what is the age of this sample?

$$t = -\ln \frac{^{129}\text{I}}{^{129}\text{I}_0} \Big/ \lambda = -\ln \frac{14000}{20000} \Big/ 4.41 \times 10^{-8} = 8{,}080{,}514\,\text{years}$$

With such assumptions (no subsurface production or addition), ^{129}I suits dating groundwaters whose age ranges from 2,400,000 to 52,000,000 years (assuming 10% and 90% of the original atoms are decayed), effectively covering the waters recharged during the entire Tertiary age. Fabryka-Martin (2000) put the upper limit of the age range at 80,000,000 years, which is based on the disintegration of 97% of the original atoms, a condition that may not be analytically possible to fulfill.

Note We can neglect the subsurface production, when and if ^{129}I concentration in the sample is below the aforementioned initial value. In the study of brines and hydrothermal fluids, subsurface production is usually assumed to be negligible. In freshwater aquifers, however, subsurface production and leaching from surrounding formation may be the dominant factors in controlling the concentration of ^{129}I. In such cases, as time goes by, the concentration of ^{129}I in the groundwater increases. Therefore, the higher the concentration of ^{129}I, the older the groundwater. To date groundwater, we would then need to calculate the annual underground production rate. To distinguish between anthropogenic sources and natural subsurface production of ^{129}I, we need to measure some other tracers such as tritium. If bomb tritium is present in the groundwater, then the high concentration of ^{129}I (above initial value of 2×10^4 atoms/L) is either due to subsurface production or a result of anthropogenic sources. If bomb tritium is absent, anthropogenic sources are ruled out and the high concentration can be confidently attributed to the subsurface production. Moran et al. (1995) attempt to produce a curve that integrates both the decay curve and the underground production curve of ^{129}I. The main problem, however, is how to satisfactorily calculate the annual production rate.

6.5.4 Advantages and Disadvantages

Advantages
1. Iodine-129 is the longest half-life isotope in the dating business; it is thus suitable to date very old groundwaters in the range of tens of millions of years and also to date very old brines and formation waters.
2. The geochemistry of iodine is simple, because it is near conservative.
3. Due to long residence time of iodine in the ocean and its long residence time in the atmosphere, pre-anthropogenic ^{129}I/I ratio for the entire hydrosphere is spatially and temporally constant and that local or short-term variations are smoothed out (Moran et al., 1995). This means that a verified universal initial ^{129}I/I (natural) ratio can be adapted for age calculations.

Disadvantages
1. Not many aquifers worldwide are that old to have the criteria to be dated by this technique.
2. Uncertainties in the initial value of ^{129}I/I because of significant temporal changes in the cosmogenic production rates and rate of release by volcanic activities (Fabryka-Martin, 2000).
3. Laboratory facilities are limited and the cost of analysis is, therefore, high.
4. Subsurface production of ^{129}I is difficult to quantify, while it is sometimes the dominant source of ^{129}I.

6.5.5 Case Studies

Nankai Hydrate Field, Japan Due to the long half-life, the majority of ^{129}I age-dating studies have dealt with pore waters associated with gas hydrates, oil field brines, and hydrothermal systems (Fehn et al., 1990, 2000; Moran et al., 1995). Fresh groundwaters have been studied to a limited extent by this technique. Here, we discuss a dating study of the pore waters sampled from two deep boreholes (BH-1 and Main) in the Nankai hydrates field, 50 km off the mouth of the Tenryu River, central Japan (Fehn et al., 2003). Gas hydrates deposits of the continental slopes contain very large quantities of hydrocarbons mainly in the form of methane. Pore waters associated with such deposits are rich in iodine. The close association of iodine with organic matter and the strong iodine enrichment found in pore waters associated with gas hydrates suggest a similar origin (organic-rich source formations) for both iodine and methane (Fehn et al., 2003). Therefore, the origin and age of the pore waters (through ^{129}I dating) would provide some clues about the origin of the hydrocarbons (methane). Table 6.5 presents the result of ^{129}I dating on 12 pore water samples, 5–15 mL in volume, squeezed from sediment cores sampled in the Nankai field. The core samples were recovered from different depths of the two boreholes. Iodine-129 ages showed that the pore waters are older than both the host sediments and the subducting marine sediments whose ages are less than 2 and 21 million years, respectively. Pore water age data, therefore, led Fehn and his colleagues to conclude that methane in the gas hydrate deposits is neither derived from the present host sediments nor from the current subducting sediments; rather it is related to the release and long-time

TABLE 6.5. The Results of ^{129}I Analysis of the Pore Waters in Nankai Hydrates Field, Japan

Hole	Depth (mbsf*)	I (µM)	^{129}I/I (×10^{15})	Age (×10^6 years)
BH-1	26	120	276 ± 52	38
BH-1	31	130	520 ± 120	24
BH-1	102	100	1,490 ± 130	0
BH-1	120	230	298 ± 50	37
BH-1	170	220	390 ± 70	31
BH-1	170	220	330 ± 120	34
BH-1	173	230	181 ± 41	48
BH-1	216	220	180 ± 100	48
BH-1	247	190	280 ± 70	38
Main	209	200	470 ± 120	26
Main	210	200	400 ± 310	30
Main	223	190	350 ± 190	33
Seawater	0	0.4	1,500 ± 150	

*: Meter below surface

Source: Reproduced with GSA permission from Fehn et al. 2003.

recycling of fluids from marine formations of early Tertiary age. In this study the ^{129}I/I marine input ratio of $1,500 \times 10^{-15}$ has been adapted as the initial value of ^{129}I to calculate the age of the samples.

Great Artesian Basin of Australia The Great Artesian Basin, or GAB, is a world-known mega-aquifer, there is no need to describe its general characteristics here once more. Iodine-129 analysis of nine water samples along two separate flow lines of the GAB were undertaken by Fabryka-Martin et al. (1985) in 1982. The water samples were collected close to the recharge area, which is characterized by low iodine and ^{129}I concentration (Group A) and from the deeper part of the aquifer, which has higher concentrations of iodine and ^{129}I (Group B). In both groups, however, samples had ^{129}I concentration higher than that of the recharge water. Iodine-129 content of Group A samples ranges from 0.5–2.7×10^5 atoms/L and Group B samples from 2.8–25×10^5 atoms/L). As pointed out earlier, normally ^{129}I concentration in the recharge water is 2×10^4 atoms/L. The conclusion was that underground production of ^{129}I is the dominant process in the GAB. Consequently, Fabryka-Martin and her colleagues were not able to deduce the age of the samples using ^{129}I isotopic decay principles. In a recent study (Lehmann et al., 2003), ^{129}I content of four water samples from the GAB were measured at 2.4, 1.5, 1.1, and 2 ($\times 10^6$) atoms/L. These measurements were used to interpret the ^{81}Kr and ^{36}Cl dates and showed that ^{129}I in the GAB originates from both the pore water of the confining layer of shale and from the atmosphere. Therefore, no ^{129}I ages were calculated.

Generally, it is safe to argue that no ^{129}I study has so far resulted in the quantitative age estimation of fresh groundwater resources. The method performance has been, however, satisfactory for brine and thermal springs. A list of aquifers and various oil and gas fields studied by the ^{129}I method is presented in Table 6.6.

Two major recent studies concerning ^{129}I are those by Moran et al. (2002b) and Cecil et al. (2003). The former discusses the concentration of ^{129}I in some major rivers in North America and Western Europe and the emission of ^{129}I by nuclear fuel reprocessing facilities. The latter determined the ^{129}I background concentration in the Snake River plain aquifer in Idaho, USA, to be 5.4×10^{-8} Curies/L, by analyzing ^{129}I content of 52 ground- and surface water samples. Such background value is needed to evaluate and quantify the impact of radioactive waste disposal practices at the Idaho National Engineering and Environmental Laboratory (INEEL), on the concentration of ^{129}I in the environment.

6.6 URANIUM DISEQUILIBRIUM SERIES

Uranium was discovered in 1789 by Martin Klaproth. Its discovery caused little interest until 1869, when Henry Becquerel discovered radioactivity

TABLE 6.6. Worldwide Case Studies Involving the ^{129}I Dating Method

Aquifer/field	Age (×10⁶ years)	Reference
Milk River aquifer of Alberta, Canada	No age obtained	Fehn et al. (1992)
Stripa groundwater, Sweden	No age obtained	Fabryka-Martin et al. (1989)
Great Artesian Basin, Australia	No age obtained	Fabryka-Martin et al. (1985)
Gulf Coast brines, Southern Louisiana, USA	$8 + \underline{4}$	Fabryka-Martin et al. (1985)
Oil field brines from the U.S. Gulf Coast basin (Texas, Louisiana, and Gulf of Mexico)	53–55	Moran et al. (1995)
Deep marine brines in the Canadian Shield	80 (minimum)	Bottomley et al. (2002)
Hydrothermal fluids of eastern Clear Lake, California, USA	0.01–0.1	Fabryka-Martin et al. (1991)
Gas hydrates of Blake Ridge in the Atlantic Ocean	55	Fehn et al. (2000)
Gas hydrates field, Nankai, Japan	24–48	Fehn et al. (2003)

in uranium salts. Since then, uranium has become an important element. Uranium oxide has been used since Roman times for yellow pigments in glass. Uranium is of great importance as a nuclear fuel. Although only mildly radioactive, uranium compounds are toxic. Much of the internal heat of the Earth results from the radioactive decay of uranium and thorium (together with potassium). Uranium's atomic weight and number are 238 and 92, respectively. Uranium has 24 isotopes, all radioactive, ranging in mass from 218 to 242, but only 9 of them have half-lives of longer than days. Naturally occurring uranium nominally contains 99.28305 by weight ^{238}U, 0.7110% ^{235}U, and 0.0054% ^{234}U. Uranium decay series include a chain of a series of atoms starting from ^{238}U (parent) to ^{206}Pb (final product). Uranuim-238 with a long half-life of 4.51×10^9 years has been used to estimate the age of igneous rocks and the Earth (Patterson, 1956). Hydrologic application of uranium disequilibrium in the groundwater has long been recognized (Osmond et al., 1968; Kaufman et al., 1969). In addition to dating groundwater, uranium disequilibrium phenomena can be used to study chemical erosion and the source of solutes in the rivers (Vigier et al., 2005), to solve the question of deep groundwater inflow into surface waters (Durand et al., 2005), to determine the source of fluvial sediments like alluvium (Yeager et al., 2002), to monitor changes to groundwater flow regime over time (Clark and Fritz, 1997), and to quantify mixing of surface water and groundwater (Briel, 1976).

Uranium-238 decays to uranium-234 through a three-step alpha and beta decay:

$$^{238}_{92}U \rightarrow {}^{234}_{90}Th + \alpha \rightarrow {}^{234}_{90}Th \rightarrow {}^{234}_{91}Pa + \beta \rightarrow {}^{234}_{91}Pa \rightarrow {}^{234}_{92}U + \beta$$

The product, uranium-234 with a half-life of 2.46×10^5 years, disintegrates also into thurium-230 by alpha decay reaction.

$$^{234}_{92}U \rightarrow {}^{230}_{90}Th + \alpha$$

Therefore, as ^{238}U atoms decays to ^{234}U, ^{234}U simultaneously disintegrates to ^{230}Th. As a consequence, on the one hand, ^{234}U atoms are increased, and on the other hand, they are decreased. So, in a closed geologic system where there is no extra source or sink for any of these two isotopes, after about 1,000,000 years, there will be a radioactive equilibrium between ^{234}U and its parent ^{238}U, i.e., their alpha activity ratio is 1.00. The reason for such conclusion is that after 10^6 years (almost four times that of the half-life of ^{234}U) almost all initially existed ^{234}U atoms are decayed and there will be no further ^{234}U atom to decay unless they are freshly produced by ^{238}U disintegration. However, if there is a source or a sink for any of these two isotopes (or the age of the system is less than 1,000,000 years), this equilibrium will not happen. This principle forms the basis of what is called *uranium disequilibrium series*, which is used for hydrological applications. A sink for ^{234}U (or if the system age is less than 10^6 years) will lead to a $^{234}U/^{238}U$ ratio of higher than unity. This is the case for most natural waters and oceans (Osmond et al., 1968). Uranium-234 is more mobile and is leached easier from where it is produced, as compared to ^{238}U. The reasons for such behavior include (1) nuclear recoil-induced bond breakage, (2) displacement of the ^{234}U within the crystal structure, and (3) oxidation of ^{234}U to 6^+ valence as a result of recoil within the lattice (Osmond et al., 1968, and various other references). Another, possibly important, factor is the direct alpha recoil of ^{234}U from the solid into the aqueous phase (Kronfeld et al., 1975).

6.6.1 Sampling, Analysis, and Reporting the Results

Groundwater samples just need to be free from particulate matters and hence may not need filtering through 0.45 micron filters (Osmond and Cowart, 2000). Collected water samples are analyzed for their uranium and $^{234}U/^{238}U$ ratio by either alpha spectrometry or more recently by mass spectrometric analysis. The former is cheaper, is time-consuming, and needs high sample volumes of up to 20 liters. The latter is quicker, expensive, and more precise. The result of analysis is reported in uranium concentration in µg/L and in the ratio of $^{234}U/^{238}U$, or simply called activity ratio, or in brief AR.

6.6.2 Dating Groundwater by UDS

Disequilibrium of $^{234}U/^{238}U$ and its variability in groundwater is due to the passage of groundwater through various geological formations and different

URANIUM DISEQUILIBRIUM SERIES

hydrogeochemical conditions. It is also partly due to the different leaching rate of ^{234}U and ^{238}U. Because ^{234}U is leached easier, groundwaters have $^{234}U/^{238}U$ ratios higher than 1. In terms of $^{234}U/^{238}U$ ratio and uranium content, three different zones can be recognized in a regional flow system (Ivanovich et al., 1991):

1. Oxidizing recharge zone where uranium content is high and $^{234}U/^{238}U$ is moderate.
2. Redox front (the point of intersection of oxidizing and reducing groundwaters) where both uranium content and $^{234}U/^{238}U$ are high.
3. Deep aquifer zone, which is characterized by constant, low-uranium concentration and steady decrease of the $^{234}U/^{238}U$ ratio.

The basis of dating groundwater by $^{234}U/^{238}U$ is the decrease in the activity ratio of $^{234}U/^{238}U$ from recharge area toward downstream (downgradient), i.e., from zone 1 to zone 3. This happens because ^{234}U decays with a faster rate than ^{238}U due to its shorter half-life. However, the two main downturns of this principle are

1. Addition of ^{234}U along the flow path (recoil of ^{234}Th downgradient may add to ^{234}U content of groundwater), and
2. The fact that age obtained through this principle is the age of uranium, not the age of groundwater because uranium travels downgradient at a slower rate than water molecules due to retardation and adsorption on the aquifer matrix (Osmond and Cowart, 2000).

To overcome these problems, a phenomenological model was developed by Ivanovich et al. (1991) to account for the evolution of uranium isotopes by taking into account dispersion, radioactive decay, and the first-order kinetic nature of sorption processes.
If a simple ^{234}U isotopic decay is assumed, the age of groundwater may be determined by Equation 6.14.

$$t = -\frac{1}{\lambda} \ln\left[\frac{C-1}{C_0-1}\right] \quad \text{(modified after Osmond and Cowart, 2000)} \quad (6.14)$$

where λ is the decay constant of ^{234}U, C is the $^{234}U/^{238}U$ ratio of a deep groundwater sample, and C_0 is the same ratio upflow near the redox front (initial value).

Example: The AR (or C) of a groundwater sample is 1.5. If the AR (or C_0) of groundwater at the redox front of the aquifer is 2.5, what is the age of this sample?

$$t = -\frac{1}{2.8 \times 10^{-6}} \ln\left[\frac{1.5-1}{2.5-1}\right] = -354,903 \times \ln 0.33 = 389,901 \text{ years}$$

The range of groundwater age datable by this method is thought to be from 10,000 years to 1.5 million years (Ivanovich et al., 1991), but this is just a rough suggestion and has yet to be proved. To be able to date groundwater successfully, one needs to know the configuration of aquifer porosity, the distribution of uranium and geochemical conditions in the aquifer, and the liquid to solid distribution coefficient of the uranium and its spatial variability in the groundwater and in the aquifer matrix. Kigoshi (1971) and Andrews et al. (1982) pro-

TABLE 6.7. Advantages and Disadvantages of Various Dating Methods

Method	Advantages	Disadvantages
^{81}Kr	No subsurface and anthropogenic sources for ^{81}Kr Limited geochemical reactions The only method to quantitatively date very old groundwaters	Limited laboratory facilities It has not been fully demonstrated as yet
^{36}Cl	Relatively large number of laboratories worldwide Well-known case studies Data can be used for other scientific applications	Subsurface sources for ^{36}Cl Initial value problem The method is not applicable to brackish and saline groundwaters
^{4}He	The method has been used in a large number of aquifers and oil well fields It covers a large age range from thousands to millions years In addition to age information, the data can be used for other scientific purposes	Calculating ^{4}He accumulation rate is a huge task Many sources for ^{4}He Lengthy and delicate sample preparation procedure
^{40}Ar	It could prove to be a supplement to the ^{4}He method ^{40}Ar data can be used in ^{4}He method for confidence building	The method has not been used extensively and is yet to be verified
^{129}I		Subsurface sources of ^{129}I that are difficult to quantify
Uranium disequlibrium series	It could be a supplementary method Measured data have applications other than dating	The method may be considered as obsolete, but may also be revived with the invent of the sophisticated analytical facilities

posed Equation 6.15 to account for some of the factors that prevent the use of simple decay equation:

$$t = -\frac{1}{\lambda} \ln\left[\frac{[U]_s \times (C - [U]_s) - 0.235[U]_r \rho S R}{(C_0 - 1)[U]_s - 0.235[U]_r \rho S R}\right]$$

modified after Metcalfe et al. (1998) (6.15)

where $[U]_s$ is the uranium content in groundwater (μg/kg), $[U]_r$ is the natural uranium content of the rock (μg/g), S is the surface area (cm²/mL), and R is the recoil range of ^{234}Th $= 3 \times 10^{-6}$. The rest of the parameters are as in the decay equation.

6.6.3 Case Studies

Dating groundwater by the ^{234}U/^{238}U method has been applied to only a limited number of aquifers worldwide. These include the Carrizo sand aquifer, Texas, USA (Cowart and Osmond, 1974), Stripa granite, Sweden (Andrews et al., 1982), Triassic Bunter sandstone aquifer of east Midland, UK (Andrews et al., 1983), and Milk River aquifer, Canada (Ivanovich et al., 1991). Here we briefly report the result of the last case. Twenty-one groundwater samples were collected from the Milk River aquifer and analyzed for uranium, ^{234}U and ^{238}U. The uranium values range from 0.008 to 14μg/kg, ^{234}U activities from 0.02 to 19 dpm/kg (disintegration per minute per kilogram), and ^{238}U activities from 0.005 to 10.3 dpm/kg. The ^{234}U/^{238}U activity ratio in the Milk River aquifer decreased from about 10.8 at the recharge zone to about 3 some 60 km downstream. Applying Equation 6.14 the age of groundwater at 60 km downstream is estimated at

$$t = -\frac{1}{2.8 \times 10^{-6}} \ln\left[\frac{3-1}{10.8-1}\right] = -354{,}903 \times \ln 0.2 = 246{,}000 \text{ years}$$

This age estimation is based on a simple decay model. However, using a phenomenological model (described above), the age of groundwater in the Milk River aquifer was determined to be younger. At the end of this subchapter we should remind the readers of the fact that dating groundwater by uranium disequilibrium series has not received any attention during the past 15 years. Therefore, it is safe to say that it can no longer be regarded as a dating method. The advent of mass spectrometric analysis of uranium isotopes, which are quicker and more precise, may, however, revive this method to some extent.

7

MODELING OF GROUNDWATER AGE AND RESIDENCE-TIME DISTRIBUTIONS

In this chapter, we try to mathematically describe how groundwater ages obtained by various methods—as described in the previous chapters—can be different from or similar to the real ages of each groundwater molecule/particle we sample in any dating study. Such a chapter is unprecedented in the groundwater textbooks, especially those that target Earth science specialists. Therefore, in contrast to all other chapters, it is necessary to start with a broad overview of the subject and to proceed step by step to help the average reader have a general impression of the content of the chapter.

The migration of groundwater particles through an aquifer is governed by the physical transport processes of advection, dispersion, and diffusion, which are described by the well-known flow and transport equations (see, e.g., Freeze and Cherry, 1979; Wang and Anderson, 1982; Huyakorn and Pinder, 1983; de Marsily, 1986; Fetter, 1999). As an intrinsic property of these moving particles, their respective age can therefore be assimilated to an extensive quantity evolving in space depending on the structure of the flow field, itself dictated by aquifer heterogeneity and recharge and discharge conditions. Hence, in any aquifer, groundwater particles move at various speeds along flow paths of various shapes and lengths. The morphology of the flow paths characterizes the geometric nature of a given flow system and determines the time water and rock are in contact. In regional systems, ages or residence times are often found between very short and very large values. The major consequence of this is that a groundwater sample taken in the aquifer, or at any of its outlets,

Groundwater Age, by Gholam A. Kazemi, Jay H. Lehr, and Pierre Perrochet
Copyright © 2006 John Wiley & Sons, Inc.

includes a range, or a distribution of particle ages. Hence, full groundwater age information at a given location must be represented by a frequency distribution, in the statistical sense, and not by a single value. This distribution may possibly exhibit a complex shape and cover a wide variety of age values depending on the aquifer nature and its hydrodispersive components. For example, multimodal, asymmetric, and skewed distributions are expected in cases of multilayer aquifers, fracture porosity, and fault systems. Figure 7.1 illustrates the residence-time variations in relation to the many possible flow paths leading to the discharge zone of a regional multilayer system. Depending on where infiltration takes place, these flow paths may reach different depths, and imply extremely different travel times to the outlet, affecting water chemistry and mineralization accordingly.

Groundwater age-dating by various tracers described in Chapters 4–6 generally gives us an average value for a given sample. This value as described in Chapter 3 is certainly very instructive and useful but gives no information at all on the actual age distribution and its possible complex features, which remain unknown. In fact, many age distributions can have the same average value (statistical mean, or first moment), and this average is sometimes far from the modes of the unknown distribution, namely the most frequent age

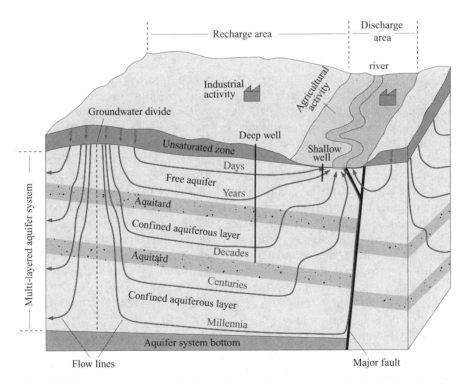

Figure 7.1. Illustration of a regional hydrogeological system with the different possible flow paths and travel times from the recharge to the discharge areas.

occurrences in the sample. Using the mean ages to evaluate aquifer characteristics such as recharge rates and flow velocities should therefore be made very carefully, and in principle only for relatively simple hydrogeological configurations for which standard age distributions can be enforced. For other, more complex flow systems, the mean age interpretation may lead to erroneous conclusions with sometimes severe consequences when water quality and water protection issues are addressed.

To date, the only ways to determine groundwater age and residence-time distributions are those of mathematical modeling based on the solution of age transport equations. To perform accurately, these approaches require full hydrodynamical description of the aquifer under investigation and the knowledge of its flow field. In general, the flow field is derived from the solution of the steady-state groundwater flow equations for calibrated hydrodynamical data sets. The calibrated flow field is then used as input, as well as other hydrodispersive parameters, to calculate the age transport processes and the resulting age distributions.

The following sections provide an overview on groundwater age modeling techniques from the 1960s up to the present state-of-the-art, and some mathematical basics on groundwater age transport, groundwater age, and residence-time distributions, as well as illustrative typical examples.

7.1 OVERVIEW AND STATE-OF-THE-ART

To quantify the distribution of ages in aquifers, several types of mathematical models have been developed during the past decades, and research in this field is progressing rapidly. The classical age modeling practice consists of using simple analytical models, commonly called lumped-parameter models, to interpret environmental tracer data. This type of approach has been extensively used and is still very popular (e.g., Maloszewski and Zuber, 1982, 1993; Campana and Simpson, 1984; Balderer, 1986; Richter et al., 1993; Amin and Campana, 1996). With lumped-parameter models, simple specific age distributions describing piston flow, exponential mixing, combined piston-exponential mixing, or dispersive mixing are assumed beforehand, sometimes very arbitrarily. The inverse problem is then solved by fitting the model response to tracer measurements by adjustment of some parameters (see Figure 7.2). This procedure calls for significant simplifications, which are often not justified, such as neglecting the reservoir structure, as well as the spatial variability of infiltration rates (Campana, 1987) and aquifer flow and transport parameters. Amin and Campana (1996) proposed to model the mixing process effects by means of a three-parameter gamma function ranging between no mixing (piston-flow model) and perfect mixing (exponential model).

Lumped-parameter models are often used to make inferences on groundwater flow velocities, aquifer turnover time, and recharge rates. However, sound verifications of the reliability of such models can hardly be found (e.g.,

OVERVIEW AND STATE-OF-THE-ART 207

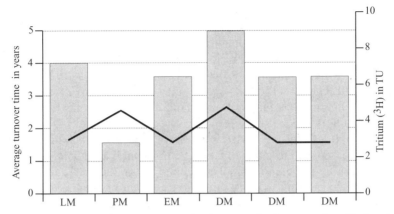

Figure 7.2. Comparison between turnover times at the Lutry spring (Switzerland), obtained with the classical linear (LM), piston-flow (PM), exponential (EM) and dispersive (DM, with three sets of parameters) models (adapted from Etcheverry, 2002). The solid line indicates the absolute difference between the modeled values and the average age measured at the spring by the ^{3}H method (in Tritium units).

see Haitjema, 1995; Luther and Haitjema, 1998; Etcheverry and Perrochet, 2000). Haitjema (1995) provided some guidelines for the use of lumped-parameter models and derived an analytical solution of the exponential model for semiconfined stratified aquifers. Luther and Haitjema (1998) claimed that the conditions for the validity of the exponential model (horizontal flow and homogeneous aquifer with respect to porosity ϕ, recharge rate i, and saturated thickness e) can be extended to configurations where the ratio $\phi e/i$ (namely the only free parameter in the exponential model) remains constant throughout the aquifer. This ratio is the system turnover time, or average residence time, which means that a water sample taken anywhere in the reservoir must give a mean age corresponding to the aquifer turnover time. Such a configuration and these uniform mean age conditions may hardly be found in nature, making the assumptions of Luther and Haitjema (1998) rather limiting. As demonstrated by Etcheverry (2001) a simple linear variation of the thickness e significantly influences the shape of the theoretical exponential age distribution. Classically, and for simplicity, age transport has also often been assumed driven by the only process of advection. In doing so, average ages can indeed be calculated easily, either analytically if the flow configuration is simple, or numerically by postprocessing the results of a groundwater flow model with particle-tracking techniques. Given a known velocity field, particle-tracking algorithms basically track down a certain number of released particles along path lines and during selected times. This can be done either forward by following the flow lines downstream or backward by following the flow lines upstream. However, the resulting advective ages ignore the effect of dilution, dispersion, and mixing on age transport (Cordes and Kinzelbach,

1992) and often reveal to be ill posed in complex heterogeneous systems (Varni and Carrera, 1998), for which the three-dimensional implementation is subject to severe technical problems. Moreover, particle-tracking techniques do not allow the calculation of residence-time distributions, since the moving groundwater volumes are not associated to the simulated ages. Nevertheless, the particle tracking method is very popular and may provide some help to validate dating methods, mainly isotopic methods, which consider the disintegration of radioactive isotopes as the age index (Smith et al., 1976).

An example of advective age simulation by particle tracking is shown in Figure 7.3(a), representing a plan view of a horizontal aquifer with a pumping well. Backward particle tracking is applied from the well, and isochrone markers at specific times (black dots) give an idea of the life expectancy of water particles in the well capture zone (i.e., the time left before absorption of these particles by the well). The advective capture zone of the well can therefore be delineated by the envelope of all isochrone markers. Following this approach, water particles are either in or out of the capture zone. However, it is clear that water particles also move laterally under dispersion and diffusion processes and, thus, depending on their distance to the well, may eventually exit or enter the capture zone. When comparisons are made between modeled and measured ages, the importance of accounting for age

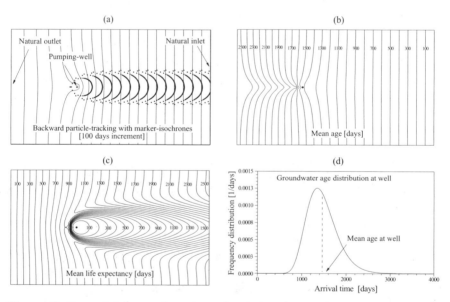

Figure 7.3. Example of groundwater age simulations in a shallow horizontal aquifer including a pumping well (modified after Cornaton, 2004). (a) Hydraulic heads and advective life expectancy calculated by the back-tracking method. (b) Advective-dispersive mean age. (c) Advective-dispersive mean life-expectancy. (d) Age distribution of the water extracted at the well.

diffusion and dispersion has been pointed out by many authors (e.g., Sudicky and Frind, 1981; Maloszewski and Zuber, 1982, 1991).

More elaborated quantitative approaches consider groundwater age as an extensive property, where transport through an aquifer can be described by volume-averaged field equations of the advective-dispersive type (e.g., Spalding, 1958; Harvey and Gorelick, 1995; Goode, 1996; Ginn, 1999). Comparing various modeling results with the experimental data of a complex system, Castro and Goblet (2005) found that the mean age field resulting from advective-dispersive age mass transport (Goode, 1996) was the most consistent. Varni and Carrera (1998) derived a set of recursive, temporal moment equations that were also compared to radiometric age measurements. According to Harvey and Gorelick (1995), the first four moments (characterizing the mean, the variance, the skewness, and the kurtosis of a breakthrough curve, respectively) may provide sufficient information to evaluate the entire age distribution at a given location. However, this finding only holds in cases where the principle of maximum entropy applies. Since many natural systems are likely to reveal multimodal age distributions of complex shapes, within the reservoir and at the discharge zones, this approach to construct entire age distributions (a priori unknown) remain somewhat uncertain, or nonunique. For instance, it is fairly obvious that calculations of mean age, or of age variance, at a given location (with first and second temporal moment equations) do not represent the full age picture since many different age distributions may exhibit the same mean and variance. Examples of mean age and mean life expectancy simulations including dispersion and diffusion effects are given in Figure 7.3(b) and (c). The mean age and mean life expectancy fields are obtained by two solutions of the advective-dispersive mean age equation (Goode, 1996), with forward and backward flow field, respectively. The mean life expectancy field in Figure 7.3(c), where the isolines indicate the time required by groundwater particles to reach one of the two outlets, can be compared to the advective backward particle-tracking solution in Figure 7.3(a), and the effects of dispersion on the size of the well capture zone can be assessed. However, none of these advective or advective-dispersive calculations provides any insight on the entire age or life expectancy distributions at a given location. For example at the extraction well, where many groundwater fractions with different ages mix, the knowledge of the entire age distribution shown in Figure 7.3(d) is of particular interest. However, approaches based on temporal moment analysis are not suitable for this purpose, and improved models must be enforced to achieve this task in a deterministic manner for any arbitrary hydrogeological configuration (Cornaton, 2004; Cornaton and Perrochet, 2005a, b).

Travel time probabilities have also been a subject of high interest in many studies characterizing solute transport in subsurface hydrology (e.g., Dagan, 1982, 1987, 1989). The travel time probability is commonly defined as the response function to an instantaneous unit mass flux impulse (Danckwerts, 1953). In their transfer function approach of contaminant transport through unsaturated soil units, Jury and Roth (1990) model tracer breakthrough curves with one-dimensional travel time probability functions. Shapiro and Cvetkovic (1988) and Dagan and Nguyen (1989) derived the forward travel time probability for a mass of solute by using the Lagrangian concept of particle displacement in porous media. Using a stochastic, longitudinal exponential model for the travel time density function, Woodbury and Rubin (2000) combined this approach with full-Bayesian hydrogeological parameter inference by inverting the travel time moments of solute breakthroughs in heterogeneous aquifers.

The derivation of forward and backward models for location and travel time probabilities has become a classical mathematical approach for contaminant transport characterization and prediction (e.g., van Herwaarden, 1994; van Kooten, 1995; Neupauer and Wilson 1999, 2001). The spreading of a contaminant mass is analyzed by following the random motion of solute particles, and, to do so, the advection-dispersion equation is assimilated to the Fokker–Planck, or forward Kolmogorov equation. The expected resident concentration of a conservative tracer is taken as the probability density function for the location of a particle, at any time after entry into the system. LaBolle et al. (1998) recall that the standard diffusion theory, which relates the dynamics of a diffusion process to Kolmogorov equations, may apply at local scale to advection-dispersion equations if the porosity and the dispersion tensor are functions varying smoothly in space. To solve these equations, random-walk procedures are often chosen as alternatives to more standard discretization methods (see, for example, Kinzelbach, 1992; LaBolle et al., 1998; Weissmann et al., 2002).

The derivation of adjoint state equations to define backward location and travel time probabilities was recently revisited by Neupauer and Wilson (1999, 2001). They derived specific forward and backward probabilities to predict contaminant sources from concentrations measured at monitoring and extraction wells. Compared to the forward modeling approach used to predict the future evolution of contamination when the source is known, the backward modeling approach reveals to be useful when the contamination has been detected and the possible sources have to be identified.

Following a different line of research, Eriksson (1961, 1971), followed by Bolin and Rhode (1973), introduced the reservoir theory in the field of environmental problems. Originating from chemical engineering, this theory considers a reservoir globally and is particularly useful to address residence-time issues. In essence, when applied to conservative particles flowing through a bounded system, the theory links the distribution of particle ages in the interior of the system to the residence-time distribution of the particle population. However, in spite of its fundamental features and capabilities, this approach has not received much attention in hydrogeology until recently (Etcheverry and Perrochet, 2000; Etcheverry, 2001). These authors proposed a method to directly calculate the groundwater residence-time distributions at the outlet of an aquifer of arbitrary heterogeneity by combining the age mass transport equation of Goode (1996) to the reservoir theory. One significant advantage of the method is that full temporal information is recovered at the outlet based on the solution of only two steady-state equations. Moreover, based on the internal organization of ages in the entire reservoir, the method minimizes the loss of age information due to mixing in the vicinity of local outlets where flow paths converge. The potential of this approach was investigated further by Cornaton (2004) and Cornaton and Perrochet (2005a, b) who generalized the reservoir theory to hydrodispersive systems with multiple inlets and outlets, and combined it to forward and backward Laplace-transformed equations for location and travel time probabilities. Achieving a particularly high computational efficiency, these recent works open a range of new research and application perspectives, not only in the field of age-dating, but also in associated contaminant hydrogeology issues involving large time scales. Facing the mathematical modelers' community, sustainable aquifer management with respect to climatic changes, accurate delineation of well-head protection zones, risk and safety assessments of nuclear and chemical waste repositories, among others, are now all 21st-century crucial issues in which groundwater age, life expectancy, and residence-time distributions play a major role in the long term.

7.2 BASICS IN GROUNDWATER AGE TRANSPORT

This section introduces the basics for modeling groundwater age and residence-time distributions by combining the reservoir theory to deterministic mathematical flow and transport models. This is made using a formalism kept as simple as possible, so that the reader can understand the principles of the method with standard mathematical knowledge and calculus techniques. However, notions on the modeling of groundwater flow and transport processes, whose presentation is not the scope of this chapter, are recommended and can be found in influential textbooks such as those by Freeze and Cherry (1979), Wang and Anderson (1982), Huyakorn and Pinder (1983), de Marsily (1986), and Fetter (1999).

7.2.1 The Reservoir Theory

The reservoir theory was introduced in environmental sciences by Eriksson (1961, 1971) to address atmospheric and surface hydrology issues. In the groundwater context, this theory can be developed from the divergence theorem applied to particle age-fluxes through a bounded domain (Etcheverry, 2001; Cornaton, 2004). This global approach enables us to quantify the intrinsic relation between age and residence-time distributions in any hydrodispersive and mixing flow regimes, and for multidimensional configurations of arbitrary heterogeneity. However, the essential features of the reservoir theory are more easily derived and interpreted for averaged steady-state flow systems, as generally assumed in groundwater modeling practices. In addition, the notion of local average ages is adopted in the following sections. This last restriction is enforced for the sake of mathematical simplicity without altering the generality and the usefulness of the approach to handle full, locally dispersed age and residence-time distributions.

Relation Between Internal Ages and Residence Times The simple system in Figure 7.4 represents a homogeneous aquifer of total pore volume V_o and total discharge rate Q_o. The various flow paths taken by water infiltrating over the inlet imply a range of residence times over the outlet and a range of ages in the aquifer. Given one particular, average age contour τ in the system, groundwater volumes can be arranged in a cumulative manner, so that the function $V(\tau)$ represents the resident volumes where ages are smaller than τ. Similarly, over the outlet, the discharge can be arranged in a cumulative manner, so that the function $Q(\tau)$ represents the volumetric rate of water leaving the system after a residence time smaller than τ. Considering now mass balance, the total recharge entering the system during the interval τ, namely the infiltrated volume $Q_o\tau$, must correspond to the sum of the groundwater volume $V(\tau)$ and the volume exfiltrated, during the interval, with a residence time smaller than τ. Hence, the relation between $Q(\tau)$ and $V(\tau)$ can be written as

$$Q_o\tau = V(\tau) + \int_0^\tau Q(t)dt \tag{7.1}$$

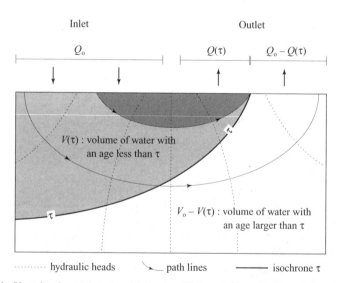

Figure 7.4. Sketch of a reservoir with hydraulic heads, flow lines, and age contour for age τ. The water in the dark gray area has an age and a residence time smaller than τ. The water in the light gray area has an age smaller than τ, but a residence time larger than τ.

BASICS IN GROUNDWATER AGE TRANSPORT

An equivalent relation based on age-mass balance over the pore volume $V(\tau)$ is

$$(Q_o - Q(\tau))\tau + \int_0^\tau t \frac{\partial Q(t)}{\partial t} dt = V(\tau) \tag{7.2}$$

where the terms in the left-hand side represent, respectively, the age-mass flux through the age contour τ and through the outlet. There is no input mass-flux through the inlet since the average age there is zero. In the above equation, the function $V(\tau)$ represents the age-mass production rate generated by the groundwater particles with age smaller than τ, and the function $\partial Q(\tau)/\partial \tau$ is the flux-average frequency distribution of the residence times over the outlet.

Using mass balance again, another way to relate the functions $V(\tau)$ and $Q(\tau)$ is to consider that the difference between the total recharge Q_o and the partial discharge $Q(\tau)$ is the discharge occurring after a residence time greater than τ (i.e., the groundwater flux through the age contour τ, or the flux of particles reaching the age τ). This corresponds to the instant rate of increase of the cumulated volume $V(\tau)$ around time τ and is simply expressed by

$$Q_o - Q(\tau) = \frac{\partial V(\tau)}{\partial \tau} \tag{7.3}$$

where $\partial V(\tau)/\partial \tau$ is the volume-averaged frequency distribution of the ages in the aquifer. It is easily seen that Equation 7.3 is the time derivative of either Equation 7.1 or 7.2. Equation 7.3 defines a general relation between the functions $V(\tau)$ or $Q(\tau)$, indicating that either one of the functions can be determined when the other is known. This relation is of a fully deterministic nature and can be expressed in a more compact form. Considering that the maximum value of these cumulated functions are $V(\tau = \infty) = V_o$ and $Q(\tau = \infty) = Q_o$, and replacing the time symbol by t, the age and residence-time probability density distributions, $\psi(t)$ and $\varphi(t)$, can be respectively defined by the standardized expressions

$$\psi(t) = \frac{1}{V_o} \frac{\partial V(t)}{\partial t}, \quad \int_0^\infty \psi(t)dt = 1 \tag{7.4}$$

and

$$\varphi(t) = \frac{1}{Q_o} \frac{\partial Q(t)}{\partial t}, \quad \int_0^\infty \varphi(t)dt = 1 \tag{7.5}$$

These two distributions describe the frequency of time occurrences with respect to fractions of aquifer pore volumes and discharge rates and, therefore, are real statistical distributions that may be used to perform probabilistic inferences. For example, in a sample collected at an aquifer outlet, the fraction of the particles with a residence time between t_1 and t_2, or, alternatively, the probability that one particle in the sample has a residence time in this interval, is

$$P(t_1 < t < t_2) = \int_{t_1}^{t_2} \varphi(t) dt \tag{7.6}$$

Differentiating Equation 7.3 with respect to time and substituting the definitions (7.4) and (7.5) yields

$$\varphi(t) = -\frac{V_o}{Q_o} \frac{\partial \psi(t)}{\partial t} \tag{7.7}$$

This equation relates the statistical distribution of groundwater ages $\psi(t)$ in the reservoir to that of the residence times at its outlet $\varphi(t)$. Replacing, in the above developments, the notion of age by the notion of life expectancy would lead to exactly the same results. Since the residence time corresponds to both the age at the outlet and the life expectancy at the inlet, it is clear that the distribution of groundwater ages in the reservoir, $\psi(t)$, is also the distribution of groundwater life expectancies in the reservoir. Equation 7.7 is the major constitutive statement of the reservoir theory and can be used to gain insights into some relevant age-related aquifer characteristics.

Characteristic Times The turnover time, τ_o, expresses the ratio of the total groundwater volume in the aquifer (or pore volume in saturated conditions) to its total discharge

$$\tau_o = \frac{V_o}{Q_o} \tag{7.8}$$

and appears as the only scaling factor in Equation 7.7. Substituting Equation 7.4 into Equation 7.3 yields the relation

$$Q_o - Q(t) = V_o \psi(t) \tag{7.9}$$

which, since $Q(t = 0) = 0$, indicates that the initial value of the age distribution, or life-expectancy distribution, is always $\psi(t = 0) = Q_o/V_o = 1/\tau_o$, namely, the inverse of the turnover time. Moreover, the cumulative functions $V(t)$ and $Q(t)$ are, by definition, nondecreasing functions of time. This implies, from Equations 7.4 and 7.5, that $\psi(t)$ and $\varphi(t)$ cannot be negative and, from Equation 7.7, that $\partial \psi(t)/\partial t$ cannot be positive. Hence, the statistical distribution of ages in an aquifer, $\psi(t)$, always starts with the value $1/\tau_o$ at $t = 0$, is never

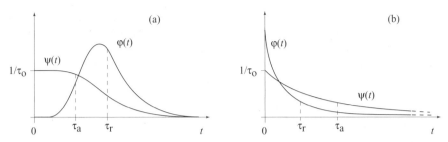

Figure 7.5. Examples of regular statistical distributions of groundwater ages in the aquifer ψ(t) and residence times at the outlet φ(t) with respective mean values. (a) $\tau_a < \tau_r$ and (b) $\tau_a > \tau_r$.

increasing, and reaches zero at $t = \infty$, or at the maximum age (necessarily located at the outlet). On the other hand, the statistical distribution of residence times at the outlet, φ(t), can have a much less regular shape governed by the evolution of ψ(t), as dictated by Equation 7.7. Depending on the geometry of the flow patterns in the aquifer, φ(t) can show virtually any initial and intermediate values, but must also vanish at $t = \infty$, or at the maximum residence time (i.e., maximum age) detected over the outlet. Two examples of simple age and residence-time distributions are given in Figure 7.5 with their respective mean values τ_a and τ_r.

The mean residence time of particles leaving the reservoir, τ_r, is a frequency-averaged value corresponding to the statistical first moment of the residence time distribution φ(t). Making use of the fundamental relation (7.7) and considering that the integral of the function ψ(t) is one by definition results, after integration by parts, in

$$\tau_r = \int_0^\infty t\varphi(t)dt = -\frac{V_o}{Q_o}\int_0^\infty t\frac{\partial \psi(t)}{\partial t}dt = -\frac{V_o}{Q_o}\left(t\psi(t)\bigg|_0^\infty - \int_0^\infty \psi(t)dt\right) = \frac{V_o}{Q_o} = \tau_o \quad (7.10)$$

Thus, the mean residence time of the particles flowing through the reservoir is identical to the turnover time. Similarly, the average age in the aquifer is the first moment of the age distribution ψ(t). Integrating by parts and making again use of the relation (7.7) results in

$$\tau_a = \int_0^\infty t\psi(t)dt = \left(\frac{t^2}{2}\psi(t)\bigg|_0^\infty - \frac{1}{2}\int_0^\infty t^2\frac{\partial \psi(t)}{\partial t}dt\right) = \frac{Q_o}{2V_o}\int_0^\infty t^2\varphi(t)dt = \frac{1}{2\tau_o}(\tau_r^2 + \sigma_r^2)$$

(7.11)

where the statistical second moment of the distribution φ(t), in the second-to-last term above, is expressed as the sum of the average residence time squared, τ_r^2, and of the variance of the residence time σ_r^2 (square of the standard deviation σ_r). It can also easily be shown with the above equation that the total age mass produced in any steady-state reservoir is always the product $V_o\tau_a$.

Substituting the result (7.10), $\tau_r = \tau_o$, in Equation 7.11 yields the interesting relation

$$\tau_a = \frac{\tau_o}{2}\left(1 + \frac{\sigma_r^2}{\tau_o^2}\right) \qquad (7.12)$$

Hence, for systems with large turnover times, or relatively low hydrodispersive properties (small variations of particles residence times at outlet, small σ_r), the average age in the reservoir may be smaller than the turnover time, or average residence time, with the lower possible value $\tau_a = \tau_o/2$ occurring in the piston-flow configuration only ($\sigma_r = 0$). For systems with small turnover time, or relatively high hydrodispersive properties, the average age in the reservoir may be much larger than the average residence time.

In cases where $\tau_a < \tau_o$, the functions $\psi(t)$ and $\varphi(t)$ have typical shapes as illustrated in Figure 7.5(a) and are characterized by the fact that no, or very few, groundwater particles leave the reservoir soon after having entered it [i.e., $\varphi(t)$ is relatively small at short times]. Reservoirs with modest transport velocities or inlet and outlet zones far apart belong to this category. Figure 7.5(b) illustrates the typical shapes of the functions $\psi(t)$ and $\varphi(t)$ for cases where $\tau_a > \tau_o$. This category is encountered when most of the water particles entering a reservoir stay there only a short period of time, making τ_o small, while the rest of the particles remain in the reservoir sufficiently long to make τ_a comparatively large.

The particular case for which $\tau_a = \tau_o$, although corresponding to very specific flow configurations, is frequently assumed in practice to interpret isotopic age data or tracer breakthroughs. As seen from Equations 7.10 and 7.11, the general condition to have $\tau_a = \tau_r = \tau_o$ is

$$\int_0^\infty t(\psi(t) - \varphi(t))dt = 0 \qquad (7.13)$$

and a sufficient condition for that is $\psi(t) = \varphi(t)$. Inserting this condition in Equation 7.7 results in

$$\varphi(t) = -\tau_o \frac{\partial \varphi(t)}{\partial t} \qquad (7.14)$$

which, after integration, yields the well-known and widely used exponential distribution

$$\varphi(t) = \psi(t) = \frac{1}{\tau_o} e^{-t/\tau_o} \qquad (7.15)$$

BASICS IN GROUNDWATER AGE TRANSPORT

Hence, if either one of the two frequency functions can be shown to have an exponential form, it follows that the other frequency function must be identical. In this very special case, both functions show equal mean and standard deviation, with $\tau_a = \tau_r = \tau_o = \sigma_r$. Exponential frequency distributions characterize reservoirs in which age occurrences in a water sample appear with equal probabilities regardless of the location. As a consequence, average age measurements at any location in the reservoir should theoretically be identical and correspond to the average residence time. One known flow configuration tending toward this special case is that of a horizontal, vertically well-mixed, shallow aquifer uniformly recharged over its whole surface.

Characteristic Volumes The specific groundwater volumes related to a given range of ages or residence times are important quantities to consider when addressing aquifer management strategies. Assessing the long-term evolution of groundwater chemistry, or defining corrective measures aiming at restoring groundwater quality after a contamination event, indeed requires appropriate age and volume-related information. Provided the residence-time distribution $\varphi(t)$ and the total discharge Q_o are known, this can be done in a relatively straightforward manner by a simple analysis of the function $Q(t)$, expressing the cumulative discharge with respect to residence time. In effect, from Equation 7.5, this function takes the integral form

$$Q(\tau) = Q_o \int_0^\tau \varphi(t) dt \tag{7.16}$$

This function is always zero at very early times and starts to increase when the fastest particles are detected at the outlet (shortest residence time, t_{min}). The total discharge Q_o is reached when the slowest particles exfiltrate (longest residence time, t_{max}).

To illustrate the principles of groundwater volume evaluation, a simple aquifer is skematized in Figure 7.6(a), with age and life expectancy isochrones, flow lines, and specific groundwater volumes. The corresponding cumulative discharge $Q(t)$ is drawn in Figure 7.6(b) with the corresponding groundwater volumes. It is seen that the volumes in the aquifer reduce to simple geometric planar forms in the diagram space, where they can be easily calculated.

Considering any given value of time, such as the value τ indicated in Figure 7.6(b), the colored zones represent the groundwater volumes in the aquifer having ages less than τ. Some of these volumes flow through the system in less than τ (V_1 and V_2) and in more than τ (V_3 and V_4). V_1 and V_3 are groundwater volumes of ages less than t_{min}, but their residence times are, respectively, smaller and larger than τ. V_2 and V_4 are groundwater volumes of ages between t_{min} and τ, but the residence times are larger than τ for V_4 and smaller than τ for V_2. The volume V_5 is the groundwater volume with ages greater than τ, i.e., the remaining white area in Figure 7.6(a). The surface below the curve $Q(t)$,

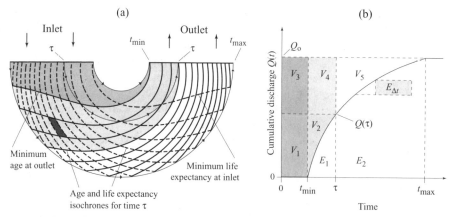

Figure 7.6. Groundwater volumes. (a) Simple flow system with age and life expectancy isochrones, flow lines, and specific volumes. (b) Plot of the cumulative discharge $Q(t)$ and evaluation of the specific volumes related to a given age τ.

between t_{min} and τ, represents the volume of exfiltrated water (E_1) having traveled from the inlet to the outlet during an observation period τ. If the observation period is t_{max}, E_2 is the volume of exfiltrated water having traveled through the entire system during the period, and in a time ranging between τ and t_{max}. E_1 and E_2 are obtained by

$$E_1 = \int_{t_{min}}^{\tau} Q(t)dt, \quad E_2 = \int_{\tau}^{t_{max}} Q(t)dt \qquad (7.17)$$

Knowing these two quantities, any of the other volumes, which can have any complex shape in a real situation, are readily determined by simple calculation of rectangular surfaces. The total surface above the curve $Q(t)$ is the total pore volume V_o in the aquifer, corresponding to

$$V_o = V(t_{max}) = Q_o t_{max} - \int_{t_{min}}^{t_{max}} Q(t)dt = Q_o t_{max} - E_1 - E_2 = V_1 + V_2 + V_3 + V_4 + V_5 \qquad (7.18)$$

as indicated by Equation 7.1 for $\tau = t_{max}$.

A large variety of other specific volumes can be defined when more refined information is required. For example, the volume of water exfiltrating at the outlet with a residence time between t_1 and t_2 during an observation period of any length Δt [the volume $E_{\Delta t}$ in Figure 7.6(b)] is simply obtained, considering Equation 7.16, by

BASICS IN GROUNDWATER AGE TRANSPORT

$$E_{\Delta t} = \Delta t Q_o \int_{t_1}^{t_2} \varphi(t)dt = \Delta t (Q(t_2) - Q(t_1)) \tag{7.19}$$

The residence time distribution $\varphi(t)$ relates to the age over the outlet with respect to discharge rates, as well as to the life expectancy over the inlet with respect to recharge rates. Similarly, the function $\psi(t)$ is the age distribution, as well as the life expectancy distribution, in the interior of the aquifer. Volumetric estimations combining these two properties can thus be made, such as finding the volume of groundwater having ages between a_1 and a_2, and life expectancies between e_1 and e_2. This volume, corresponding to the little red zone in Figure 7.6(a), is

$$V_{(a_1 < a < a_2, e_1 < e < e_2)} = \int_{a_1+e_2}^{a_2+e_2} Q(t)dt - \int_{a_1+e_1}^{a_2+e_1} Q(t)dt \tag{7.20}$$

and can also be easily calculated from the cumulated discharge diagram $Q(t)$ in Figure 7.6(b).

Reservoirs with Multiple Inlets and Outlets The theory presented and illustrated above on the basis of a very simple system applies to steady-state flow systems of arbitrary geometry and heterogeneity. For a flow configuration including several inlets and outlets, the age or life expectancy distribution $\psi(t)$ is relative to the pore volume of the whole system V_o, and the residence-time distribution $\varphi(t)$ is that of the total discharge (or recharge) Q_o including all outlets (or inlets). If a residence-time distribution is required for analysis at only one of the outlets, for example at a given well or spring, then all the calculations presented above must be performed on the subsystem related to that outlet. Several mathematical techniques can be applied to delineate the subsystems contributing to one given outlet, or, alternatively, the subsystems invaded by the water infiltrated over one particular inlet. A particularly efficient way to achieve this task in a general hydrodispersive context, and, at the same time, perform the specific calculations required by the reservoir theory, is to combine the analysis of reverse flow-field transport equations with the concepts of fractional, or probabilistic, capture zones (Cornaton and Perrochet, 2005b). However, when the outlets are sufficiently apart from each other, or when hydrodispersive and mixing processes are relatively small, the detailed analysis of individual subsystems is not needed. For such cases, individual outlet signatures with their specific mode and nonoverlapping bounds may indeed be distinctly identified from the residence-time distribution calculated globally for the total discharge. As illustrated in Figure 7.7, a global residence-time distribution $\varphi(t)$ can be expressed as

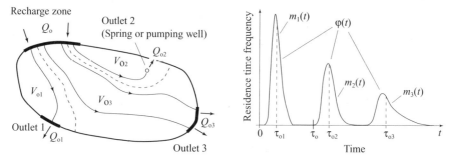

Figure 7.7. Illustration of a flow system with multiple outlets and subsystems. Global residence time distribution $\varphi(t)$ and signatures of individual outlet subsystems $m_i(t)$.

$$\varphi(t) = \sum_{i}^{n} m_i(t) \qquad (7.21)$$

where the terms $m_i(t)$ are the signatures (distinct in the present example, but often overlapping in general cases) of each individual outlet i, and n is the total number of oulets of the global system. The integral of an individual signature is

$$\int_0^\infty m_i(t)dt = \frac{Q_{oi}}{Q_o} \qquad (7.22)$$

expressing the ratio between the discharge at the corresponding outlet, Q_{oi}, and the total discharge of the system Q_o. Scaling $m_i(t)$ by this ratio and taking its first moment results in

$$\frac{Q_o}{Q_{oi}}\int_0^\infty tm_i(t)dt = \int_0^\infty t\varphi_i(t)dt = \tau_{oi}, \qquad \varphi_i(t) = \frac{Q_o}{Q_{oi}}m_i(t) \qquad (7.23)$$

where $\varphi_i(t)$ is the scaled residence-time distribution at the particular outlet, and τ_{oi} is the turnover time, or average residence time in the subsystem contributing to that outlet. The pore volume of the subsystem i, or the groundwater volume contributing to the outlet i, is therefore $V_{oi} = Q_{oi}\tau_{oi}$. For a system with n outlets, it is then clear that

$$\sum_{i}^{n}Q_{oi}\tau_{oi} = \sum_{i}^{n}Q_{oi}\int_0^\infty t\varphi_i(t)dt = Q_o\int_0^\infty t\sum_{i}^{n}m_i(t)dt = Q_o\int_0^\infty t\varphi(t)dt = Q_o\tau_o = V_o \qquad (7.24)$$

where V_o is the total pore volume of the system. It also follows that the average turnover time of the system is the discharge weight average of the individual turnover times

BASICS IN GROUNDWATER AGE TRANSPORT

$$\tau_o = \frac{V_o}{Q_o} = \frac{\sum_{}^{n} Q_{oi}\tau_{oi}}{\sum_{}^{n} Q_{oi}} \qquad (7.25)$$

Further manipulations can be made with the age or life expectancy distributions $\psi_i(t)$, and it can be demonstrated that Equations 7.1 to 7.20 also hold for any aquifer subsystem. For instance, the important relation, in Equation 7.7, between the age and residence-time distributions in a global system, becomes

$$\varphi_i(t) = -\frac{V_{oi}}{Q_{oi}}\frac{\partial \psi_i(t)}{\partial t} = -\tau_{oi}\frac{\partial \psi_i(t)}{\partial t} \qquad (7.26)$$

in the subsystems contributing to the discharge at a given outlet i.

Hence, all the features of the reservoir theory discussed, illustrated, and interpreted previously for a simple aquifer with single inlet and outlet zones similarly apply to systems with multiple inlet and outlet zones, either in individual subsystems or in any combination thereof.

Analogies with a Human Population   The age-related concepts introduced in the above sections apply to any kind of reservoir, including that of the human population, or of a population subgroup. If, over a total number of people V_o, $V(\tau)$ is the number of people of an age less than τ, measured in years, $V_o\psi(\tau)$ is then the population age distribution measured as the number of people in each year class. Q_o is the birthrate, measured in unit per year, and $Q(\tau)$ is the yearly number of deaths among the part of the population that is younger than τ. The relation (7.9), $Q_o - Q(\tau) = V_o\psi(\tau)$, expresses here the fact that, in a steady-state population, the number of deaths among people older than τ must be balanced by the number of persons reaching the age τ each year.

The characteristic times τ_o and τ_a defined in Equations 7.10 and 7.11 can also be easily understood when representing the average lifetime and the average age of a population, respectively. The average lifetime, or turnover of the population, is the total number of people divided by the birthrate. In most industrialized countries, due to a relatively low rate of infant mortality, the life expectancy, or average lifetime of a newborn is about 75 years (τ_o), whereas the average age of the living people is about 40 years (τ_a). This is typical of the case where $\tau_a < \tau_o$. The reverse is true in other countries where a high rate of infant mortality does not prevent a significant number of people from reaching large ages before dying. Moreover, and although birth and death are the overall single inlet and outlet of a human reservoir, the relative importance of various fatal diseases or accidents can be more specifically analyzed by application of the theory to the corresponding population subgroups or individual subsystems.

Interpreting Mean Age in Large Aquifer-Aquitard Systems Age measurements often result in an average value that represents the various ages of the particles in a given water sample. When the sample is collected at a point in the aquifer, this value corresponds to the average time needed for the particles to reach the sampling location after having infiltrated the aquifer. When the sample is taken at an aquifer outlet, and includes a mixture of particles having converged toward this outlet along different flow paths, the measured average age value is the mean residence time in the aquifer subsystem contributing to that outlet. The mean age, or mean residence time, is certainly a first instructive quantity that may give some insights into global, age-related aquifer characteristics, such as turnover time and qualitative age classification of the water produced at a given location. Mean age data are also often used in practice to evaluate hydrodynamical aquifer parameters such as recharge rates and flow velocities. However, to do so, working assumptions have to be enforced as to the aquifer structure and flow patterns. These assumptions often result in oversimplified conceptual flow models, yielding thus a considerable uncertainty about the results. Typical, simplified conceptual flow models frequently used for mean age interpretation are the piston-flow model and the exponential model. These models characterize only very few specific flow configurations and they owe their popularity much more to their simplicity and ease of use, rather than to their hydrogeological relevance. As a consequence, enforcing such simplified models to interpret mean age data when the real hydrogeological settings are unknown may often lead to severe errors, as illustrated below.

Figure 7.8 illustrates a simple conceptual model of an aquifer-aquitard system. The aquifer is assumed horizontal and homogeneous with high permeability K, thickness e, and porosity ϕ. The underlying aquitard is also assumed homogeneous with low permeability K_a, thickness e_a, and porosity ϕ_a. The inlet and outlet zones are separated by the distance L, and the total steady-state discharge rate through the system is Q_o. If the average residence time is measured at the outlet in the prospect of estimating the horizontal flow

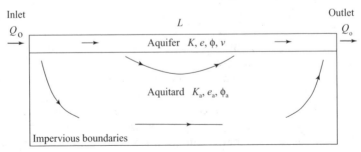

Figure 7.8. Aquifer-aquitard system with single local inlet and outlet. Simple conceptual model for the interpretation of average ages and residence time.

BASICS IN GROUNDWATER AGE TRANSPORT

velocity v in the aquifer, then the conceptual piston-flow model will undoubtedly be chosen to interpret the data. Given this very simple configuration, and the fact that the permeability contrast implies a constant and horizontal velocity in the aquifer, one may indeed expect that average groundwater ages in the aquifer grow linearly with distance between the inlet and outlet zones. Accordingly, age growth with distance would therefore simply be expressed by

$$A(x) = \frac{x}{v} \qquad (7.27)$$

where x is the horizontal distance counted from the inlet ($x = 0$).

Hence, the average residence time τ_o measured at the outlet, namely the age of groundwater at the distance $x = L$, can be substituted in the above piston-flow model $[A(L) = \tau_o = L/v]$ and the pore velocity $v = L/\tau_o$ is obtained. Knowing the aquifer porosity ϕ, the pore velocity can then be transformed into a Darcy velocity $q = \phi v$, which, in turn, can be used to estimate the assumed homogeneous aquifer permeability if the overall hydraulic gradient is known. The standard piston-flow model in Equation 7.27 is, however, not suitable in the present case because it ignores the age-mass exchanges occurring at the interface aquifer-aquitard. Since the porosity, and thus the permeability of the aquitard, are not zero, age-mass fluxes do exist between the aquifer and the aquitard where an enormous amount of age mass is generated.

In effect, considering the general result of the reservoir theory derived in Equation 7.10, the average residence time is the turnover time $\tau_o = V_o/Q_o$. In a two-component aquifer-aquitard system, the total pore volume V_o is the sum of the pore volume of the aquifer and that of the aquitard. Hence, one can write

$$\tau_o = \frac{V_o}{Q_o} = \frac{V_{aquifer} + V_{aquitard}}{Q_o} = \frac{V_{aquifer}}{Q_o}\left(1 + \frac{V_{aquitard}}{V_{aquifer}}\right) \qquad (7.28)$$

This relation clearly indicates the contribution of the aquitard pore volume to the global aging process of the system expressed in the average residence time. Since aquitards occupy very significant volumes in most hydrogeological settings, the ratio between aquitard and aquifer pore volumes in the above equation can be relatively large. Beyond a few units, this ratio practically acts as a multiplier of the sole quantity $V_{aquifer}/Q_o$ usually, and wrongly, taken as the turnover time.

Applying this result to the example in Figure 7.8, in which $Q_o = eq = e\phi v$, yields

$$\tau_o = \frac{L}{v}\left(1 + \frac{e_a \phi_a}{e \phi}\right) = \frac{L}{v} R \qquad (7.29)$$

as the correct piston-flow model to enforce in this type of configuration. The required correction R due to the aquitard volume clearly appears here as a retardation factor. This factor is similar to the retardation factor characterizing the transport of sorbing solutes, and making the apparent transport velocity smaller than the real pore velocity of the water particles. Since, in the piston-flow model, average age is used as an advective tracer, neglecting this correction by setting $R = 1$ in the above equation is analogous to analyzing the advective migration of a sorbing tracer, but without accounting for the sorption processes that delay breakthrough times. This common, but erroneous, interpretation of average residence times systematically results in velocity underestimations often up to one or two orders of magnitude.

When groundwater age measurements are made in the aquifer between the inlet and the outlet, a reasonable, piston-flow estimation of the pore velocity between two observation points can be made with the formula

$$v = \frac{x_2 - x_1}{\tau_2 - \tau_1} R, \qquad R = 1 + \frac{e_a \phi_a}{e \phi} \qquad (7.30)$$

where $x_2 - x_1$ is the investigated distance and $\tau_2 - \tau_1$ is the travel time (difference between the measured ages at the locations x_2 and x_1). However, this estimation is not entirely correct because the age isochrones in the aquifer are not vertical. This is due to the massive production of age mass in the aquitard and to gradual age-mass fluxes through the aquifer–aquitard interface. Hence, significant and nonuniform vertical age gradients exist in the aquifer, making the measurements τ_1 and τ_2 depth-dependent, and, therefore, not necessarily representative of the average age at the locations x_1 and x_2.

This last point is highlighted in Figure 7.9, showing the results obtained by a full solution of the age transport equation. The average age contours in Figure 7.9(a) indicate how age mass is produced in the aquitard and how it is gradually released into the aquifer. Due to age-mass continuity at the aquifer–aquitard interface, the age contours are far from being vertical in the aquifer, although most of the discharge rate Q_o flows through it. Average age values measured at the observation points x_1 to x_4 in the aquifer are, therefore, local values representative of their sampling depth only. Thus, equal sampling depths would at least be required to interpret these data in terms of horizontal flow velocities by means of a piston-flow model.

Figure 7.9(b) illustrates the full age distributions modeled at the locations x_3 and x_4. In the linear-scale plot, two modes, or two signatures, are clearly identified at each of the two sampling points, revealing the presence of two contributing bodies. The magnitudes of these signatures indicate the respective, volumetric contributions of each of the bodies to the total age mass detected at these points. These contributions necessarily evolve with distance according to specific age mass increases in both the aquifer and the aquitard, and to the rate of age-mass exchanges at the interface. Given the size of the aquitard, and the downstream location of the sampling points x_3 and x_4,

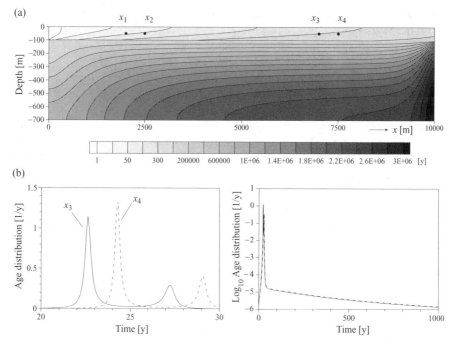

Figure 7.9. Aquifer-aquitard system with single local inlet and outlet (modified after Cornaton, 2004). (a) Mean age contours; (b) Age frequency distributions at locations x_3 and x_4 in linear and log scales.

particles with ages reaching thousands of years are included in the samples, albeit in much smaller proportions. This is particularly apparent in the log-scale plot where the age distribution exhibits a long, and slowly decreasing tail. Therefore, the average age at these sampling locations may well be far above the time required by the water particles in the aquifer to travel from the inlet to the locations x_3 and x_4.

The average ages calculated at the four observation points (first moment of the modeled age distributions) are now assimilated to a set of data available to evaluate flow velocities in the aquifer. In doing so, the relevance of the piston-flow approach can be directly assessed by comparison of the inferred velocities with the true model velocity. In the example of Figure 7.9, age transport was modeled with standard hydrodispersive properties and with the parameters $K = 10^{-3}$ m/s, $e = 100$ m, and $\phi = 0.1$ for the aquifer, and $K_a = 10^{-8}$ m/s, $e_a = 600$ m, and $\phi_a = 0.3$ for the aquitard. A hydraulic head gradient of 10^{-3} was specified between the inlet and outlet zones, separated by the distance $L = 10,000$ m. With these conditions, and given the permeability contrast, the model pore velocities turned out to be completely uniform over the aquifer thickness at $v = 315.4$ m/yr ($v = 10^{-5}$ m/s), yielding the discharge rate $Q_o = 3,154$ m³/yr per unit width. With the total pore volume $V_o = 1.9 \cdot 10^6$ m³ per unit width, the turnover time, or average residence time of this aquifer–aquitard system is $\tau_o = 602.4\,\text{yr}$, and the retardation factor in Equation 7.29 is $R = 19$. The horizontal locations of the four observation points are $x_1 = 2,000\,\text{m}$, $x_2 = 2,500\,\text{m}$, $x_3 = 7,000\,\text{m}$, and $x_4 = 7,500\,\text{m}$, each at the depth of 50 m, and the average ages calculated at these locations are $\tau_1 = 27.1\,\text{yr}$, $\tau_2 = 44.6\,\text{yr}$, $\tau_3 = 278.3\,\text{yr}$, and $\tau_4 = 307.4\,\text{yr}$. Combining these data and assuming piston flow, the evaluation of the pore velocity v can be done by means of Equation 7.30, without and with retardation correction.

Considering the model velocity $v = 315.4\,\text{m/yr}$ as the correct theoretical result, it is seen, in Table 7.1, that severe underestimations occur when the aquitard is neglected ($R = 1$). The error grows with the distance from the inlet according to growing age-mass releases from the aquifer, which are not accounted for. Using the retardation factor $R = 19$ provides overestimated, but more realistic values (except in the upstream part of the aquifer), with accuracy increasing with the distance from the inlet. Thus, with minimum additional geological information, such as the thickness and porosity of the underlaying aquitard in the above example, data interpretation could reach a more reasonable accuracy, particularly when several measurements can be combined in a systematical manner. The comparative analysis performed above can be applied to other flow configurations and used to assess the validity of the exponential, piston-exponential, or other simplified interpretation models, resulting in similar restrictions and relevance level. The fundamental problem with standard interpretation models is that the average age is not necessarily the appropriate input quantity, but is nevertheless assumed as such because it is the only data available, or measurable. In fact, for a straight and accurate evaluation of aquifer velocities, the appropriate value to consider is the time corresponding to the first peak of the full age distribution, namely the most frequent breakthrough time, or age mode, of the water particles. Considering

TABLE 7.1. Aquifer Pore-Velocity Estimation Based on Average Age Data and Piston-Flow Assumptions, Without ($R = 1$) and With ($R = 19$) Retardation Effects Due to the Aquitard

Distance [m]	Average age (yr)	Velocity (m/yr), $R = 1$	Velocity (m/yr), $R = 19$
$x_1 = 2,000$	$\tau_1 = 27.1$	73.8	1,402.2
$x_2 = 2,500$	$\tau_2 = 44.6$	56.1	1,065.0
$x_3 = 7,000$	$\tau_3 = 278.3$	25.2	477.9
$x_4 = 7,500$	$\tau_4 = 307.4$	24.4	463.6
$x_2 - x_1 = 500$	$\tau_2 - \tau_1 = 17.5$	28.6	542.9
$x_3 - x_1 = 5,000$	$\tau_3 - \tau_1 = 251.2$	19.9	378.2
$x_4 - x_1 = 5,500$	$\tau_4 - \tau_1 = 280.3$	19.6	372.8
$x_3 - x_2 = 4,500$	$\tau_3 - \tau_2 = 233.7$	19.3	365.9
$x_4 - x_2 = 5,000$	$\tau_4 - \tau_2 = 262.8$	19.0	361.5
$x_4 - x_3 = 500$	$\tau_4 - \tau_3 = 29.1$	17.2	326.5
$L = 10,000$	$\tau_0 = 602.4$	16.6	**315.4**

TABLE 7.2. Aquifer Pore-Velocity Estimation Based on Age Mode and Piston Flow

Distance [m]	Age mode (y)	Velocity (m/y)
$x_3 = 7{,}000$	$\tau_3 = 22.6$	309.7
$x_4 = 7{,}500$	$\tau_4 = 24.2$	309.9
$x_4 - x_3 = 500$	$\tau_4 - \tau_3 = 1.6$	312.5
Theoretical velocity		**315.4**

Figure 7.9(b) and the age distributions modeled at the locations x_3 and x_4, the age modes at these two locations can be evaluated to $\tau_3 = 22.6$ yr and $\tau_4 = 24.2$ yr (more than 10 times smaller than the average ages at these points), yielding the difference $\Delta\tau = 1.6$ yr over the distance $\Delta x = 500$ m. As shown in Table 7.2, average pore velocities obtained with these age modes and the formula $v = x/\tau$ are indeed in excellent agreement with the theoretical value. However, direct age mode measurements do not exist and an approximation of the latter can only be attempted by means of selected tracing experiments, not always successful at large scales. Classical interpretation of groundwater ages in terms of velocities in large, real systems remains therefore a difficult task when geological information is not sufficient. Unless more detailed flow models can be constructed and calibrated on the basis of measured age values, results of this type often remain crude and relatively inaccurate approximations.

7.2.2 Determination of Age and Residence-Time Distributions

The determination of the age and residence-time distributions, $\psi(t)$ and $\varphi(t)$, can be achieved in several ways depending on the flow structure in the aquifer, the type of inlet and outlet zones, and the level of resolution at which the age transport processes are analyzed. In some instances, it is sometimes easier to calculate first the residence-time distribution at a given outlet and deduct the age distribution in the aquifer from the relation given by the reservoir theory. In some other highly heterogeneous situations, evaluating first the age distribution in the aquifer provides more accurate residence-time distributions at localized outlets, where mixing perturbations can be very important.

Age transport modeling principles are presented in the following sections. Particular emphasis is laid upon advection-dominated systems, which make the analysis much more straightforward. This is due to the fact that, whatever the flow configuration, the advective age at a point in the aquifer is a single value corresponding to the average age at that point. The methodology is also briefly described for general hydrodispersive systems, and the notions of forward and backward modeling techniques are introduced in relation to age and life expectancy calculations.

Advective Systems A few classical, homogeneous flow configurations are first examined. For these simple cases the problem of age transport can be explic-

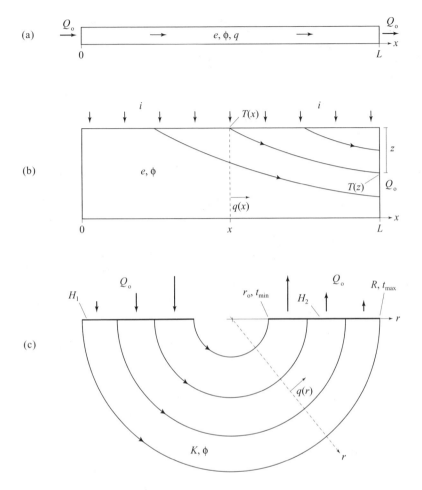

Figure 7.10. Flow configuration sketches for analytical modeling of residence time distributions. (a) piston-flow, (b) exponential, and (c) piston-hyperbolic models.

itly solved by closed-form analytical solutions. Figure 7.10 illustrates three typical, ideal situations for which the residence time distributions, shown in Figure 7.11(a), (b), and (c), are singular, exponential, and hyperbolic, respectively.

In the first case [Figure 7.10(a)], an aquifer of length L is assumed with uniform thickness e, porosity ϕ, and Darcy velocity q. There is no recharge between the inlet ($x = 0$) and the outlet ($x = L$) and, therefore, groundwater particles penetrate the system in a piston-like manner and age growth is characterized by $A(x) = \phi x/q = x/v$, where v is the pore velocity. This means that the residence time is $\tau_o = L/v$ for all particles, and that particle ages in the range zero to τ_o are represented in equal proportions in the aquifer. This is expressed by the age and residence-time statistical distributions

BASICS IN GROUNDWATER AGE TRANSPORT

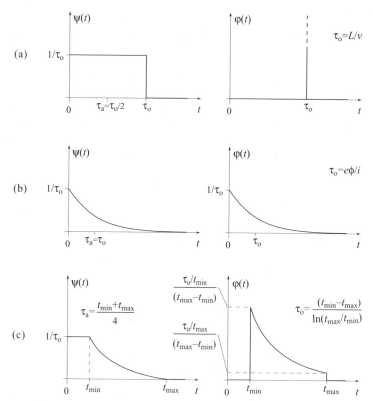

Figure 7.11. Analytical age and residence time distributions $\psi(t)$ and $\varphi(t)$ and characteristic times. (a) piston-flow, (b) exponential, and (c) piston-hyperbolic models.

$$\psi(t) = \frac{1}{\tau_o} H(\tau_o - t), \qquad \varphi(t) = \delta(\tau_o - t), \qquad \tau_o = \frac{L}{v} \qquad (7.31)$$

where H(u) is the Heaviside step function [H(u) = 1 for u > 0, H(u) = 0 for u < 0] and its derivative is $\delta(u)$, the Dirac delta function [$\delta(u) = \infty$ for u = 0, $\delta(u) = 0$ for u ≠ 0]. Therefore, as illustrated in Figure 7.11(a), ages in the aquifer are statistically described by the uniform distribution $\psi(t)$, yielding $\tau_a = \tau_o/2$, and this implies, after the general rule (7.7) of the reservoir theory, a singular distribution of the residence times at the outlet [$\partial\psi(t)/\partial t = -\infty$ at $t = \tau_o$, $\partial\psi(t)/\partial t = 0$ for $t \neq \tau_o$].

The second case [Figure 7.10(b)] is that of a horizontal aquifer, with thickness e and porosity ϕ, recharged by a uniform surface infiltration rate i. The flow rate through any vertical section at a location x must be equal to the amount of effective infiltration over the area upstream from x. Assuming that the Dupuit assumption holds in the present situation (horizontal velocity constant over depth), the Darcy flux at any location is therefore $q(x) = ix/e$. Hence, during their travel in the aquifer, the particles infiltrated at the location x are submitted to the accelerated pore velocity $v(x) = q(x)/\phi$, and the travel time $T(x)$ needed to reach the location $x = L$ is

$$T(x) = \int_x^L \frac{dx}{v(x)} = \frac{e\phi}{i} \int_x^L \frac{dx}{x} = \frac{e\phi}{i} \ln\left(\frac{L}{x}\right) \tag{7.32}$$

At the distance L, the depth z at which this travel time is detected must be such that the total flow rate over that depth must be equal to the total recharge over the horizontal distance $L - x$. This is expressed by

$$q(L)z = \frac{iL}{e} z = i(L-x), \qquad x = \left(1 - \frac{z}{e}\right)L \tag{7.33}$$

The relation, between the infiltration location x and the depth z at the distance L, can then be substituted into Equation 7.32 to yield the depth-dependent age evolution

$$T(z) = -\frac{e\phi}{i} \ln\left(1 - \frac{z}{e}\right), \qquad \lim_{e \to \infty} T(z) = \frac{\phi}{i} z \tag{7.34}$$

tending to vertical, piston-flow age for aquifer of very large thicknesses. Since this equation depends on neither x nor L, it implies that groundwater ages theoretically range between zero (at $z = 0$) and infinity (at $z = e$), at any location. Moreover, it is easily shown that averaging the travel times over the length L is equivalent to averaging the ages over the depth e. In effect, the results

$$\overline{T(x)} = \frac{1}{L} \int_0^L \frac{e\phi}{i} \ln\left(\frac{L}{x}\right) dx = \frac{e\phi}{i}, \qquad \overline{T(z)} = -\frac{1}{e} \int_0^e \frac{e\phi}{i} \ln\left(1 - \frac{z}{e}\right) dz = \frac{e\phi}{i} \tag{7.35}$$

indicate that both average values correspond to the ratio $e\phi/i$, for any value of L. Hence, the depth-averaged age is the same at any location, and the depth z^* at which this average value is encountered is readily deducted from Equation 7.34 by the solution of $\ln(1 - z^*/e) = -1$, yielding $z^* = (1 - e^{-1})e = 0.632e$. The interpretation of average age data in terms of upstream infiltration rates should always be made on samples taken at this depth. Indeed, the average value in a sampling profile should always be close to the depth z^*, regardless of the hydrodispersive mixing effects between flow paths that may render Equation 7.34 locally inaccurate. If an outlet is located at $x = L$, both functions $T(x)$, in Equation 7.32, and $T(z)$, in Equation 7.34, express particle residence times (i.e., life expectancy over the inlet or age over the outlet), and the average residence time at the outlet, or system turnover time $\tau_o = V_o/Q_o$ is again the ratio $e\phi/i$.

BASICS IN GROUNDWATER AGE TRANSPORT

The statistical distribution $\varphi(t)$, characterizing the residence times in a water sample collected at the outlet, can now easily be determined based on Equation 7.34. Reversing the latter to explicit the depth as a function of the residence time yields

$$z(t) = \left(1 - e^{-t/\tau_o}\right)e, \qquad \tau_o = \frac{e\phi}{i} \qquad (7.36)$$

At the outlet, the Darcy flux is $q = iL/e$ at any depth and the cumulative discharge with respect to depth is $Q(z) = qz$. Given that the residence time monotonically increases with depth, Equation 7.36 can be directly substituted and the cumulative discharge with respect to time becomes

$$Q(t) = qz(t) = iL(1 - e^{-t/\tau_o}), \qquad Q(\infty) = Q_o = iL \qquad (7.37)$$

Considering the definition (7.5), the occurrence of the residence times in the sample is therefore described by the exponential distribution

$$\varphi(t) = \frac{1}{Q_o}\frac{\partial Q(t)}{\partial t} = \frac{1}{\tau_o}e^{-t/\tau_o} \qquad (7.38)$$

implying $\psi(t) = \varphi(t)$ and $\tau_a = \tau_o$, as discussed previously.

Replacing the linear outlet of this example by a well, and repeating the above analytical development for the horizontal-radial flow configuration, results in exactly the same distribution. In fact, the exponential model may apply in all kinds of flow and outlet configurations, as long as the aquifer is horizontal and satisfies the Dupuit assumption, is solely recharged by precipitations, and is relatively shallow with a uniform ratio $e\phi/i$.

The third case illustrated in Figure 7.10(c) represents a flow configuration with deep, half-circular flow paths. The aquifer is homogeneous with respect to hydraulic conductivity K and porosity ϕ, and a hydraulic head difference $\Delta H = H_1 - H_2$ is specified between the inlet and outlet zones. The system can be described by means of the radial distance r, with r_o and R as inner and outer radii, respectively. A given flow path at a radius r has thus a length πr. Along this path, between the inlet and outlet zones, the constant Darcy flux $q(r)$ and the travel time $t(r)$ are therefore the monotonic radial functions

$$q(r) = K\frac{\Delta H}{\pi r}, \qquad t(r) = \frac{\phi\pi r}{q(r)} = \frac{\phi\pi^2 r^2}{K\Delta H} \qquad (7.39)$$

with the minimum and maximum travel times $t_{min} = t(r_o)$ and $t_{max} = t(R)$. The above function $t(r)$ indifferently represents, at the distance r from the origin, the life expectancy of the particles when they infiltrate over the inlet and their age, or residence time, when they exfiltrate over the outlet.

The total pore volume of the aquifer is $V_o = \phi\pi(R^2 - r_o^2)/2$ and the discharge at the outlet, cumulated with respect to r, is

$$Q(r) = \int_{r_o}^{r} q(r)dr = \frac{K\Delta H}{\pi}\int_{r_o}^{r}\frac{dr}{r} = \frac{K\Delta H}{\pi}\ln\left(\frac{r}{r_o}\right), \quad Q_o = Q(R) = \frac{K\Delta H}{\pi}\ln\left(\frac{R}{r_o}\right) \tag{7.40}$$

Expressing r as a function of t in Equation 7.39, accounting for the definitions of $t_{\min}$ and $t_{\max}$, and substituting into Equation 7.40 yield

$$Q(t) = \frac{K\Delta H}{2\pi}\ln\left(\frac{K\Delta H t}{\phi\pi^2 r_o^2}\right) = \frac{K\Delta H}{2\pi}\ln\left(\frac{t}{t_{\min}}\right), \quad Q_o = \frac{K\Delta H}{2\pi}\ln\left(\frac{t_{\max}}{t_{\min}}\right) \tag{7.41}$$

and the statistical distribution of the residence times at the outlet is

$$\varphi(t) = \frac{1}{Q_o}\frac{\partial Q(t)}{\partial t} = \frac{1}{\ln(t_{\max}/t_{\min})}\frac{1}{t}, \quad t_{\min} \le t \le t_{\max} \tag{7.42}$$

This function is of a hyperbolic nature between $t_{\min}$ and $t_{\max}$ and vanishes for $t < t_{\min}$ and $t > t_{\max}$. Its first moment, namely the turnover time, is $\tau_o = (t_{\max} - t_{\min})/\ln(t_{\max}/t_{\min})$.

The age distribution in the aquifer can be obtained from the general rule (7.7) of the reservoir theory, and it follows that

$$\psi(t) = \frac{1}{t_{\max} - t_{\min}}\ln(\frac{t_{\max}}{t_{\min}}) = \frac{1}{\tau_o}, \quad 0 \le t < t_{\min} \tag{7.43}$$

and that

$$\psi(t) = \frac{1}{t_{\max} - t_{\min}}\ln\left(\frac{t_{\max}}{t}\right), \quad t_{\min} \le t \le t_{\max} \tag{7.44}$$

with $\psi(t) = 0$ for $t > t_{\max}$. The first moment of this distribution is the average age in the aquifer $\tau_a = (t_{\min} + t_{\max})/4$, greater than τ_o when $t_{\max}/t_{\min} > 2e^\pi$. The two functions $\varphi(t)$ and $\psi(t)$ are illustrated in Figure 7.11(c).

The analytical modeling approach described above can be applied to other flow configurations with simple geometry and parametric fields, resulting in a number of straightforward formulas. Although exact in the mathematical sense, these practical formulas are of course restricted to the specific hydrogeological situations where the underlying simplifying assumptions can reasonably be justified.

For more complex flow configurations, or more detailed hydrodynamic characterization, the age and residence-time distributions cannot be found

BASICS IN GROUNDWATER AGE TRANSPORT

analytically anymore, and discrete numerical modeling approaches are required for this purpose. Finite difference or finite element techniques are efficient tools to discretize a given system, accounting for all relevant structural information, parametric details, and flow conditions. The flow and age-mass transport processes are then simulated by discrete solutions of their governing partial differential equations, and the derivation of age-related quantities is obtained by subsequent regionalization and analysis of the results.

At any point of spatial coordinates x, y, and z in an arbitrary hydrogeological system, the fluid flux is defined by Darcy's law in the vector form

$$\mathbf{q} = -\mathbf{K}\nabla H \tag{7.45}$$

where ∇H is the local hydraulic head gradient and $\mathbf{K}$ is the local permeability tensor. These two functions vary in space and determine the local direction and magnitude of the flux vector $\mathbf{q}$. In a general case, the latter has three components expressing the flux in each of the space dimensions. In a steady-state, divergence-free flow regime this flux vector is submitted to the local mass balance condition

$$\nabla \cdot \mathbf{q} = 0 \tag{7.46}$$

where ∇ is the partial differential gradient operator and the dot denotes a scalar product ($\nabla \cdot$ is the divergence operator). The equation governing steady-state groundwater flow is therefore

$$\nabla \cdot (-\mathbf{K}\nabla H) = 0 \tag{7.47}$$

and its solution for specific boundary conditions (specified fluxes or hydraulic heads on inlet and outlet boundaries) yields the spatial distribution of the hydraulic heads $H(x, y, z)$ in the system. The flow velocity field $\mathbf{q}$ is then derived by application of Equation 7.45 and can be used as input in an age transport equation.

Under advective conditions, the age-mass flux at a point is submitted to the local age-mass balance condition

$$\nabla \cdot (\mathbf{q}A) = \phi \tag{7.48}$$

in which the porosity ϕ is the volumetric age-mass production rate. Expanding the left-hand side of this equation in $A\nabla \cdot \mathbf{q} + \mathbf{q} \cdot \nabla A$, and inserting the mass balance requirement (7.46), results in

$$\mathbf{q} \cdot \nabla A = \phi \tag{7.49}$$

This equation governs steady-state advective age transport. It states that the scalar product of the Darcy flux vector and the age gradient at a point must always be equal to the porosity of the medium at that point. Since the latter is never zero, it follows that flux vectors and age gradients can never be locally perpendicular. An alternative form of this equation in terms of pore velocity is $\mathbf{v} \cdot \nabla A = 1$, indicating that the magnitude of the age gradient along a given flow path is exactly the inverse of the pore velocity magnitude, and that the aging rate of flowing particles is necessarily one (i.e., one additional unit of age per unit of elapsed flow time).

Enforcing the required condition $A = 0$ on all inlet boundaries, the solution of Equation 7.49 yields the spatial age field $A(x, y, z)$ in the system. Including all relevant internal age information, this spatial age distribution can then be mapped for various interpretation purposes, and more specifically analyzed in the context of the reservoir theory. On the system outlets, the distribution $A(x, y, z)$ only indicates the average age values, or average residence times resulting from the mechanical mixing of numerous converging flow paths. Thus, the full statistical residence time distributions at these outlets remain apparently unknown at this stage, and this may look somehow logical since only steady-state equations have been solved. However, the desired transient information can indeed be recovered when considering the intrinsic statistical relation between the outlet residence time and internal age distributions. The latter, $\psi(t)$, can actually be constructed based on a classification of the spatial age data $A(x, y, z)$, from which the former, $\varphi(t)$, can be derived.

In effect, recalling the general rule (7.7) of the reservoir theory and substituting in it the definition (7.4), the residence time distribution can also be expressed by

$$\varphi(t) = -\frac{1}{Q_o} \frac{\partial^2 V(t)}{\partial t^2} \qquad (7.50)$$

where it appears as the second derivative of the function $V(t)$, scaled by the total system discharge. The function $V(t)$, as defined previously, represents the cumulated resident water volumes having ages A smaller than a given value of time. Knowing the spatial distribution of internal ages, this cumulated function is readily obtained by computing the integral

$$V(t) = \int_\Omega \phi H(t - A(x, y, z)) d\Omega \qquad (7.51)$$

over the model domain Ω. This integration is performed for a range of increasing discrete times, starting from zero and ending at t_{max} at which $V(t_{max}) = V_o$. In the above integral, the value of the Heaviside function is 1 for the elementary water volumes $\phi d\Omega$ having ages smaller than the working value of t, and is 0 otherwise. The resulting discrete function is then differentiated twice by means of an appropriate finite difference scheme, and the residence-time distribution is found according to Equation 7.50. The size of the time increment used in the calculation plays of course an important role in the final resolution of the function $\varphi(t)$.

In spite of its simplicity, this modeling procedure achieves good levels of accuracy, particularly in heterogeneous systems. The fact that the cumulative function $V(t)$ includes the largest age in the reservoir, ensures that this age is accounted for in the residence time distribution, minimizing thus the loss of information due to mixing in the outlet vicinity. In addition, requiring only the numerical solution of two steady-state equations, this approach is computationally very efficient for the analysis of large systems.

Forward and Backward Flow Fields The modeling notions introduced above associate the age transport processes to any given groundwater flow field governed by Darcy's law and mass conservation. Dictated by the boundary conditions specified on the inlet and outlet zones, these flow fields always indicate downstream particle motion and are thus called forward flow fields. Age transport calculations performed in these conditions, with $A = 0$ on inlet zones, always result in increasing age values along the various flow paths until an outlet is reached. Assuming now that a given flow field could be uniformly and instantaneously reversed, the water particles at a given outlet would just start to flow back into the aquifer, and, following exactly the same flow paths in the reverse direction, would make their way upstream to the corresponding inlet zones. The situation would then be similar to that of playing a movie backward and, thus, of going back in time. Such a temporarily reversed, or backward, flow field would certainly be used by hydrogeologists for tracing the water particles back to their origin. In the process, valuable new insights would be gained into the extension and overall transport properties of the aquifer subsystems contributing to any given outlet.

Unrealizable by nature, these fictive physical prospects can, however, be straightforwardly simulated and analyzed by simply removing the negative sign in Darcy's law (7.45). In doing so, the resulting flux vector at a point keeps its original orientation, but points in the opposite direction toward higher hydraulic heads. Hence, backward age transport calculations provide information on the time required by groundwater particles to reach a given location in the aquifer when flowing back from a given outlet. In this case, the backward "age" calculated at a point is the life expectancy of the groundwater particles, namely the time that these particles will spend in the system when flowing in the real, forward flow field before they reach the outlet. Therefore, when a classical forward flow field $\mathbf{q}$ is known, the equation governing the transport of groundwater life expectancy E in the system is simply

$$-\mathbf{q} \cdot \nabla E = \phi \qquad (7.52)$$

This equation only differs from the age transport Equation 7.49 by the negative sign and by the required boundary conditions for its solution. In this case, in effect, the condition $E = 0$ has to be enforced on all outlet boundaries. The resulting life expectancy field $E(x, y, z)$ can then be mapped and interpreted to address numerous water-quality issues, such as those related to the design of well heads protection zones or to the contamination risks generated by waste repositories.

Moreover, given the similar and complementary nature of the age and life expectancy, it can be shown that all the elements of the reservoir theory, essentially discussed with respect to groundwater age so far, are also fully valid when applied to life expectancy. For instance, the procedure described previously to determine the residence-time distribution at an outlet from the age field $A(x, y, z)$ produces exactly the same results when the field $E(x, y, z)$ is considered, and this is true for any arbitrary flow system or subsystem (Etcheverry, 2001; Cornaton, 2004). This is due to the fact that forward and backward flow fields generate exactly the same discharge and the same flow paths, but in reversed directions. Consequently, age-mass fluxes at the outlets are exactly balanced by backward life-expectancy-mass fluxes at the inlets. Indeed, the cumulated function $V(t)$ computed in Equation 7.51 can as well be obtained with the domain integral

$$V(t) = \int_\Omega \phi H(t - E(x, y, z)) d\Omega \qquad (7.53)$$

and the resulting residence-time distribution $\varphi(t)$ in Equation 7.50 is identical. This also implies that the internal age and life expectancy distributions are identical to the function $\psi(t)$.

At a given point along a particular flow path, the age of groundwater particles represents the time that has elapsed since the particles entered the system. At the same point, life expectancy represents the time left before the particles exfiltrate the system. At that point, the sum $T = A + E$ represents thus the travel time of the particles between the inlet and the outlet, also called transit time. This transit time T is obviously constant along the flow path and can be associated to it as an invariant characteristic. The transit time, as defined here, is therefore also the residence time of the particles that could be monitored at the end of the flow path when the outlet is reached. When both spatial distributions $A(x, y, z)$ and $E(x, y, z)$ are simulated, a very interesting operation is to add them up and analyze the resulting transit time spatial distribution. Mapping the field $T(x, y, z)$ produces isoline contours exactly matching the corresponding flow paths throughout the system. Subtracting now the two Equations 7.49 and 7.52 governing the differential evolution of the fields $A(x, y, z)$ and $E(x, y, z)$ yields the local condition

$$\mathbf{q} \cdot \nabla T = 0 \qquad (7.54)$$

stating that the flux vector and the transit time gradient have to be always perpendicular. In effect, this is required to keep transit times, and transit-time-

BASICS IN GROUNDWATER AGE TRANSPORT   237

mass fluxes, constant along the flow paths. The above condition also indicates that there is no transit-time-mass production in the system.

Extended at the scale of the reservoir, these considerations may be expressed in a global form by analyzing the cumulated pore volume $V_T(\tau)$ in the reservoir, where the groundwater transit times, or residence times, are smaller than a given value τ. Recalling the characteristic volumes defined in Figure 7.6, this volume $V_T(\tau)$ corresponds to the volume $V_1(\tau) + V_2(\tau)$, which flows through the system in a time less than τ. Calculating the corresponding surface on the diagram in Figure 7.6(b) results in the mass balance condition

$$V_T(\tau) = V_1(\tau) + V_2(\tau) = Q(\tau)\tau - \int_0^\tau Q(t)dt \qquad (7.55)$$

where $Q(\tau)$ is the cumulated discharge exfiltrating over the outlet with a residence time smaller than τ, as defined previously. As stated by the above condition, the groundwater volume $V_T(\tau)$ must correspond to the flow rate $Q(\tau)$ applied to it during the period τ minus the quantity of water exfiltrated during that period with a residence time less than τ.

Differentiating Equation 7.55 with respect to time, replacing the time variable by t, and accounting for the statistical definition of the residence-time distribution $\varphi(t)$ (7.5) yield

$$\frac{\partial V_T(t)}{\partial t} = t\frac{\partial Q(t)}{\partial t} = tQ_o\varphi(t) \qquad (7.56)$$

Expressing now the statistical description of the internal spatial field $T(x, y, z)$ by the classical definition

$$\psi_T(t) = \frac{1}{V_o}\frac{\partial V_T(t)}{\partial t}, \qquad \int_0^\infty \psi_T(t)dt = 1 \qquad (7.57)$$

and substituting this function into Equation 7.56 results in

$$\varphi(t) = \frac{\tau_o}{t}\psi_T(t), \qquad \tau_o = \frac{V_o}{Q_o} \qquad (7.58)$$

This fundamental relation initially formulated by Cornaton and Perrochet (2005a) includes all the features of the reservoir theory in the most compact form. Combining the outcomes of forward and backward transport models, it relates the outlet residence-time distribution $\varphi(t)$ to its internal counterpart $\psi_T(t)$. Both functions relate to residence times and provide complementary information. The distribution $\varphi(t)$ characterizes the residence times with respect to discharge rates at the outlets, while the distribution $\psi_T(t)$

characterizes the residence times with respect to groundwater volumes in the aquifer.

Compared to the standard rule (7.7) based on the internal age or life expectancy distribution $\psi(t)$, the above formulation is a great improvement. It is simpler and provides residence-time distribution with much higher resolution and accuracy. This is due to the fact that the spatial field $T(x, y, z)$ is always more regular than the fields $A(x, y, z)$ or $E(x, y, z)$, particularly in the vicinity of outlets, and that there is no time differentiation between $\psi_T(t)$ and $\varphi(t)$. The cumulated function $V_T(t)$ obtained by the domain integral

$$V_T(t) = \int_\Omega \phi H(t - T(x, y, z)) d\Omega \tag{7.59}$$

is thus relatively smoother, and the substitution of Equation 7.57 into 7.58, reading

$$\varphi(t) = \frac{1}{tQ_o} \frac{\partial V_T(t)}{\partial t} \tag{7.60}$$

indicates that $V_T(t)$ needs to be differentiated only once with respect to time to yield the residence time distribution.

From Equation 7.58, the internal transit time distribution can be written $\psi_T(t) = t\varphi(t)/\tau_o$ and its first moment is

$$\tau_T = \int_0^\infty t\psi_T(t)dt = \frac{1}{\tau_o}\int_0^\infty t^2\varphi(t)dt = \frac{1}{\tau_o}(\tau_r^2 + \sigma_r^2) \tag{7.61}$$

where τ_r and σ_r are, respectively, the mean (always equal to τ_o) and the standard deviation of the residence-time distribution $\varphi(t)$. Combining this result to the mean age in the reservoir τ_a in Equation 7.11 or 7.12 shows that $\tau_T = 2\tau_a$. The mean transit time in the reservoir is therefore always twice larger than the mean age (or the mean life expectancy) in the reservoir and, according to Equation 7.61, is always greater than the turnover time, except in the particular piston-flow case for which $\tau_T = \tau_o$.

When combined to the reservoir theory, the transport processes simulated with forward and backward flow fields provide all the required information to efficiently determine the distributions $\varphi(t)$, $\psi(t)$, and $\psi_T(t)$ for any arbitrary system. Knowing these functions, various global properties of a given aquifer system can be analyzed. For instance, the discharge rate exfiltrating from the system with a residence time less than τ is

$$Q(\tau) = Q_o \int_0^\tau \varphi(t)dt \tag{7.62}$$

BASICS IN GROUNDWATER AGE TRANSPORT   239

The volume of groundwater in the system having ages, or life expectancies, ranging between τ_1 and τ_2 is the volume delineated by the isochrones (isolines or isosurfaces) of the spatial fields $A(x, y, z)$ or $E(x, y, z)$, and is

$$V(\tau_2) - V(\tau_1) = V_o \int_{\tau_1}^{\tau_2} \psi(t) dt \qquad (7.63)$$

and the volume of groundwater flowing through the system in a time smaller than τ (volume $V_1 + V_2$ in Figure 7.6) is

$$V_T(\tau) = V_o \int_0^{\tau} \psi_T(t) dt \qquad (7.64)$$

Other numerous statistical and probabilistic inferences can be made in a similar manner. For example, the probability that a particle in a water sample collected at the system outlet has a residence time ranging between τ_1 and τ_2 is

$$P(\tau_1 < t < \tau_2) = \int_{\tau_1}^{\tau_2} \varphi(t) dt = -\tau_o \int_{\tau_1}^{\tau_2} \frac{\partial \psi(t)}{\partial t} dt = -\tau_o (\psi(\tau_2) - \psi(\tau_1)) \qquad (7.65)$$

and is $P(t < \tau_2) = 1 - \tau_o \psi(\tau_2)$ if $\tau_1 = 0$.

The probability that a particle in the reservoir has, simultaneously, an age ranging between a_1 and a_2, and a life expectancy ranging between e_1 and e_2, is

$$P(a_1 < t < a_2, e_1 < t < e_2) = \int_{a_1+e_1}^{a_2+e_1} \psi(t) dt - \int_{a_1+e_2}^{a_2+e_2} \psi(t) dt \qquad (7.66)$$

where $a_1 + e_1$ and $a_2 + e_2$ mark the range of possible transit times for that particle. Replacing the internal distributions $\psi(t)$ in this equation by its intrinsic relation with the outlet residence-time distribution $\varphi(t)$, multiplying by the pore volume of the system V_o, and expanding the calculation, results in Equation 7.20, counterpart of the above equation for the volumetric estimations.

When the groundwater flow model of a given real situation can be constructed and calibrated based on relevant hydrogeological data, the resulting age-related, spatial, and statistical distributions are easily obtained with very modest additional work. These distributions can of course also help in the calibration phase of the groundwater model, when age field data are available. The approach can also be used to simulate hypothetical hydrogeological setups and verify the validity and limitations of some assumed, simplified age and residence-time model distributions.

Advective-Dispersive Systems In general hydrodispersive contexts, the migration of water particles is governed by the processes of advection, dispersion, and diffusion. These processes imply that the age at a given location is not the single value corresponding to the advective age anymore. Instead, a range of particle ages has to be considered. Advective-dispersive ages at any point in the system must therefore be described by a local age distribution, denoted $C_A(\mathbf{x}, t)$, where $\mathbf{x}$ is the location vector. Given that this local age distribution describes the resident age fractions, it can be seen as an age concentration.

For a given steady-state, divergence-free flow field ($\nabla \cdot \mathbf{q} = 0$) the evolution of the local age distribution C_A through the system is governed by the advective-dispersive transport equation

$$\frac{\partial \phi C_A}{\partial t} = -\nabla \cdot (\mathbf{q} C_A - \mathbf{D}\nabla C_A) \qquad (7.67)$$

where fractional age-mass fluxes include components of both advective ($\mathbf{q}C_A$) and dispersive ($-\mathbf{D}\nabla C_A$) natures. The latter is submitted to Fick's law in which $\mathbf{D}$ is the standard hydrodynamic diffusion-dispersion tensor characterizing particle dispersion with respect to the flow direction. When the local distribution C_A is known at every point of a given flow domain Ω, the global frequency distribution of ages $\psi(t)$ in the system is obtained by the domain integral

$$\psi(t) = \frac{1}{V_o} \int_\Omega \phi C_A(\mathbf{x}, t) d\Omega \qquad (7.68)$$

Differentiating this global distribution with respect to time yields

$$\frac{\partial \psi(t)}{\partial t} = \frac{1}{V_o} \int_\Omega \frac{\partial (\phi C_A)}{\partial t} d\Omega \qquad (7.69)$$

which corresponds to the domain integral of the left-hand side of the transport Equation 7.67, scaled by the total pore volume of the system. The above equation may therefore also be expressed by

$$\frac{\partial \psi(t)}{\partial t} = -\frac{1}{V_o} \int_\Omega \nabla \cdot (\mathbf{q} C_A - \mathbf{D}\nabla C_A) d\Omega \qquad (7.70)$$

or, after application of the divergence theorem, by the boundary integral

$$\frac{\partial \psi(t)}{\partial t} = -\frac{1}{V_o} \int_\Gamma (\mathbf{q} C_A - \mathbf{D}\nabla C_A) \cdot \mathbf{n} d\Gamma \qquad (7.71)$$

where $\mathbf{n}$ is a unit vector normal to the domain boundary Γ. This equation relates the global internal age distribution $\psi(t)$ to the fractional age-mass boundary fluxes. The domain boundary Γ is the union of all no-flow boundaries (Γ°), as well as all inlet (Γ^-) and outlet (Γ^+) boundaries.

Hence, the knowledge of the distribution C_A at every point of the flow domain requires the solution of the transport Equation 7.67 under a set of boundary conditions specific to the age and residence-time problem. These conditions have to preserve the continuity of age-mass fluxes at the outlet boundaries, and to ensure that these fluxes may correctly be interpreted at outlets in terms of residence-time distributions.

As demonstrated by Cornaton et al. (2004) and Cornaton and Perrochet (2005a), the flux continuity requirement at the oulet boundaries Γ^+ is satisfied by an implicit Neumann condition, replacing the classical homogeneous Neumann condition ($\mathbf{D}\nabla C_A \cdot \mathbf{n} = 0$), and allowing the development of unperturbed, natural age gradients across any outlet. In addition, to make the solution C_A and the outlet fractional fluxes correspond, respectively, to local age and outlet residence-time distributions, an instantaneous age-mass, pulse-injection Cauchy condition has to be enforced on the inlet boundaries Γ^-. With these two conditions, and given that there is no age-mass flux through the no-flow boundaries Γ°, Equation 7.71 becomes

$$-V_o \frac{\partial \psi(t)}{\partial t} = \int_{\Gamma^-} \mathbf{q} \cdot \mathbf{n}\delta(t)d\Gamma + \int_{\Gamma^+} (\mathbf{q}C_A - \mathbf{D}\nabla C_A) \cdot \mathbf{n}d\Gamma \qquad (7.72)$$

where the Dirac function $\delta(t)$ is used to instantaneously mark the particle fluxes recharging the system at $t = 0$. Since the integral of the Darcy fluxes on the inlet boundary is the total, steady-state recharge rate Q_o, the first term on the right-hand side of the above equation is $-Q_o\delta(t)$ (the negative sign indicating that the vectors $\mathbf{q}$ and $\mathbf{n}$ point in opposite directions on Γ^-), while the second term fractionates the discharge rate Q_o with respect to advective-dispersive arrival times at the outlet boundary Γ^+. Scaling Equation 7.72 by Q_o and rearranging the terms results then in

$$\frac{1}{Q_o}\int_{\Gamma^+} (\mathbf{q}C_A - \mathbf{D}\nabla C_A) \cdot \mathbf{n}d\Gamma = -\frac{V_o}{Q_o}\frac{\partial \psi(t)}{\partial t} + \delta(t) \qquad (7.73)$$

where the left-hand side is the global outlet residence-time distribution $\varphi(t)$. Given that $\delta(t) = 0$ for $t > 0$, this equation finally reduces to

$$\varphi(t) = -\tau_o \frac{\partial \psi(t)}{\partial t}, \quad \tau_o = \frac{V_o}{Q_o} \qquad (7.74)$$

which confirms that the major rule of the reservoir theory—derived previously in Equation 7.7 following a different approach—also applies to advective-dispersive systems. By extension, the evaluation of characteristic times and volumes based on the global distributions $\varphi(t)$ and $\psi(t)$ (Section 7.2.1) also applies in such systems, as well as the modeling strategies based on forward and backward flow fields discussed previously. The considerations made above for the local age distribution $C_A(\mathbf{x}, t)$ can indeed be similarly performed for the local life expectancy distributions $C_E(\mathbf{x}, t)$, based on the reversed flow-field transport equation

$$\frac{\partial \phi C_E}{\partial t} = \nabla \cdot (\mathbf{q} C_E + \mathbf{D} \nabla C_E) \qquad (7.75)$$

and appropriate boundary conditions.

The general procedure required to accurately model outlet residence-time distributions is to solve first the transport Equation 7.67 for the local resident age fractions $C_A(\mathbf{x}, t)$, then calculate the global internal age distribution $\psi(t)$ by the domain integral (7.68), and finally apply Equation 7.74 to find $\varphi(t)$. In doing so, the outlet residence-time distribution is fully defined as a property of the reservoir internal structure and hydrodispersive properties. Following this procedure, there is no need to record or count the arrival times of water particles at the outlet, or perform the boundary integrals of age-mass fluxes. Such operations may not only be sometimes difficult to implement technically, but they are also at the origin of significant resolution losses.

In effect, at a system outlet, the age fractions C_A are distributed along the discharge boundary, implying mixing and superposition of the information transmitted by each breakthrough curves, and the true minimum and maximum residence times are often diluted. These effects particularly affect outlets of relatively small sizes where the capture of individual breakthrough curves or arrival times is easily ill-posed because of the numerical/physical mixing of converging flow patterns. In other words, the temporal resolution of a residence time distribution evaluated directly at an outlet is strongly dependent on both the local spatial resolution (mesh refinement), and the way the boundary fractional age-mass fluxes are integrated (line or surface integrals). At equal temporal resolution, numerical methods based on this approach require a much higher level of refinement in the neighborhood of the integration boundaries than the global approach presented here.

Compared to standard approaches, this modeling strategy achieves thus a far more accurate evaluation of residence time distributions by capturing inside the system the age information that is missing at its boundaries. Moreover, computational efficiency is another significant advantage of the proposed modeling strategy, and this efficiency can further, and dramatically be improved when transforming all working equations from the time domain onto the Laplace space.

Additional fundamental findings concerning the modeling of age, life expectancy, and residence-time distributions can be found in Cornaton and Perrochet (2005a, 2005b). Based on forward and backward Laplace-transformed equations for location and travel time probabilities, these recent works use the concepts of fractional capture zones to generalize the reservoir theory to multidimensional hydrodispersive flow fields, with sinks and sources ($\nabla \cdot \mathbf{q} \neq 0$), and with multiple inlet and outlet zones.

Simple manipulations of the transport Equation (7.67) can further be performed to analyze and regionalize various statistical, age-related quantities, such as average values, variances and other higher moments of the local distributions in advective-dispersive conditions (Harvey and Gorelick, 1995; Goode, 1996; Varni and Carrera, 1998). Concerning average values, the local age distribution $C_A(\mathbf{x}, t)$ describes the statistical occurence of ages at a given location and, by definition, its first moment is the local average age

$$A(\mathbf{x}) = \int_0^\infty t C_A(\mathbf{x}, t) dt, \quad \int_0^\infty C_A(\mathbf{x}, t) dt = 1 \qquad (7.76)$$

When dispersion-diffusion effects are neglected, this local average age is equivalent to the advective age discussed in the preceding sections, in which case the local distribution is the singular Dirac function $C_A(\mathbf{x}, t) = \delta(t - A(\mathbf{x}))$.

When diffusive-dispersive effects are taken into account, the spatial evolution of the average field $A(\mathbf{x})$ in the system is governed by a specific equation obtained by taking the first moment of the transient Equation (7.67) and of its boundary conditions. Multiplying each terms in Equation 7.67 by t and integrating over all times result in

$$\int_0^\infty t \frac{\partial \phi C_A}{\partial t} dt = -\nabla \cdot \left(\mathbf{q} \int_0^\infty t C_A dt - \mathbf{D} \nabla \int_0^\infty t C_A dt \right) \qquad (7.77)$$

Using the definitions in (7.76), the left-hand side of this equation is integrated by parts to give

$$\int_0^\infty t \frac{\partial \phi C_A}{\partial t} dt = \phi t C_A \Big|_0^\infty - \phi \int_0^\infty C_A dt = -\phi \qquad (7.78)$$

while the first moment of the distribution C_A on the right-hand side can directly be replaced by the mean value A. The resulting advective-dispersive, age-mass-transport equation is

$$\nabla \cdot (\mathbf{q} A - \mathbf{D} \nabla A) = \phi \qquad (7.79)$$

which simplifies in Equation 7.48 when dispersion is neglected. To model the proper average field $A(\mathbf{x})$ with this equation, the implicit Neumann condition (Cornaton et al., 2004) has to be applied on the outlet boundaries Γ^+ to ensure the continuity of the age-mass fluxes exiting the system, and a zero age-mass,

Cauchy input flux [i.e., the first moment of the quantity $\mathbf{q} \cdot \mathbf{n}\delta(t)$ in Equation 7.72] must condition the inlet boundaries Γ^-.

Similarly, the advective-dispersive average life expectancy field $E(\mathbf{x})$ can be calculated by the backward flow-field equation

$$-\nabla \cdot (\mathbf{q}E + \mathbf{D}\nabla E) = \phi \qquad (7.80)$$

and both fields $A(\mathbf{x})$ and $E(\mathbf{x})$ can be mapped and post-processed following the procedure described for advective systems, and according to Equations 7.50, 7.51, and 7.53 to approximate the global distributions $\varphi(t)$ and $\psi(t)$ based on these averaged quantities.

Finally, integrating either Equation 7.79 or 7.80 over the flow domain, applying the divergence theorem, and accounting for the boundary conditions on Γ^- and Γ^+ yield the time-average form of Equation 7.72, or age-mass conservation equation,

$$\int_\Omega \phi d\Omega = \int_{\Gamma^+} (\mathbf{q}A - \mathbf{D}\nabla A) \cdot \mathbf{n} d\Gamma \qquad (7.81)$$

in which the left-hand side represents the total pore volume of the system V_o, or total internal age-mass production rate, while the right-hand side is the total advective-dispersive age-mass flux at the outlet. Dividing this total age-mass flux by the discharge Q_o yields, therefore, the average age over the outlet, or average residence time τ_r defined in Equation 7.10 by means of the reservoir theory. Scaling Equation 7.81 by Q_o yields then the average form of Equation 7.73, namely,

$$\frac{V_o}{Q_o} = \frac{1}{Q_o} \int_{\Gamma^+} (\mathbf{q}A - \mathbf{D}\nabla A) \cdot \mathbf{n} d\Gamma = \tau_r \qquad (7.82)$$

which confirms the well-known definition of the turnover time, $\tau_o = \tau_r$, and extends the notion to advective-dispersive systems.

7.3 SELECTED TYPICAL EXAMPLES

The theoretical age modeling elements and age modeling strategies presented in Sections 7.2.1 and 7.2.2 are applied in the following simulation examples, in which age and residence-time distributions result from the combination of the reservoir theory to numerical solution of age transport equations. These examples aim at showing the large variety of age and residence-time distributions that can be found in natural conditions. The response of these distributions and of the corresponding characteristic groundwater volumes is illustrated and analyzed in terms of sensitivity to reservoir hydrodispersive properties and boundary conditions.

SELECTED TYPICAL EXAMPLES

7.3.1 Aquifer with Uniform and Localized Recharge

The behavior of an aquifer with good turnover properties, uniformly recharged from the surface, is compared to that of an aquifer with distant inlet and outlet zones in Figure 7.12. Both aquifers have the same homogeneous structure and size, and diffusive-dispersive effects are ignored.

For the uniform recharge condition [Figure 7.12(a)], the internal age and outlet residence time distributions in Figure 7.12(c) are quasi-exponential functions, with more or less equal average values $\tau_a \approx \tau_o$, and, therefore, $\varphi(t) \approx \psi(t)$ as discussed in Section 7.2.1. In this case, the characteristic groundwater volumes $V_3(t) + V_4(t)$ (see Figure 7.6 and related comments for definitions) diverges rapidly from $V(t)$, indicating that the minimum residence time at the outlet approaches zero. When the recharge is localized far upstream of the outlet [Figure 7.12(b)], the residence-time distribution $\varphi(t)$ in Figure 7.12(c) is nil and the age distribution $\psi(t)$ is constant until the minimum residence time is reached. In this case the average residence time τ_o is slightly less than twice

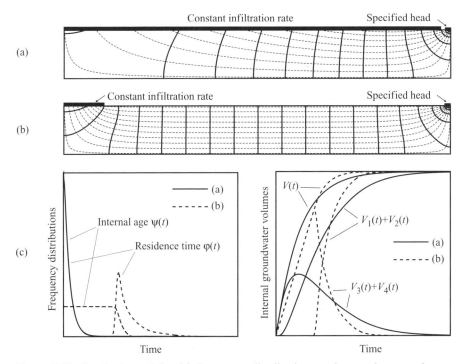

Figure 7.12. Synthetic models with frequency distributions and groundwater volumes as indicators of recharge conditions (modified after Cornaton, 2004). Hydraulic heads and flow paths for (a) the uniformly recharged aquifer, (b) the aquifer with distant inlet and outlet zones. (c) internal age and outlet residence-time distributions, internal cumulated groundwater volumes ($V = V_1 + V_2 + V_3 + V_4$).

the internal average age τ_a (a factor of two corresponds to pistonflow). This is the situation $\tau_a < \tau_o$ discussed in Section 7.2.1, for which the groundwater volumes $V_3(t) + V_4(t)$ and $V(t)$ remain identical until the minimum residence time is reached. Past this time, the decrease of $V_3(t) + V_4(t)$ is immediate and reflects the fact that most of the discharge rate is exfiltrated over a relatively narrow residence time span, related to the good uniformity of the flow paths in the aquifer. A narrow triangular-shaped function $V_3(t) + V_4(t)$ is typical of aquifers with significant minimum residence times and bad turnover properties. For aquifer with good turnover properties this function shows a much longer tail, according to the broader range of travel distances from the recharge areas to the outlet.

7.3.2 Hydrodispersive Multilayer Aquifer

The dispersive properties of a heterogeneous system are evaluated for the four-layer-aquifer shown in Figure 7.13. The four layers are homogeneous with thicknesses decreasing with depth, and permeabilities and porosities set in a way to generate contrasted pore velocities [velocities v_i in Figure 7.13(a)] and transit times in each layer. Specifying layers with decreasing thicknesses and increasing pore velocities is meant to create flux-averaged arrival-time peaks of comparable magnitudes at the outlet. The later is located at the surface over a relatively narrow zone where high mixing of arrival times is expected due to flux convergence. This is illustrated in Figure 7.13(b), showing a linear evolution of the mean age contours in each of the layer, as well as high-density isochrones separating the four distinct age categories of the water converging toward the outlet. Age and residence-time distributions are simulated for both advection-dominated and dispersion-dominated conditions, enforcing strong hydrodynamic dispersion parameters in the second case.

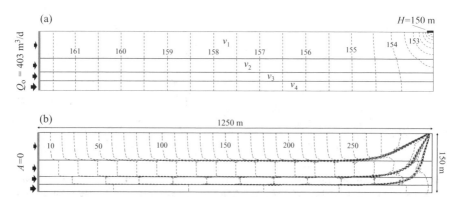

Figure 7.13. Synthetic multilayer aquifer (modified after Cornaton, 2004). (a) Boundary conditions and hydraulic head contours (H in m), (b) mean age contours (A in d). Average pore velocities in m/d are $v_1 = 4$, $v_2 = 6.5$, $v_3 = 10.5$, and $v_4 = 23.5$.

With a total pore volume $V_o = 65{,}312\,\text{m}^3$ and a total flow rate $Q_o = 403\,\text{m}^3/\text{d}$, the turnover time is $\tau_o = 162\,\text{d}$ for the two cases, and the average age in the reservoir shows only little sensitivity to dispersion (about 5%, around $\tau_a = 100\,\text{d}$). This is due to the fact that, even if the velocities are well contrasted between the four layers, flow is generally fast in the whole system analyzed here. In principle, dispersion acting parallel to the flow direction tends to make the system older by generating residence-time distributions with longer tails, and, therefore, with larger standard deviation σ_r. As stated theoretically in Equation 7.12, this has a direct consequence on the average age τ_a. In an inverse manner, dispersion acting perpendicularly to the flow direction has an age-homogenizing effect by mixing water particles of adjacent flow paths. This effect reduces the tailing (and the standard deviation) of the residence-time distributions and tends to make the system younger.

The outlet residence-time distribution $\varphi(t)$ in Figure 7.14(a), and the internal residence time distribution $\psi_T(t)$ in Figure 7.14(b) are of particular interest for the characterization of the internal flow dynamics. When the age transport process is dominated by advection, the two functions clearly exhibit

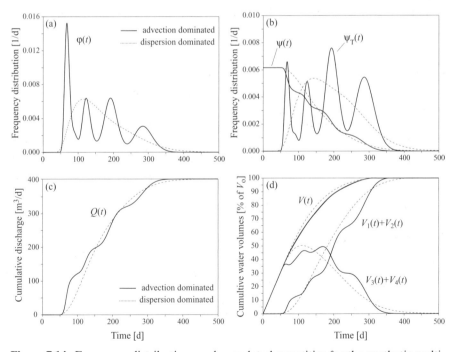

Figure 7.14. Frequency distributions and cumulated quantities for the synthetic multilayer aquifer (modified after Cornaton, 2004). Sensitivity to dispersion. (a) Outlet residence-time distribution, (b) internal age and residence-time distributions, (c) cumulative discharge at the outlet, (d) relative cumulated groundwater volumes $V = V_1 + V_2 + V_3 + V_4$.

as many peaks as the number of contributing layers. However, the magnitudes of the peaks are different and discern the residence-time frequencies of the outlet discharge from those of the system pore volume. It is also noted on these figures that the turnover time $\tau_o = 162\,d$, which is also the average outlet residence time, has one of the weakest frequencies of occurrence. This characteristic value is therefore poorly representative because there are only very few groundwater particles with the residence time τ_o at the outlet. The fact that the residence-time frequencies do not vanish between the four modes is due to mild fractional age-mass exchanges between the layers and to the delayed transition of the deeper waters through the upper layers in the vicinity of the outlet. In this example, the first mode of $\varphi(t)$, around 70 d, expresses the most frequent residence time with respect to discharge rate fractions, and is representative of the flow in the fourth, bottom layer where pore velocity is the highest. The contribution of this layer to the cumulative discharge at the outlet can be estimated in Figure 7.14(c) at about $125\,m^3/s$, from the cumulative function $Q(t)$ at about $t = 100\,d$. However, it is the third mode of $\psi_T(t)$, around 200 d and corresponding to the second layer of the system, that shows the most frequent residence time with respect to pore volume fractions, because the pore volume of this layer is relatively high. The latter can be estimated to about 40% of V_o on Figure 7.14(d), by the increase of the cumulated function $V_1(t) + V_2(t)$ between the bounds of the third mode signature (about 150 d and 250 d).

Using high diffusive-dispersive parameters results in important fractional age-mass fluxes between the layers. Since there are no real aquitard or impervious formations in this example, both functions $\varphi(t)$ and $\psi_T(t)$ show a much smoother aspect, making it impossible to distinctly identify the contributions of the different layers. The cumulative discharge at the outlet and the cumulative groundwater volumes are moderately affected by these internal mixing processes, particularly the total groundwater volume $V(t)$ cumulated with respect to ages in the system. This simple principle example points to the complex nature of age and residence-time distributions in hydrogeological systems, and calls for a very careful use of mean age data when the system dynamics and its vulnerability have to be assessed. Generally, hydrodynamic dispersion, heterogeneity, and long-distance induced mixing are assumed to be the main sources of uncertainties for the interpretation of environmental tracer data and age-dating methods. As shown here, the geological structure and the hydraulic configuration of the system boundaries are indeed also at the origin of very specific outcomes, for which mean ages may not be representative, even in advection-dominated conditions.

7.3.3 The Seeland Phreatic Aquifer

The Seeland aquifer is located in the northwestern part of Switzerland, in Quaternary deposits on the molassic plateau in the region of Bienne (Figure 7.15). This alluvial, horizontal, and unconfined aquifer covers approximately $70\,km^2$

SELECTED TYPICAL EXAMPLES

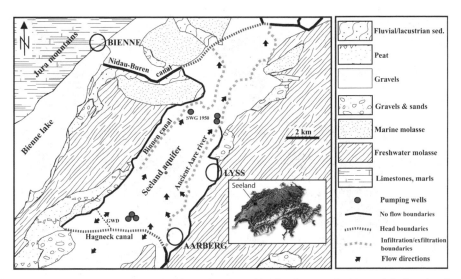

Figure 7.15. Situation map of the Seeland aquifer in the North-West of Switzerland with hydrogeological information (after Cornaton, 2004).

and consists of a heterogeneous set of well sorted gravels, sandy gravels, and silty sands. The average thickness is about 20 to 30 m, with a maximum at about 50 m. This aquifer is recharged by precipitations and by bordering canals and rivers, which also drain the system downstream. It is intensely exploited by several pumping wells, and its surface is essentially devoted to dense agricultural activities.

A groundwater flow model of this region was elaborated and calibrated under both steady-state and transient flow regimes (Jordan, 2000) to support the design of the protection zones and the production of vulnerability maps for the pumping well SWG 1950, an important public water supply. Among the various scenarios enforced to achieve this task, two were analyzed for average and low-water conditions with, respectively, surface infiltrations of 1 mm/d (typical hydrological conditions) and 0.1 mm/d (low-water conditions). Using the calibrated permeability field and transport parameters, and applying the methodology described in Cornaton and Perrochet (2005b), the fractional capture zone of the pumping well (Figure 7.16) and the corresponding residence-time information (Figure 7.17) were computed and analyzed in terms of sensitivity to recharge conditions.

The fractional, or probabilistic, capture zones shown in Figure 7.16 indicate the local groundwater fractions actually reaching the well under investigation. Dispersive-diffusive processes may indeed divert groundwater particles across various flow paths and, thus, inside or outside the purely advective capture zone of the well. Particles located on advective flow paths leading to the well have, therefore, a capture probability that decreases with the distance from

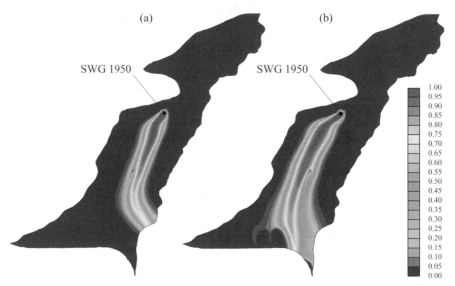

Figure 7.16. Fractional capture zone (P-field) for the SWG 1950 pumping well for surface infiltration rates of (a) 1 mm/d (average scenario), and (b) 0.1 mm/d (low-water scenario). The color fringes indicate local volumetric water fractions reaching the well (after Cornaton, 2004).

the well. Inversely, the distance from the well increases the capture probability of particles not initially located on advective flow paths leading to the well. Because they indicate capture probabilities between zero and one, fractional capture zones generally look much wider than the classical advective capture zones.

Knowing the capture probability field $P(\mathbf{x})$ of a given well, the relative pluvial and canal/river contributions to the well discharge rate Q_{op} can be evaluated, as well as the pore volume V_{op} of the subsystem contributing to the well. In the case of the present horizontal, unconfined aquifer with uniform surface infiltration rate i, the pluvial contribution Q_{Ip} is

$$Q_{Ip} = i \int_\Omega P(\mathbf{x}) d\Omega = i S_{op} \tag{7.83}$$

where S_{op} (domain integral of the probability field) is the effective surface of the well capture zone. With spatially varying saturated thicknesses and porosities, $e(\mathbf{x})$ and $\phi(\mathbf{x})$, the pore volume of the well subsystem is

$$V_{op} = \int_\Omega P(\mathbf{x})\phi(\mathbf{x})e(\mathbf{x}) d\Omega \tag{7.84}$$

According to Equations 7.21 to 7.26 for systems with multiple inlets and outlets, the results of the reservoir theory apply in this subsystem, for which the age and residence-time distributions, $\varphi_p(t)$ and $\psi_p(t)$, are related by

SELECTED TYPICAL EXAMPLES 251

$$\varphi_p(t) = -\frac{V_{op}}{Q_{op}}\frac{\partial \psi_p(t)}{\partial t} = -\tau_{op}\frac{\partial \psi_p(t)}{\partial t} \qquad (7.85)$$

where $\tau_{op} = V_{op}/Q_{op}$ is the subsystem turnover time.

As shown in Figure 7.16 for the constant pumping rate $Q_{op} = 3{,}600\,\text{m}^3/\text{d}$, decreasing the infiltration rate increases the lateral extent of the capture zone, as would be expected. The consequence of this is an increase of the contributions of infiltrating boundaries. In Figure 7.16(a), for the average hydrologic conditions, the capture zone is relatively narrow and clearly shows the portion of the ancient Aare river, on which P varies in the range 0–0.65, indicating that only a fraction of the infiltrated waters reaches the well. In this case the effective surface of the capture zone in Equation 7.83 is $S_{op} = 2.85\,\text{km}^2$, and the pluvial contribution is $Q_{Ip} = 2{,}850\,\text{m}^3/\text{d}$, namely 79% of the extraction rate at the well. This pluvial contribution is thus not enough to satisfy the demand at the well and the remaining 21% ($750\,\text{m}^3/\text{d}$) represents therefore groundwaters originating from river infiltrations. The pore volume of the well subsystem in Equation 7.84 is $V_{op} = 1.04 \cdot 10^7\,\text{m}^3$, yielding the turnover time $\tau_{op} = 7.9\,\text{y}$.

Taking a sensibly different orientation for the low-water conditions in Figure 7.16(b), the capture zone is wider and intersects inflowing portions of the ancient Aare river, as well as some portions of the Binnen and Hagneck canals (see Figure 7.14 for their location). In this case, the values $S_{op} = 5.04\,\text{km}^2$ and $V_{op} = 2.85 \cdot 10^7\,\text{m}^3$ indicate the pluvial contribution $Q_{Ip} = 504\,\text{m}^3/\text{d}$ and the subsystem turnover time $\tau_{op} = 21.7\,\text{yr}$. In such hydrological conditions, the precipitations satisfy only up to 14% of the water demand at the well. The remaining 86% must come from the river and the canals, and this explains the significant turnover time increase. These basic information are important to address water-quality issues and to anticipate the changes in water chemistry that could result from a degradation of the recharge conditions.

Fractional capture zones such as those illustrated in Figure 7.16 for the pumping well SWG 1950 could be obtained for all the other wells implemented in the area, yielding the required individual information, and accounting for possible overlaps of the P-fields. For example, the three wells located near the Hagneck canal generate their own capture zones with relatively high P-values, and this necessarily implies an abrupt drop in the SWG 1950 P-field in this area, as shown in Figure 7.16(b). The effect of another well, with a much smaller extraction rate, can also be seen in the form of a very local, but steep depression in the plotted capture zone for both scenarios.

For the average conditions, the information given by the well residence-time distribution $\varphi_p(t)$ in Figure 7.17(a) confirms a strong pluvial contribution to the pumping well until about 6 yr, with particularly high frequencies from the origin up to about 3 yr. The rest of the curve results from the mixture of pluvial and river infiltrations and the third peak, at about 15 yr, corresponds to the mean arrival time of the Aare river waters. The standard deviation around the mean residence time $\tau_{op} = 7.9\,\text{yr}$ is $\sigma_{rp} = 6.5\,\text{yr}$, yielding, from Equation 7.61, the volume averaged mean residence time $\tau_{Tp} = 13.2\,\text{yr}$. For the low-water conditions, the residence-time distribution is clearly bimodal, attesting

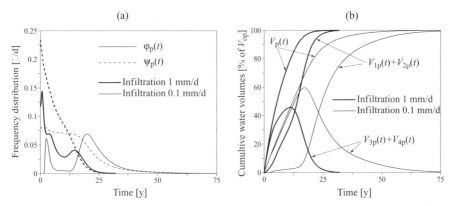

Figure 7.17. Frequency distributions and characteristic groundwater volumes for the Seeland aquifer (after Cornaton, 2004). Sensitivity to surface infiltration rates. (a) Residence-time distribution $\varphi_p(t)$ at the pumping well SWG 1950 and internal age distribution $\psi_p(t)$ (upscaled by a factor 1.8 for clarity) in the contributing pore volume, (b) cumulated groundwater volumes $V_p = V_{1p} + V_{2p} + V_{3p} + V_{4p}$ expressed as percentages of V_{op} (here the subscript p refers to the subsystem corresponding to the fractional capture zone of the pumping well).

of the main contributions of the canals and the river. The first peak is representative of the infiltrations from the Binnen canal, while the second results from the mixture of the Aare River and Hagneck canal infiltrations in the southeastern end of the domain. In this case, $\tau_{op} = 21.7\,\text{yr}$, $\sigma_{rp} = 11.3\,\text{yr}$, and $\tau_{Tp} = 27.6\,\text{yr}$.

The effects of the recharge conditions are also quite apparent in the age distributions $\psi_p(t)$, which show the typical responses of a system where uniform pluvial dilution dominates ($i = 1\,\text{mm/d}$), and of a system where recharge mainly occurs at distant locations ($i = 0.1\,\text{mm/d}$). In the contributing pore volume V_{op}, the reduction of the infiltration rate by a factor of 10 makes the mean age value increase from $\tau_{ap} = 6.6\,\text{yr}$ to $\tau_{ap} = 13.8\,\text{yr}$. The comparison of these values to the respective turnover times indicates that $\tau_{ap} < \tau_{op}$ and that the ratio τ_{ap}/τ_{op} decreases from 0.84 to 0.64, according to the change in aquifer recharge type. Improving the mixing, or the dilution, of groundwaters with young surface waters always tends to equalize τ_{ap} and τ_{op} ($\tau_{ap}/\tau_{op} = 1$ for perfect mixing properties, exponential model), while τ_{ap} tends toward the half of τ_{op} for poor mixing properties ($\tau_{ap}/\tau_{op} = 1/2$ in the case of perfect piston flow) as discussed in Section 7.2.1. In both average and low-water conditions, the tail of the functions $\varphi_p(t)$ and $\psi_p(t)$ is important and is due to the combined effect of dilution and dispersion. These effects are also apparent in the characteristic groundwater volumes shown in Figure 7.17(b). The curves $V_p(t)$ and $V_{3p}(t) + V_{4p}(t)$ diverge at early times because of the specified surface infiltrations, which imply short residence times in the vicinity of the well. In average conditions, the function $V_{3p}(t) + V_{4p}(t)$ does not show important tailing patterns

SELECTED TYPICAL EXAMPLES   253

because the origin of the boundary infiltrations is restricted to a relatively narrow portion of the Aare River. This is not the case in low-water conditions, in which the long tail of the curve $V_{3p}(t) + V_{4p}(t)$ is related to the large extension of the fractional capture zone collecting a variety of groundwater particles with very different infiltration locations.

In Figure 7.17(b), the curves $V_{1p}(t) + V_{2p}(t)$ also indicate that the groundwater volumes reaching the well in less than about 30 yr range between 100% and 75% depending on the infiltration rate. For any intermediate hydrological scenario in the range 0.1–1 mm/d, this means that groundwater occupying up to 25% of the pore volume contributing to the pumping well need more than 30 yr to reach the well. Given that the order of magnitude of V_{op} is of the order of 10^7m^3, the corresponding groundwater volumes may be enormous, and this has a direct implication in the time required to restore water quality after a regional, diffuse contamination event.

The groundwater of the Seeland aquifer suffers from such diffuse contaminations. For instance, the threshold nitrate concentration of 40 mg/l is exceeded in many parts of the system, including the SWG 1950 pumping well. Assuming that future hydrologic conditions may regularly lead to average surface infiltrations ranging between the two simulated values, the residence-time distributions at this well suggest that significant, late-time frequencies could be found in the interval 25 to 50 yr. Taking the indicative value of 30 yr discussed above, and considering that nitrates and other fertilizers were massively used up to the 1990s, before being gradually reduced or replaced, one may indeed conclude to a large amount of residual polluted groundwaters. Significantly declining nitrate concentrations at the well may therefore not occur before a decade, when the relatively slow dilution mechanisms will be combined to flushing boundary inflows accelerating the restoration process, which may take another decade to reach a satisfactory level, and a few more others to be complete.

8

ISSUES AND THOUGHTS IN GROUNDWATER DATING SCIENCE

This chapter deals with some new and important topics that have not been covered in the previous chapters due to the specialized nature of each chapter. First we describe the need to have more user-friendly groundwater dating methods and list a range of potential dating techniques that might be developed further and employed to supplement those discussed earlier. How to translate the results of age modeling exercises into real-world, practical applications is dealt with next. We then propose some new concepts and measures that might be considered as the sign of the maturity of groundwater age science. The future of groundwater dating will also be dealt with in the context of the "Future of Hydrogeology," i.e., whether groundwater age-dating is as mature as hydrogeology or is a branch for which major advancement has still to come. The Authors' concluding remarks about the science of groundwater age-dating are included in this chapter.

8.1 THE NEED FOR MORE DATING METHODS AND THE CURRENTLY PROPOSED POTENTIAL METHODS

In this subchapter, we briefly point to some of the less used and specific groundwater dating methods that may find wider applications in the future. The general consensus among scientists is that the number of age-dating methods is quite limited (Loosli et al., 2000). Since 50 years ago, some tech-

---

Groundwater Age, by Gholam A. Kazemi, Jay H. Lehr, and Pierre Perrochet
Copyright © 2006 John Wiley & Sons, Inc.

niques for dating groundwaters have been developed, with some already losing their credibility. Tritium and silicon-32 methods, for instance, are disappearing, and Krypton-85, CFCs, and SF_6 have recently come into play. Iodine-129 and ^{36}Cl have proved to suffer from some serious drawbacks, but the new ATTA technique to measure ^{81}Kr content of groundwater samples have created serious hopes in dating very old groundwaters by this isotope. Further, ^{4}He and ^{40}Ar techniques have been used to supplement very old dating methods. In general, there are many more dating methods now than there were 20 years ago, and there could be many more 20 years from now. The motivation for the development in the field of groundwater dating is the serious need, felt by the international scientific community, to know the age of groundwater. From the viewpoint of the aquifer management, it is of crucial value to have more widely and easily applicable cheap dating methods. It is also necessary to extend the application of age data and to simplify the interpretation of the presently costly research-level-only dating methods. We need to have reasonably precise age indicators for various age categories. For example, we need methods for dating 0–100 years old, 100–1,000 years old, 1,000–10,000-year-old waters, and so on. The ideal situation would be to have a set of methods to enable groundwater professionals to date each and every single molecule of water. One such technique should, for instance, be able to reveal the very last moment that the groundwater molecule, the age of which is of interest, was subjected to the sunlight. In other words, we should be able to find out when the last time that the molecule of interest saw the light of the sun. In this regard, researchers should concentrate to find out whether sunlight leaves any recognizable impact on the water molecule. This is to some extent similar to thermoluminocene dating of archaeological materials. Now we turn to briefly discuss some of the potential dating methods that have been either proposed or applied to only a minor extent.

Radon-222 Method As stated in Chapter 4, ^{222}Rn method can be grouped along with other methods for dating young groundwaters, but it has been separated because the timescale (age) involved is much shorter than the other isotopes. Radon-222 is mostly used to determine the quantity and location of river discharge to streams (Cook et al., 2003), to delineate groundwater recharge and flow paths, and to estimate groundwater apparent ages over short timescales (Cecil and Green, 2000) and as a natural tracer to monitor the remediation of NAPL contamination in the subsurface. A good discussion about the hydrogeology of radon in an area near Conifer, Colorado, is presented by Lawrence et al. (1991), where they measured the concentration of ^{222}Rn in 46 water wells and one spring.

Radon-222 with a half life of 3.8 days is a product in the decay chain of uranium-238. It is generated by the disintegration of radium-226 (the immediate precursor to radon in the decay chain with a half-life of 1,620 years) through alpha–particle emission and it decays to polonium-218 by the same alpha-particle emission reaction:

$$^{226}_{88}Ra \rightarrow\, ^{222}_{86}Rn + \alpha, \qquad ^{222}_{86}Rn \rightarrow\, ^{218}_{84}Po + \alpha$$

There are two sources for ^{222}Rn in groundwater including (1) radioactive disintegration of dissolved radium and (2) direct release of radon from minerals containing members of the uranium/thorium decay series. The short half-life of ^{222}Rn makes it suitable to date very young groundwaters whose age ranges from 1 to 20 days. Groundwaters have higher concentration of ^{222}Rn as compared to surface waters because ^{222}Rn content of surface waters escapes into the atmosphere. This forms the base of dating groundwater by ^{222}Rn. Recharge waters with minimum or no ^{222}Rn concentration start to acquire ^{222}Rn from the time they enter aquifer. This is called radon "ingrowth." After equilibrating with the ambient groundwater, which happens in a period of about 20 days, it is no longer possible to distinguish between newly recharged water and the ambient groundwater on the basis of ^{222}Rn concentration. Therefore, in dating groundwaters with ^{222}Rn, only zones of active groundwater circulation are of interest. In such a situation, we need to measure the concentration of ^{222}Rn in the ambient groundwater at a distance far from the recharge area. We then compare the concentration of ^{222}Rn in the sample (to be dated) with that of the ambient groundwater through the following radioactive decay equation:

$$C = C_e\left(1 - e^{-\lambda t}\right) \qquad (8.1)$$

where C is the concentration of ^{222}Rn in the sample, C_e is the concentration of ^{222}Rn in ambient groundwater, λ is the decay constant of ^{222}Rn (0.18 day^{-1}), and t is the age of groundwater.

Hoehn and von Gunten (1989) measured the age of infiltrated river water in three unconfined aquifer in various locations in Switzerland by measuring the concentration of ^{222}Rn. The contamination of groundwater as a result of river recharge is a concern in the majority of European countries, including Switzerland. Therefore, it is important to know the velocity of the groundwater to have an estimation of the migration rate of the contaminants. Hohen and von Gunten showed that it takes River Glatt water 15 days to travel to a distance of 26 meters (from river bank) in an unconfined aquifer in River Glatt Valley, Switerland. On the basis of ^{222}Rn groundwater ages at various distances from the River Glatt, the average groundwater flow velocity was estimated at 4.6 m/day.

Atmospheric Noble Gas Signals Noble gas solubility in atmospheric precipitation depends on the air temperature; the higher the temperature, the lower the solubility. Infiltrating rainfall carries the dissolved noble gas into the unsaturated zone and then to the saturated zone where it can be sampled and its concentration analyzed and interpreted. It is possible to estimate groundwater recharge temperature by measuring the concentration of dissolved noble gases. If the original atmospheric noble gas signal is preserved in the aquifer, it is possible to constrain water residence times by comparing the recharge paleotemperatures with independently derived paleotemperatures (Pinti et al., 1997). The method works better at mid- and high latitudes, where the temperature increase from Pleistocene to Holocene has been greater (Gonfiantini et al., 1998). However, the main problem that limits and complicates this particular application of noble gasses is the "excess air" or "supersaturation" problem, air entrapped by the infiltrating rainwater molecules in the unsaturated zone (Heaton and Vogel, 1981).

Chemical Timescale Chemicals dissolved in water may provide some clues about the age and residence time of groundwater. In some circumstances, a high concentration of chemicals is a rough indication of long ages. This has been referred to as chemical timescale or time information derived by studying the chemistry of the groundwater (Edmunds and Smedley, 2000). In addition to age information, the chemical timescale, supplemented by inert and other quantitative indicators, provides a mean to evaluate aquifer homogeneity and/or stratification unrelated to the present-day flow regime (Edmunds and Smedley, 2000). This means that a careful study of the chemical composition of groundwater may have dual use and replace some of the costly age-dating methods that are not available in many countries.

Anthropogenic Iodine-129 Iodine-129 is a tracer to date very old groundwaters as described in Chapter 6. However, a very high concentration of this isotope in shallow groundwaters is associated with anthropogenic sources such as radioactive waste disposal facilities. In such circumstances, groundwater containing excessive ^{129}I is thought to have recharged in modern times. For instance, Schwehr et al. (2005) have shown that the concentration of ^{129}I in shallow groundwater in Orange County aquifer, USA, is decreased with the increase in the age of groundwater. Ages, measured by ^{3}H/^{3}He method and ^{129}I/^{127}I ratio of nine samples were as follows:

^{129}I/^{127}I × 10^{-11}	63.8	44.5	29	14.6	9	16	26.6	24.5	29.7	7.1
Age (year)	6.8	1.9	1.2	0.2	0.1	0.25	0.08	0.55	6	25.3

This data indicates that the source of ^{129}I in the Orange County aquifer is anthropogenic, which is why its concentration in older groundwaters is less.

Sulfur-35 Sulfur-35, a radioactive isotope of sulfur with a half-life of 87.4 days, is formed in the atmosphere from the cosmic-ray spallation of ^{40}Ar. It reaches the ground surface through wet and dry deposition. Cooper et al. (1991), in a study on Imnavait Creek, Alaska, have shown that deposited ^{35}S is mostly adsorbed within the watershed and does not enter the hydrologic cycle. This is the main problem in using ^{35}S as a hydrologic tracer. However, Michel (2000) argues that ^{35}S is a useful tracer to date less than one-year-old groundwaters if the conditions—minimal biological uptake, and no water–rock interaction—are met. So far the three successful studies involving ^{35}S in dating very young groundwaters are those by Michel and Turk (1996) on Flat Top Wilderness region, Colorado, Sueker et al. (1999) on Colorado Front Range, and Plummer et al. (2001) on Shenandoah National Park, Blue Ridge Mountains, Virginia.

Strontium Isotopes Collerson et al. (1988) have shown that the ratio of $^{87}Sr/^{86}Sr$ isotopes in groundwater in the Jurassic aquifer of the Great Artesian Basin of Australia increases linearly with the distance from the recharge area and age of groundwater (from 0.7045 to 0.7118). Such an increase was attributed to the interaction between groundwater and igneous lithologies or allochthonous material derived by dissolution of rocks having low Rb/Sr ratios. On the basis of this research, we argue that the ratio of strontium isotopes could be used as an age indicator, but this needs research and further ratification.

Dating Groundwater by Studying Aquifer Matrix In a study by Zwingmann et al. (1998), the age of formation fluids was determined by dating hydrothermally formed illites in the Rotliegend sandstone reservoirs of northern Germany. The basis of such kind of dating is the difference between the ages of illites (which were formed by fluids) in different parts of the mentioned reservoirs. Illite ages decrease from an average of 198 million years in the vicinity of the fault system (as the source and conduit for fluids) to 177 million years some 4 km away from the fault zone. This means that the fluids traveled 4 km in 21 million years or 0.2 mm per year, or simply the age of the fluids was calculated at 21 million years. In short, this method of dating is based on the imprints of the moving fluids in different parts of the reservoir.

Graphical Analysis of Conductivity Time Series This method, which is based on measurement of simple parameters, was applied by Hofer et al. (2005) to date young groundwaters in a shallow aquifer in the Thur Valley, Switzerland. The producing unconfined aquifer is hydraulically connected to, and qualitatively affected by, the quality of the River Thur. Time series of electrical conductivity values in the River Thur were correlated with the respective time series in the shallow aquifer. From the resulting diagrams, the travel time of the infiltrating river water to the pumping wells tapping shallow aquifer was calculated at 20 ± 10 days. However, higher-resolution time series of electrical conductivity values (hourly values during 10 months) resulted in different ages of 34 and 16 days, depending on the weighting of short-term fluctuations in the river-water conductivity. The large discrepancy between the two sets of results clearly indicates that this technique is prone to high uncertainty. This is why Hofer et al. (2005) confess that it is mostly to help decide if the application of other sophisticated tracer methods is needed and which tracers are most suitable to study the local groundwater dynamics.

Geologic History There are some groundwater age indicators that are referred to as "Geologic history method of dating" (Cornman and Marine, dateless). These include obtaining information about previous ocean or sea water level, information about past distinct climates (e.g., hot, dry, wet, etc.), or information about a specific ion in groundwater that might have existed at the Earth's surface during a very special and short time period. Any one of these factors may leave a recognizable signature on the groundwater which could be attributed to a specific timeframe.

Mineralogical Techniques for Dating Quaternary Groundwater Flow The following paragraph extracted from Metcalfe et al. (1998) describes how mineralogical techniques may show the existence of groundwater flow during Quaternary at a particular location. Mineralogical evidence such as mineral distribution, mineral chemical variations, and mineral textures are used to provide a record of groundwater flow at a particular locality after the groundwaters themselves have left the area. The basis of mineralogical dating technique is twofold: (1) precipitation of minerals as calcite, quartz, and anhydrite is time-dependent and (2) Quaternary groundwater flow ought to have resulted in recognizable mineralogical changes. For example, reactions involving sulfide, oxide, and carbonate are predicted to occur in response to groundwater movements. Also, periods of enhanced groundwater recharge, either during glacial periods or during more temperate episodes, are expected to carry oxidizing waters to greater depth than in periods of more subdued recharge. Studies at Sellafield, England, show that mineralogical information when used in conjunction with hydrogeochemical evidence has considerable potential to provide a more continuous record of groundwater flow events during the Quaternary geological period. We must emphasize that such technique is solely to show that a groundwater flow occurred at some locations during some period of time. It does not provide us with the age of a groundwater sample as we described before.

Other Methods A host of other potential methods are proposed throughout the literature. These include agrochemicals, bacteriophages, HCFCs, Xe, ^{21}Ne, $^{13}C/^{12}C$, $^{34}S/^{32}S$, etc.

8.2 TRANSLATING SIMULATION OF GROUNDWATER AGES TECHNIQUES INTO PRACTICE—MORE APPLICATIONS FOR AGE DATA

A number of researches have been devoted to mathematically simulate the distribution of groundwater ages and residence times. These models are the only means to obtain the distribution of various ages in a groundwater sample or in a spring water. The link between these efforts and isotopic dating is missing. There is a need to combine these two fields and to try to isotopically date those aquifers whose residence time and ages have been modeled. This will help to verify the model results and also to interpret the isotopic ages more confidently. The final practical outcome would be a refined age concept and a more robust use of it. There is also a need to find out more usages for groundwater age modeling exercises such as remediation of contaminated aquifers, installation of monitoring networks, etc.

8.3 WORLDWIDE PRACTICES OF GROUNDWATER AGE-DATING

Table 8.1 lists countries, mostly those with over one million in population, in which age-dating has been applied to at least one aquifer. It demonstrates that fewer than 40 countries, out of 140 countries, have undertaken groundwater dating. This statistic, which could suffer from some errors due to the difficulty in accessing all published materials from all countries, clearly suggests that groundwater dating is an unknown field in about 75% of the world countries. Further, the table shows that only a few countries in Asia, Africa, and South America have undertaken dating studies, in contrast to Western Europe and North American countries in which age-dating is a common exercise. Therefore, action is needed to publicize this discipline and to make this science available to all countries, especially those that rely heavily on groundwater.

8.4 PROPOSAL FOR GROUNDWATER AGE MAP—WORLDWIDE GROUNDWATER AGE MAPS

Geological maps portray various rock types, unconsolidated sediments, and some other geological features of the Earth's surface, and the corresponding ages of each. These maps usually include considerable information about the local stratigraphic column as a means to constrain the age and the lithology of various subsurface formations. Presently, geological maps are available for most parts of the Earth's surface, which means that the relative ages and to some extent the absolute ages of the worldwide geological formations are now well documented. In a smaller scale, quite a few other types of specific maps have been developed within various subdisciplines of geology with themes such as engineering geological maps, hydrogeological maps, environmental geological maps, etc. In even smaller scales, within each of these subdisciplines, other types of even more-detailed maps are being developed and prepared. For instance, slope stability maps, groundwater vulnerability maps, and transmissivity maps all cover a very narrow field or a particular parameter. As the sciences of hydrogeology and groundwater evolve and the need for precise and detailed information about the groundwater resource becomes imminent, more maps are to be produced to depict various angles of these invaluable resources.

It is the purpose of this section of the book to propose one new type of map, groundwater age map, on which the ages of the groundwater resources are depicted. The preliminary steps to construct such a map include

TABLE 8.1. List of World Countries with One or More Groundwater Dating Studies (Countries Whose Population is Higher Than One Million Have Been Investigated)

Country	Continent		Country	Continent	
Afghanistan	Asia		Lebanon	Middle East	
Albania	East Europe		Libya	Africa	√
Algeria	Africa		Lithuania	East Europe	
Angola	Africa		Madagascar	Africa	
Argentina	S & C America		Malawi	Africa	
Australia	Pacific	√	Liberia	Africa	
Austria	Europe	√	Malaysia	Asia	
Bangladesh	Asia		Maldives	Asia	
Belgium	Europe		Malta	Europe	
Benin	Africa		Mauritius	Africa	
Bermuda	Caraibes		Mexico	S & C America	
Bolivia	S & C America	√	Mongolia	Asia	
Botswana	Africa		Malaysia	Asia	
Brazil	S & C America		Morocco	Africa	
Bulgaria	East Europe		Mozambique	Africa	
Burkina Faso	Africa		Myanmar	Asia	
Burundi	Africa		Namibia	Africa	√
C. African R.	Africa		Nepal	Asia	√
Cambodia	Asia		Netherlands	Europe	√
Cameroon	Africa		New Zealand	Pacific	√
Canada	North America	√	Nicaragua	S & C America	
Chad	Africa		Niger	Africa	√
Chile	S & C America		Nigeria	Africa	
China	Asia	√	North Korea	Asia	
Colombia	S & C America		Norway	Europe	
Congo	Africa		Oman	Middle East	√
Cook Islands	Pacific		Pakistan	Asia	√
Costa Rica	S & C America		Panama	Pacific	
Croatia	East Europe		P. N. Guinea	Asia	
Cuba	Caraibes		Paraguay	S & C America	
Cyprus	Europe	√	Peru	S & C America	
Czech R	East Europe	√	Philippines	Asia	
Denmark	Europe		Poland	East Europe	
Dominican R	Caraibes		Portugal	Europe	
East Timor	Asia		Puerto Rico	Caraibes	
Ecuador	S & C America		Saudi Arabia	Middle East	√
Egypt	Middle East	√	Romania	East Europe	
El Salvador	S & C America		Russia	East Europe	√
Estonia	East Europe		Senegal	Africa	
Ethiopia	Africa		Sierra Leone	Africa	
Finland	Europe	√	Singapore	Asia	
France	Europe	√	Slovakia	East Europe	
Gabon	Africa		Somalia	Africa	
Gambia	Africa		South Africa	Africa	√
Germany	Europe	√	South Korea	Asia	

TABLE 8.1. Continued

Country	Continent		Country	Continent	
Ghana	Africa		Spain	Europe	√
Gibraltar	Europe		Sri Lanka	Asia	
Greece	Europe	√	Sudan	Africa	
Greenland	North America		Sudan	Africa	
Guadaloupe	Caraibes		Suriname	S & C America	
Guam (US)	Pacific		Swaziland	Africa	
Guatemala	S & C America		Sweden	Europe	√
Guinea	Africa		Switzerland	Europe	√
Guinea Bis.	Africa		Syria	Middle East	√
Honduras	S & C America		Taiwan	Asia	
Hungary	East Europe		Tanzania	Africa	
India	Asia	√	Thailand	Asia	√
Indonesia	Asia		Togo	Africa	
Iran	Middle East		Trinid. & Tob.	Caraibes	
Iraq	Middle East		Tunisia	Africa	
Ireland	Europe	√	Turkey	East Europe	√
Israel	Middle East		UAE	Middle East	
Italy	Europe	√	Uganda	Africa	
Ivory Coast	Africa		UK	Europe	√
Jamaica	Caraibes		Ukraine	East Europe	
Japan	Asia	√	United States	North America	√
Jordan	Middle East	√	Uruguay	S & C America	
Kazakhstan	Asia		Venezuela	S & C America	√
Kenya	Africa		Vietnam	Asia	
Kuwait	Middle East	√	Yugoslavia	East Europe	
Laos	Asia		Zambia	Africa	
Latvia	East Europe		Zimbabwe	Africa	

1. At a very general scale, there is a need to have a world map to show the countries that have attempted dating of one or more of their aquifers. The raw data for such a map could derive, for example, from Table 8.1. This map will help potential international researchers or international institutions to direct their efforts to those countries that are in most need of such practices and that can also provide potential pristine research sites.
2. In a slightly more detailed fashion, we have to have a single map or a set of maps for each country that shows the aquifers that have been subjected to dating exercises.
3. Finally, we have to construct a groundwater age map for each important aquifer which could vary in detail as described in the next paragraph.

Practically, we should first of all attempt to classify aquifers, for example, all aquifers in a country, into young, old, and very old. This will be useful for the

ENHANCE GROUNDWATER AGE SCIENCE 263

management of groundwater resources in at least two respects, renewability of the aquifers and their susceptibility to pollution. Such a map will illustrate a country-wide picture of the recharging (renewable) and nonrecharging (nonrenewable) freshwater resources, and susceptible and non-/less susceptible resources. The second step should be to determine the range of age of each aquifer, for example, from 2,000 to 25,000 years or from 10–25 years. The third stage would be a more detailed age measurement across the aquifers, which would include age measurements in the recharge area, transmission zone, and discharge area of the aquifers. The last and final step would be to build an isochrone map of the important aquifers, which should take advantage of both hydraulic ages and tracer determined ages.

8.5 WORKS THAT CAN AND NEED TO BE DONE TO ENHANCE GROUNDWATER AGE SCIENCE

A number of initiatives are thought to result in a wider recognition and appreciation of groundwater age science. These are outlined below.

- Advancement in the collection and measurements techniques for dissolved noble gas isotopes such as ^{39}Ar, ^{85}Kr, and ^{85}Kr.
- Enhancement of the worldwide laboratory facilities especially in less developed countries.
- Establishment of regional central laboratories.
- Establishment of an IAH working group.
- Arrangement of groundwater age conferences and forums. The details of first such conferences which was held in 2005 in Tahoe City, California, is given below.

  The Fall 2005 Theis Conference

  "Groundwater Age: Estimation, Modeling, and Water Quality Sustainability"

  Granlibakken Conference Lodge Center, Tahoe City, California, September 23–26, 2005.

  However, this was a conference of local significance attended by U.S. scientists only. Larger national and international forums need to be organized to boost the hydrologists' and geologists' knowledge about the groundwater age subject.
- Inclusion of groundwater age in the curriculum of graduate courses in isotope hydrology and geochemistry.
- Creating an association for groundwater age specialists.
- Construction of the world groundwater age map.
- Writing a book or books about "groundwater dating methods." In this book we have tried to cover the entire subjects associated with ground-

water age. As a result, we have reduced our coverage of the dating methods to avoid enlargement of the book. These is a need to devote a full book entirely to dating techniques in order to allocate one chapter to each method. Also, other chapters of the present book may need a more thorough discussion to form a book of its own.

8.6 MAJOR PROBLEMS FACING GROUNDWATER AGE DISCIPLINE

Apart from minor, easily solvable technical issues, groundwater dating business faces three major problems that could be addressed only if an international consensus and agreement among the scientists is achieved.

1. Dating laboratories are limited and the cost of analysis is often high.
2. The methods are generally complicated and they require sophisticated interpretations as well as considerable extra information in order to yield a reliable age.
3. Dating techniques are limited with some techniques being useful only in special rare situations.

Unless all of the above difficulties are either fully addressed or alleviated, groundwater dating will remain confined to a few countries and it will not gain widespread recognition.

8.7 SOME THOUGHTFUL QUESTIONS—CONCLUDING REMARKS AND FUTURE OF GROUNDWATER DATING

Here we raise some questions that need careful thoughts:

1. Groundwater dating was started 50 years ago, but it has not received considerable attention and still remains relatively unknown. Do you agree with such statement? If yes: What could be the cause/s? Is the subject too detailed? Is it because it is not an applied science, i.e., too academic? Are the dating methods too imprecise? Is it because of the expenses involved?
2. Do we really need to know the age of a groundwater sample? Can we really know the age of a groundwater sample? If we are to delete the parameter of "age of groundwater," is there another parameter to replace it successfully? What would it be?
3. If you were to set up an age-dating laboratory, which dating method would you feature and what are the criteria? Do you use a noble gas laboratory, a radiocarbon laboratory, or a ^{36}Cl? What would be your crite-

ria, budgetary constraints, type of aquifers in your country, or the characteristics of the dating method, i.e., its precision, complexity, range of age, etc.?
4. If you are in charge of a world organization such as the Hydrology section of the IAEA, which country would you help to establish a dating laboratory? If you are in charge of a major dating laboratory, how would you determine the cost of your services? Do you attempt to provide cheap, affordable analysis to publicize the age concept and to make other scientists aware of the benefit of this special piece of the scientific sphere?
5. What is the application of groundwater age modeling practices? Do they have any real-world applications, or are they just for academic hobbies?

Recently, a theme issue of the *Hydrogeology Journal* was devoted to the future of hydrogeology. The overall conclusion was that hydrogeology is not precisely a quantitative science (Voss, 2005). One might ask the same question here about the future of groundwater dating? Will dating groundwater be a widely exercised practice for the next 20 years? Will it grow and expand or decline and shrink? Will scientists pay more attention to this field, or will it be replaced by other disciplines? Will new dating laboratories be built, or will the present ones remain unchanged or even reduced? Will new methods be invented?

Groundwater age-dating discipline was born 50 years ago. It has gone through various cycles and has benefited from the talent of many scientists worldwide. A number of applications are achieved with this discipline, but there are a number of deficiencies, some rectifiable, some not. If one is to list the fundamental difficulties that the age-dating concept has faced and is facing, for sure the problems of "mixing" and "initial value" are ranked high above. These two problems are an integral part of the concept, and it may not be an easy task to remove them. This book is intended to present this discipline to a wide range of audiences in the geo-environmental fields and to collect and to disseminate all the information that is already available in the world literature in a single volume. It aims to create a driving force behind this subject in order to enhance its usefulness to all interested parties.

REFERENCES

Aeschbach-Hertig, W., Schlosser, P., Stute, M., Simpson, H. J., Ludin, A., and Clark, J. F. (1998) A ^{3}H/^{3}He study of ground water flow in a fractured bedrock aquifer. *Ground Water*, 34:661–670.

Aggarwal, P. K. (2002) Isotope hydrology at the International Atomic Energy Agency. *Hydrological Processes*, 16:2257–2259.

Aggarwal, P. K., Gat, J. R., and Froehlich, K. F. (eds.) (2005) *Isotopes in the Water Cycle: Past, Present and Future of a Developing Science*, Springer, Not yet published.

Albright, D., Berkhout, F., and Walker, W. (1993) *World Inventory of Plutonium and Highly Enriched Uranium*, 1992. Oxford University Press, Oxford.

Alburger, D. E., Harbottle, G., and Norton, E. F. (1986) Half life of ^{32}Si. *Earth Planet Sci. Lett.*, 78:168–176.

Aller, L., Bennett, T., Lehr, J. H., and Petty, R. J. (1985) DRASTIC—A standardized system for evaluating ground water pollution potential using hydrogeologic settings, US EPA, Robert S. Kerr Environmental Research Laboratory, Office of Research and Development, EPA/600/2–85/018, 163 p.

Amin, I. E., and Campana, M. E. (1996) A general lumped parameter model for the interpretation of tracer data and transit time calculation in hydrologic systems. *J. Hydrol.*, 179:1–21.

Andre, L. (2002) Contribution de la géochimie à la connaissance des écoulements souterrains profonds: Application à l'aquifère des Sables Infra-Mollasiques du Bassin Aquitain. Ph.D. thesis, University of Bordeaux 3, France, 230 p.

Groundwater Age, by Gholam A. Kazemi, Jay H. Lehr, and Pierre Perrochet
Copyright © 2006 John Wiley & Sons, Inc.

REFERENCES

Andrews, J. N., Edmunds, W. M., Smedley, P. L., Fontes, J. Ch., Fifield, L. K., and Allan, G. L. (1994) Chlorine-36 in groundwater as a palaeoclimatic indicator: The East Midlands Triassic sandstone aquifer (UK). *Earth Planet Sci Lett.*, 122:159–171.

Andrews, J. N., Drimmie, R. J., Loosli, H. H., and Hendry, M. J. (1991) Dissolved gases in the Milk River aquifer, Canada. *Applied Geochemistry*, 6:393–403.

Andrews, J. N., Hussain, N., and Youngman, M. J. (1989) Atmospheric and radiogenic gases in groundwaters from the Stripa Granite. *Geochimica et Cosmochimica Acta*, 53:1831–1841.

Andrews, J. N., Davis, S. N., Fabryka-Martin, J., Fontes, J. Ch., Lehman, B. E., Loosli, H. H., Michelot, J. L., Moser, H., Smith, B., and Wolf, M. (1989) The in situ production of radioisotopes in rock matrices with particular reference to the Stripa granite. *Geochimica et Cosmochimica Acta*, 53:1803–1815.

Andrews, J. N., Fontes, J. Ch., Michelot, J-L., and Elmore, D. (1986) In situ ^{36}Cl production and ground water evolution in crystalline rocks at Stripa, Sweden. *Earth Planet Sci Lett.*, 77:49–58.

Andrews, J. N., Goldbrunner, J. E., Darling, W. G., Hooker, P., Wilson, G. B., Youngman, M. J., Eichinger, L., Rauert, W., and Stichler, W. (1985) A radiochemical, hydrochemical and dissolved gas study of groundwaters in the Molasse basin of Upper Austria. *Earth Planet Sci Lett.*, 73:317–332.

Andrews, J. N., Balderer, W., Bath, A., Clausen, H. B., Evans, G. V., Florowski, T., Goldbrunner, J. E., Ivanovich, M., Loosli, H., and Zojer, H. (1983) Environmental isotope studies in two aquifer systems. In: Isotope Hydrology 1983, IAEA, Vienna, 535–576.

Andrews, J. N., and Kay, R. L. F. (1982) Natural production of tritium in permeable rocks. *Nature*, 298:361–363.

Andrews, J. N., Giles, I. S., Kay, R. L. F., Lee, D. J., Osmond, J. K., Cowart, J. B., Fritz, P., Baker, J. F., and Gale, J. (1982) Radioelements, radiogenic helium and age relationships for groundwaters from the granites at Stripa, Sweden. *Geochimica et Cosmochimica Acta*, 46:1533–1543.

Andrews, J. N., and Lee, D. J. (1979) Inert gases in groundwater from the Bunter Sandstone of England as indicators of age and paleoclimatic trends. *J. Hydrol.*, 41:233–252.

Appelo, C. A. J., and Postma, D. (1999) *Geochemistry, Groundwater and Pollution*. Balkema, Rotterdam, 536 p.

Arnold, J. R., and Libby, W. F. (1951) Radiocarbon dates. *Science*, 113:111–120.

Arumugum, V. (1994) Site characterization for location of radioactive waste repository: A case study. Ph.D. thesis, Indian Institute of Technology Bombay, India.

Aryal, S. K., O'Loughlin, E. M., and Mein, R. G. (2005) A similarity approach to determine response times to steady-state saturation in landscapes. *Advances in Water Resources*, 28(2):99–115.

Attendorn, H. G., and Bowen, R. N. C. (1997) *Radioactive and Stable Isotope Geology*. Chapman & Hall, London, 522 p.

Back, W. (1994) Hydrologic time and sustainability of shallow aquifers. In: *Proc. of International conf. Water Down Under '94*. Adelaide, Australia, 331–335.

Badr, O., Probert, D. S., and O'Callaghan, P. W. (1990) Chlorofluorocarbons and the environment: Scientific, economic, social, and political issues. *Applied Energy*, 37:247–327.

Balderer, W. (1986) Signification de l'âge moyen de l'eau souterraine donné par les isotopes radioactifs. *Bulletin du Centre d'Hydrogéologie de l'Université de Neuchâtel*, 6:43–66.

Ballentine, C. J., and Sherwood-Lollar, B. (2002) Regional groundwater focusing of nitrogen and noble gases into the Hugoton-Panhandle giant gas field, USA. *Geochimica et Cosmochimica Acta*, 66(14):2483–2497.

Ballentine, C. J., and Burnard, P. G. (2002) Production, release and transport of noble gases in the continental crust. In: Porcelli, D., Ballentine, C. J., and Wieler, R. (eds.) *Noble Gases in Geochemistry and Cosmochemistry*. The Mineralogical Society of America, Washington, 481–538.

Barrett, E. W., and Huebner, L. (1960) Atmospheric tritium analysis: Chicago University technical progress report No. 2, Contract AT (11-1)-636, February 16.

Bauer, S., Fulda, C., and Schäfer, W. (2001) A multi-tracer study in a shallow aquifer using age dating tracers ^{3}H, ^{85}Kr, CFC-113 and SF_6—indication for retarded transport of CFC-113. *J. Hydrol.*, 248:14–34.

Bayari, S. (2002) TRACER: An EXCEL workbook to calculate mean residence time in groundwater by use of tracers CFC-11, CFC-12 and tritium. *Computers & Geosciences*, 28:621–630.

Bayer, R., Schlosser, P., Bönisch, G., Rupp, H., Zaucker, F., and Zimmek, G. (1989) Performance and blank components of a mass spectrometric system for routine measurement of helium isotopes and tritium by the ^{3}He ingrowth method. *Sitzungsberichte der Heidelberger Akademie der Wissenschaften, Mathematisch-naturwissenschaftliche Klasse*, 5:241–279, Springer-Verlag, Heidelberg.

Begemann, F., and Libby, F. W. (1957) Continental water balance, groundwater inventory and storage times, surface ocean mixing rates, and worldwide water circulation patterns from cosmic ray and bomb tritium. *Geochimica et Cosmochimica Acta*, 12:227–296.

Benson, B. B., and Krause, D., Jr. (1980) Isotopic fractionation of helium during solution. A probe for the liquid state. *Journal of Solution Chemistry*, 9:895–909.

Bentley, H. W., Phillips, F. M., and Davis, S. N. (1986a) Chlorine-36 in the terresterial environment. In: Fritz, P., and Fontes, J. Ch. (eds.), *Handbook of Environmental Isotope Geochemistry*. Elsevier, New York, 422–475.

Bentley, H. W., Phillips, F. M., Davis, S. N., Habermehl, M. A., Airey, Calf, G. E., P. l., Elmore, D., Gove, H. E. and Torgersen, R. T. (1986b) Chlorine-36 dating of very old groundwater 1. The Great Artesian Basin, Australia. *WRR*, 22:1991–2001.

Bentley, H. W., Phillips, F. M., Davis, S. N., Gifford, S., Elmore, D., Tubbs, L. E., and Gove, H. E. (1982) Thermonuclear ^{36}Cl pulse in natural water. *Nature*, 300:737–740.

Bertleff, B., Watzel, R., Eichinger, L., Heidinger, M., Schneider, K., Loosli, H. H., and Stichler, W. (1998) The use of isotope based modeling techniques for groundwater management in a Quaternary aquifer. In: Isotope techniques in the study of past and current environmental changes in the hydrosphere and the atmosphere, IAEA, Vienna, 437–452.

Bethke, C. M., and Johnson, T. M. (2002) Paradox of groundwater age: Correction. *Geology*, 30(4):385–388.

Bethke, C. M., Torgersen, T., and Park, J. (2000) The age of very old groundwater: Insights from reactive transport models. *J. of Geochemical Exploration*, 69–70:1–4.

Beyerle, U., Aeschbach-Hertig, W., Peeters, F., and Kipfer, R. (2000) Accumulation rates of radiogenic noble gases and noble gas temperatures deduced from the Great Artesian Basin, Australia. *Beyond 2000—New Frontiers in Isotope Geoscience*, 1:21.

Beyerle, U., Purtschert, R., Aeschbach-Hertig, W., Imboden, D. M., Loosli, H. H., Wieler, R., and Kipfer, R. (1998) Climate and groundwater recharge during the last glaciation in an ice-covered region. *Science*, 282:731–734.

Bockgard, N., Rodhe, A., and Olsson, K. A. (2004) Accuracy of CFC groundwater dating in a crystalline bedrock aquifer: Data from a site in southern Sweden. *Hydrogeology Journal*, 12:171–183.

Böhlke, J. K., and Denvor, J. M. (1995) Combined use of groundwater dating, chemical and isotopic analyses to resolve the history and fate of nitrate contamination in two agricultural watersheds, Atlantic Coastal Plain, Maryland. *WRR*, 31:2319–2339.

Bolin, B., and Rodhe, H. (1973) A note on the concepts of age distribution and transit time in natural reservoirs. *Tellus*, 25(1):58–62.

Boronina, A., Renard, P., Balderer, W., and Stichler, W. (2005) Application of tritium in precipitation and in groundwater of the Kouris catchment (Cyprus) for description of the regional groundwater flow. *Applied Geochemistry*, 20:1292–1308.

Bottomley, D. J., Gascoyne, J., and Kamineni, D. C. (1990) The geochemistry, age and origin of groundwater in a mafic pluton, East Bull Lake, Ontario, Canada. *Geochimica et Cosmochimica Acta*, 54:993–1008.

Bottomley, D. J., Ross, J. D., and Clarke, W. B. (1984) Helium and neon isotope geochemistry of some groundwaters from the Canadian Precambrian Shield. *Geochimica et Cosmochimica Acta*, 48:1973–1985.

Bowen, R. (1988) *Isotopes in the Earth Sciences*. Elsevier, London and New York, 647 p.

Brauer, F. (1974) Environmental ^{129}I measurements: Pacific Northwest Laboratory Report BNWL-SA-4983, 11 p.

Brawley, J. W., Collins, G., Kremer, J. N., and Sham, C. H. (2000) A time-dependent model of nitrogen loading to estuaries from coastal watersheds. *J. Environ. Qual.*, 29:1448–1461.

Briel, L. I. (1976) An investigation of the ^{234}U/^{238}U disequiblirium in the natural waters of the Santa Fe river basin of north central Florida. Ph.D. thesis, Florida State University.

Brinkmann, R., Münnich, J. C., and Vogel, J. C. (1959) C^{14}-Altersbestimmung von grundwasser. *Naturewissenchaften*, 46:10–12.

Broad, W. J. (2005) With a push from the U.N., Water reveals its secrets. *New York Times*, July 26, 2005.

Broers, H. P. (2004) The spatial distribution of groundwater age for different geohydrological situations in the Netherlands: implications for groundwater quality monitoring at the regional scale. *J. Hydrol.*, 299:84–106.

Brown, J. D. (1980) Evaluation of fluorocarbon compounds as groundwater tracers: Soil column studies. M.Sc. thesis, University of Arizona, Tucson.

Brown, R. H. (1964) Hydrologic factors pertinent to groundwater contamination. *Ground Water*, 2:5–12.

Bu, X., and Warner, M. J. (1995) Solubility of chlorofluorocarbon 113 in water and seawater. *Deep-sea research*, 42:1151–1161.

Burgess, A. B., Grainger, R. G., Dudhia, A., Payne, V. H., and Jay, V. L. (2004) MIPAS measurement of sulphur hexafluoride (SF_6)-art. No. L05112. *Geophysical Research Lett.*, 31:5112.

Busenberg, E., and Plummer, L. N. (2000) Dating young ground water with sulfur hexafluoride—Natural and anthropogenic sources of sulfur hexafluoride. *WRR*, 36:3011–3030.

Busenberg, E., and Plummer, L. N. (1997) Use of sulfur hexafluoride as a dating tool and as a tracer of igneous and volcanic fluids in ground water. *Abst. Geol. Soc. Amer.*, 1997 Annual Meeting, GSA Abstracts with Programs, v. 29, no. 6, p. A-78.

Busenberg, E., and Plummer, L. N. (1996) Concentrations of chlorofluorocarbons and other gases in ground water at Mirror Lake, New Hampshire, in Morganwalp, D. W., and Aronson, D. A. (eds.) USGS Toxic substances hydrology program. In: *Proc. of the Technical Meeting*, Colorado Springs, Colorado, September 20–24, 1993: USGS WRIR 94-4014, 151–158.

Busenberg, E., and Plummer, L. N. (1992) Use of chlorofluorocarbons (CCl_3F and CCl_2F2 as hydrologic tracers and age-dating tools: The alluvium and terrace system of central Oklahoma. *WRR*, 9:2257–2283.

Calf, G. E., Bird, J. R., Kellet, J. R., and Evans, W. R. (1998) Origins of chlorine variation in the Murray Basin—Using environmental chlorine-36. Australian Bureau of Mineral Resources, Geology and Geophysics Record 1988/43, Groundwater series No. 16, 28–30.

Cameron, A. C. W. (1973) Abundance of elements in the solar system. *Space Science Review*, 15:121–146.

Campana, M. E. (1987) Generation of ground-water age distributions. *Ground Water*, 25:51–58.

Campana, M. E., and Simpson, E. (1984) Groundwater residence times and recharge rates using a discrete-state compartment model and ^{14}C data. *J. Hydrol.*, 72:171–185.

Caplow, T., Schlosser, P., Ho, D. T., and Santella, N. (2003) Transport dynamics in a sheltered estuary and connecting tidal straits: SF6 tracer study in New York Harbor. *Environmental Science & Technology*, 37(22):5116–5126.

Carrillo-Rivera, J., Clark, I. D., and Fritz, P. (1992) Investigating recharge of shallow and paleogroundwaters in the Villa De Reyes basin, SLP, Mexico, with environmental isotopes. *Hydrogeology Journal*, 1:35–48.

Castro, M. C., and Goblet, P. (2005) Calculaation of groundwater ages—A comparative analysis. *Ground Water*, 43:368–380.

Cecil, L. D., Hall, L. F., and Green, J. R. (2003) Re-evaluation of background iodine-129 concentrations in water from the snake river plain aquifer, Idaho, U.S. Geological Survey Water-Resources Investigations Report 03-4106. Idaho Falls, Idaho, DOE/ID-22186.

Cecil, L. D., and Green, J. R. (2000) Radon-222. In: Cook, P. G., and Herczeg, A. L. (eds.), *Environmental Tracers in Subsurface Hydrology*. Kluwer Academic Publishers, Boston, 176–194.

Cecil, L. D., Welhan, J. A., Green, J. R., Frape, S. K., and Sudicky, E. R. (2000) Use of chlorine-36 to determine regional scale aquifer dispersivity, eastern Snake River Plain aquifer, Idaho/USA. *Nucl. Instrum. Methods Phys. Res.*, B172:679–687.

Cecil, L. D., Beasley, T. M., Pittman, J. R., Michel, R. L., Kubik, P. W., Sharma, P., Fehn, U., and Gove, H. E. (1992) Water infiltration rate in the unsaturated zone at the Idaho National Engineering Laboratory estimated from Chlorine-36 and tritium profiles and neutron logging. In: Kharaka, Y. F., and Maest, A. S. (eds.), *Water Rock Interaction*, Vol. 1., Balkema, Rotterdam, 709–714.

Ceric, A., and Haitjema, H. (2005) On using simple time-of-travel capture zone delineation methods. *Ground Water*, 43:408–412.

Chappelle, F. Z., Zelibor, J. L., Grimes, D. J., and Knobel, L. L. (1987) Bacteria in deep coastal plain sediments of Maryland: a possible source of CO_2 to groundwater. *WRR*, 23:1625–1632.

Chen, C.-Y., Li, Y. M., Bailey, K., O'Conner, T., Young, L., and Lu, Z.-T. (1999) Ultrasensitive isotope trace analysis with a magneto-optical trap. *Science*, 286: 1139–1141.

Chesnaux, R., Molson, J. W., and Chapuis, R. P. (2005) An analytical solution for groundwater transit time through unconfined aquifers. *Ground Water*, 43:511–517.

Clark, I. D., Douglas, M., Raven, K., and Bottomley, D. J. (2000) Recharge and preservation of Laurentide glacial melt water in the Canadian Shield. *Ground Water*, 38:735–742.

Clark, I. D., and Fritz, P. (1997) *Environmental Isotopes in Hydrogeology*. Lewis Publishers, Boca Raton, FL, 328 p.

Clark, J. F., Schlosser, P., Stute, M. and Simpson, H. J. (1996) SF_6-He-3 tracer release experiment: A new method of determining longitudinal dispersion coefficients in large rivers. *Environmental Science & Technology*, 30:1527–1532.

Clarke, D. D., Meerschaert, M. M., and Wheatcraft, S. W. (2005) Fractal travel time estimates for dispersive contaminants. *Ground Water*, 43:401–407.

Clebsch, A. Jr. (1961) Tritium age of groundwater at the Nevada test site, Nye County, Nevada. USGS Prof. Paper 424-C, 122–125.

Clerck, T. D., Poffijn, A., and Eggermont, G. (2002) Measurement of atmospheric [85]Kr in Gent, Belgium: History and perspectives. Poster presented at "L'évaluation et la surveillance des rejets radioactifs des installations nucléaires" 13–14 November 2002, Strasbourg, France.

Collerson, K. D., Ullman, W. J., and Torgersen, T. (1998) Ground waters with unradiogenic $^{87}Sr/^{86}Sr$ ratios in the Great Artesian Basin, Australia. *Geology*, 16:59–63.

Collon, P., Bichler, M., Caggiano, J., Cecil, L. D., Masri, Y. E., Golser, R. et al. (2004) Development of an AMS method to study oceanic circulation characteristics using cosmogenic [39]Ar. *Nuclear Instruments and Methods in Physics Research Section B: Beam Interactions with Materials and Atoms*, 223–224:428–434.

Collon, P., Kutschera, W., Loosli, H. H., Lehmann, B. E., Purtschert, R., Love, A., Sampson, L., Anthony, D., Cole, D., Davids, B., Morrissey, D. J., Sherrill, B. M., Steiner, M., Pardo, R. C., and Paul, M. (2000) [81]Kr in the Great Artesian Basin, Australia: A new method for dating very old groundwater. *Earth Planet Sci Lett.*, 182:103–113.

Cook, P. G. (2003) Groundwater ages in fractured rock aquifers. In: Krasny-Hrkal-Bruthans (eds.), *Proce. International Conf. Groundwater in Fractured Rocks*, PRAGUE, IHP-VI, series on groundwater no. 7, 139–140.

Cook, P. G., Love, A. J., Robinson, N. I., and Simmons, C. T. (2005) Groundwater ages in fractured rock aquifers. *J. Hydrol.*, 308:284–301.

Cook, P. G., Favreau, G., Dighton, J. C., and Tickell, S. (2003) Determining natural groundwater influx to a tropical river using radon, chlorofluorocarbons and ionic environmental tracers. *J. Hydrol.*, 277:74–88.

Cook, P. G., and Robinson, N. I. (2002) Estimating groundwater recharge in fractured rock from environmental ^{3}H and ^{36}Cl, Clare Valley, South Australia. *WRR*, 38:1136, 11-1–11-13. Doi: 10.1029/2001 WR000772.

Cook, P. G., and Böhlke, J. K. (2000) Determining timescales for groundwater flow and solute transport. In: Cook, P. G., and Herczeg, A. L. (eds.), *Environmental Tracers in Subsurface Hydrology*. Kluwer Academic Publishers, Boston, 1–30.

Cook, P. G., and Herczeg, A. L. (1998) Groundwater chemical methods for recharge studies. Part 2 of the *Basics of Recharge and Discharge* (Ed. L. Zhang), CSIRO Publishing, CSIRO Australia.

Cook, P. G., and Solomon, D. K. (1997) Recent advances in dating young groundwater: Chlorofluorocarbons, ^{3}H/^{3}He, and ^{85}Kr. *J. Hydrol.*, 191, 245–265.

Cook, P. G., Love, A., and Dowie, J. (1996) Recharge estimation of shallow groundwater using CFC age dating. *Mineral and Energy of South Australia Journal*, 3:32–33.

Cook, P. G., and Solomon, D. K. (1995) Transport of atmospheric trace gases to the water table: Implications for groundwater dating with chlorofluorocarbons and krypton 85. *WRR*, 31:263–270.

Cook, P. G., Solomon, D. K., Plummer, L. N., Busenberg, E., and Schiff, S. L. (1995) Chlorofluorocarbons as tracers of groundwater transport processes in a shallow, silty sand aquifer. *WRR*, 31:425–434.

Cooper, L. W., Olsen, C. R., Solomon, D. K., Larsen, I. L., Cook, R. B., and Grebmeier, J. M. (1991) Stable isotopes of oxygen and natural and fallout radionuclides used for tracing runoff during snowmelt in an Arctic watershed. *WRR*, 27:2171–2179.

Coplen, T. B. (1993) Uses of environmental isotopes. In: Alley, W. M. (ed.), *Regional Ground-Water Quality*, Van Nostrand Reinhold, New York, 255–293.

Corcho Alvarado, J. A., Purtschert, R., Hinsby, K., Troldborg, L., Hofer, M., Kipfer, R., Aeschbach-Hertig, W., and Arno-Synal, H. (2005) ^{36}Cl in modern groundwater dated by a multi-tracer approach (^{3}H/^{3}He, SF$_6$, CFC-12 and ^{85}Kr): A case study in quaternary sand aquifers in the Odense Pilot River Basin, Denmark. *Applied Geochemistry*, 20:599–609.

Corcho Alvarado, J. A., Purtschert, R., Barbecot, F., Chabault, C., Rueedi, J., Schneider, V., Aeschbach-Hertig, W., Kipfer, R., and Loosli, H. H. (2004) Groundwater age distribution in the unconfined Fontainebleau sands aquifer of the Paris Basin (France): A multi tracer study including ^{39}Ar measurements. Submitted to *Geochimica et Cosmochimica Acta*.

Cordes, C., and Kinzelbach, W. (1992) Continuous groundwater velocity fields and path lines in linear, bilinear, and trilinear finite elements. *WRR*, 28:2903–2911.

Cornaton, F. (2004) Deterministic models of groundwater age, life expectancy and transit time distribution in advective-dispersive systems. Ph.D. thesis, University of Neuchatel, Switzerland. 147 p.

Cornaton, F., and Perrochet, P. (2005a) Groundwater age, life expectancy and transit time distributions in advective-dispersive systems; 1. Generalized Reservoir Theory. *Advances in Water Resources*, in press.

Cornaton, F., and Perrochet, P. (2005b) Groundwater age, life expectancy and transit time distributions in advective-dispersive systems; 2. Reservoir theory for arbitrary aquifers, and its implications for water resources protection. *Advances in Water Resources*, in press.

Cornaton, F., Perrochet, P., and Diersch, H. J. (2004) A finite element formulation of the outlet gradient boundary condition for convective-diffusive transport problems. *International Journal for Numerical Methods in Engineering*, 61:2716–2732.

Cornett, R. J., Andrews, H. R., Chant, L. A., Davies, W. G., Greiner, B. F., Imahori, Y., Koslowsky, V. T., Kotzer, T., Milton, J. C. D., and Milton, G. M. (1997) Is ^{36}Cl from weapons' test fallout still cycling in the atmosphere? *Nuclear Instruments and Methods in Physics Research*, B123:378–381.

Cornett, R. J., Cramer, J., Andrews, H. R., Chant, L. A., Davies, W., Greiner, B. F., Imahori, Y., Koslowsky, V., McKay, J., Milton, G. M., and Milton, J. C. D. (1996) In situ production of ^{36}Cl in uranium ore: A hydrogeological assessment tool. *WRR*, 32:1511–1518.

Cornman, W. R., and Marine, I. W. (dateless) The potential for dating groundwater using radiogenic noble gases. Accessed through Internet.

Cosgrove, B. A., and Walkley, J. (1981) Solubilities of gases in H_2O and ^2H_2O. *Journal of Chromatography*, 216:161–167.

Cowart, J. B., and Osmond, J. K. (1974) ^{234}U and ^{238}U in the Carrizo sandstone aquifer of south Texas. In: *Isotope Techniques in Groundwater Hydrology*, IAEA, Vienna, Volume II, 131–149.

Cresswell, R., Wischusen, J., Jacobson, G., and Fifield, K. (1999) Assessment of recharge to groundwater systems in the arid southwestern part of Northern Territory, Australia, using chlorine-36. *Hydrogeology Journal*, 7:393–404.

Cresswell, R., Fifield, K., Wischusen, J., and Jacobson, G. (1998) Groundwater sustainability in Central Australia studied using Chlorine-36. In: Weaver, T. R., and Lawrence C. R. (eds.) *Proc. of International Groundwater Conference*. New Generation Print and Copy, Victoria, Australia, 25–30.

Dagan, G. (1989) *Flow and Transport in Porous Formations*. Springer, New York, 465 p.

Dagan, G. (1987) Theory of solute transport by groundwater. *Annual Review of Fluid Mechanics*, 19:183–215.

Dagan, G. (1982) Stochastic modeling of groundwater flow by unconditional and conditional probabilities, 2, The solute transport. *WRR*, 18:835–848.

Dagan, G., and Nguyen, V. (1989) A comparison of travel time and concentration approaches to modelling transport by groundwater. *Journal of Contaminant Hydrology*, 4:79–91.

Dalrymple, G. B. (1991) *The Age of the Earth*, Stanford University Press, Stanford, California, 474 p.

Danckwerts, P. V. (1958) The effect of incomplete mixing on homogeneous reactions. *Chem. Eng. Sci.*, 8(1):93–102.

Danckwerts, P. V. (1953) Continuous flow systems: distribution of residence times. *Chemical Engineering Science*, 2:1–13.

Dassargues, A. (ed.) (2000) *Tracers and Modelling in Hydrogeology*. IAHS publication no. 262. IAHS Press, Oxfordshire, UK, 571 p.

Davie, R. F., Kellet, J. R., Fifield, L. K., Evans, W. R., Calf, G. E., Bird, J. R., Topham, S., and Ophel, T. R. (1989) Chlorine-36 measurements in the Murray Basin—Preliminary results from the Victorian and South Australian Mallee region. *BMR Journal of Australian Geology and Geophysics*, 11:261–272.

Davie, R. F., and Schaeffer, O. A. (1955) Chlorine-36 in nature. *Annals of the New York Academy of Sciences*, 62:105–122.

Davis, G. H., and Coplen, T. B. (1987) Stable isotopic composition of groundwater of central California as an indicator of Mid-Pleistocene tectonic evolution. In: *Isotope Techniques in Water Resources Development and Management*, IAEA, Vienna, 225–239.

Davis, S. N., Moysey, S., Cecil, L. D., and Zreda, M. (2003) Chlorine-36 in groundwater of the united states: Empirical data. *Hydrogeology Journal*, 11:217–227.

Davis, S. N., Cecil, L. D., Zreda, M., and Moysey, S. (2001) Chlorine-36, bromide, and the origin of spring water. *Chemical Geology*, 179:3–16.

Davis, S. N., Campbell, D. J., Bentley, H. W., and Flynn, T. J. (1985) *Groundwater Tracers*. Worthington, Ohio, National Water Well Association Press, 200 p.

Davis, S. N., and Bentley, H. W. (1982) Dating groundwater: A short review. In: *Nuclear and Chemical Dating Techniques* (ed. Lloyd, A. C.), American Chemical Society Symposium Series, Washington, D.C., 187–222.

Davis, S. N., and DeWiest, R. J. M. (1966) *Hydrogeology*. John Wiley, New York, 463 p.

de Marsily, G. (1986) *Quantitative Hydrogeology*. Academic Press, New York, 464 p.

Deak, J., and Coplen, T. B. (1996) Identification of Holocene and Pleistocene groundwaters in Hungary using oxygen and hydrogen isotopic ratios. In: *Isotopes in Water Resources Management*, Vol. 1, IAEA, Vienna, 438.

Delcore, M. R. (1986) Groundwater tracing using environmental tritium: Application to recharge rate determination. In: Morfis, A., and Paraskevoponlou, P. (eds.), *Proc. of 5th Intl. Symp. on Underground Water Testing*, Institute of geology and mineral exploration, Athens, Greece, 45–50.

Dickin, A. P. (1995) *Radiogenic Isotope Geology*. Cambridge University Press, Cambridge, 490 p.

Dincer, T., AL Mugrin, W., and Zimmermann, V. (1974) Study of the infiltration and recharge through the sand dunes in arid zones with special references to the stable isotopes and thermonuclear tritium. *J. Hydrol.*, 23:79–109.

Dissanayake, C. (2005) Of stones and health: Medical geology in Sri Lanka. *Science*, 309:883–885.

Divine, C. E., and Humphrey, J. D. (2005) Groundwater dating with H-He. In: Lehr, J. H., and Keeley, J. (eds.), *Water Encyclopedia: Groundwater*. John Wiley and Sons, Hoboken, New Jersey, 65–69.

Divine, C. E., and Thyne, G. (2005) Detecting modern groundwaters with ^{85}Kr. In: Lehr, J. H., and Keeley, J. (eds.), *Water Encyclopedia: Groundwater*. John Wiley and Sons, Hoboken, New Jersey, 248–249.

Doerfliger, N., Jeannin, P.-Y., and Zwahlen, F. (1999) Water vulnerability assessment in karst environments: A new method of defining protection areas using a multi-attribute approach & GIS tools (EPIK method), *Environmental Geology*, 39:165–176.

Domenico, P. A., and Schwartz, F. W. (1998) *Physical and Chemical Hydrogeology*. John Wiley and Sons, New York. 2nd ed., 824 p.

Dörr, H., Schlosser, P., Stute, M., and Sonntag, C. (1992) Tritium and ^{3}He measurements as Calibration data for groundwater transport models. In: *Progress in Hydrogeochemistry*, Mattess et al. (eds.), Springer, Berlin, 461–466.

Dowgiallo, J., Nowicki, Z., Beer, J., Bonani, G., Suter, M., Synal, H. A., and Wolfli, W. (1990) ^{36}Cl in ground water of the Mazowsze Basin (Poland). *J. Hydrol.*, 18:373–385.

Dowling, C. B., Poreda, R. J., and Basu, A. R. (2003) The groundwater geochemistry of the Bengal Basin: Weathering, chemsorption, and trace metal flux to the oceans. *Geochimica et Cosmochimica Acta*, 67:2117–2136.

Du, X., Purtschert, R., Bailey, K., Lehmann, B. E., Lorenzo, R., Lu, Z.-T., Mueller, P., O'Connor, T. P., Sturchio, N. C., and Young, L. (2003) A new method of measuring ^{81}Kr and ^{85}Kr abundances in environmental samples. *Geophysical Research Lett.*, 30:2068, DOI:10.1029.

Dunai, T. J., and Porcelli, D. (2002) Storage and transport of noble gases in the subcontinental lithospehere. In: Porcelli, D., Ballentine, C. J., and Wieler, R. (eds.) *Reviews in Mineralogy and Geochemistry*, Volume 47: Noble gases in geochemistry and cosmochemistry. The Mineralogical Society of America, Washington, 371–409.

Durand, S., Chabaux, F., Rihs, S., Duringer, P., and Elsass, P. (2005) U isotope ratios as tracers of groundwater inputs into surface waters: Example of the Upper Rhine hydrosystem. *Chemical Geology*, 220:1–19.

Dutton, A. R., and Simpkins, W. W. (1989) Isotopic evidence for paleohydrologic evolution of groundwater flow-paths, Southern Great Plains, United States. *Geology*, 17:653–656.

Edmunds, W. M., and Smedley, P. L. (2000) Residence time indicators in groundwater: The East Midlands Triassic sandstone aquifer. *Applied Geochemistry*, 15:737–752.

Edwards, R. R. (1962) Iodine-129: Its occurrence in nature and its utility as a tracer. *Science*, 137:851–853.

Eichinger, L. (1983) A contribution to the interpretation of ^{14}C groundwater ages considering the example of a partially confined sandstone aquifer. *Radiocarbon*, 25:347–356.

Einloth, S., Ekwurzel, B., Eastoe, C., and Lal, D. (2003) Si-32 as a potential tracer for century-scale recharge in semi-arid regions. *Geophysical Research Abstracts*, 5:13164.

Einloth, S., Ekwurzel, B., Lal, D., Long, A., Eastoe, C., and Phillips, F. (2001) ^{32}Si: A new method for estimating semi-arid vadose zone recharge. Paper presented at the Annual Meeting of SAHRA, held at Tucson, Arizona, 22–23 Feb. 2001.

Ekwurzel, B., Schlosser, P., Smethie, Jr, W. M., Plummer, L. N., Busenberg, E., Michel, R. L., Weppernig R., and Stute, M. (1994) Dating of shallow groundwater: Comparison of the transient tracers ^{3}H/^{3}He, CFCs and ^{85}Kr. *WRR*, 30:1693–1708.

Elmore, D., Tubbs, L. E., Newman, D., Ma, X. Z., Finkel, R., Nishiizumo, K., Beer, J., Oeschger, H., and Andree, M. (1982) ^{36}Cl bomb pulse measured in a shallow ice core from Dye 3, Greenland. *Nature*, 300:735–737.

Elmore, D., Fulton, B. R., Clover, M. R., Marsden, G. R., Gove, H. E., Naylor, H., Purser, K. H., Kilius, L. R., Beukens, R. P., and Litherland, A. E. (1979) Analysis of ^{36}Cl in environmental water samples using an electrostatic accelerator. *Nature*, 272:22–25, errata 246.

Eriksson, E. (1971) Compartment models and reservoir theory. *Annual Review of Ecology and Systematics*, 2:67–84.

Eriksson, E. (1961) Natural reservoirs and their characteristics. *Geofisica International*, 1:27–43.

Eriksson, E. (1958) The possible use of tritium for estimating groundwater storage. *Tellus*, 10:472–478.

Etcheverry, D. (2002) Valorisation des méthodes isotopiques pour les questions pratiques liées aux eaux souterraines. Isotopes de l'oxygène et de l'hydrogène. Rapports de l'Office fédéral des eaux et de la géologie, Série géologie, No. 2, Bern, Switzerland, 71 p.

Etcheverry, D. (2001) Une approche déterministe des distributions des temps de transit de l'eau souterraine par la théorie des réservoirs. Ph.D. Thesis, Centre of Hydrogeology, University of Neuchâtel, Switzerland.

Etcheverry, D., and Perrochet, P. (2000) Direct simulation of groundwater transit time distributions using the reservoir theory. *Hydrogeology Journal*, 8:200–208.

Etcheverry, D., and Perrochet, P. (1999) Reservoir theory, groundwater transit time distributions and lumped parameter models. In: Isotope techniques in water resources development and management, IAEA, Vienna, (CD-ROM).

Fabryka-Martin, J. (2000) Iodine-129 as a groundwater tracer. In: Cook, P., and Herczeg, A. (eds.), *Environmental Tracers in Subsurface Hydrology*. Kluwer Academic Publishers, Boston, 504–510.

Fabryka-Martin, J., Whittemore, D. O., Davis, S. N., Kubik, P. W. and Sharma, P. (1991) Geochemistry of halogens in the Milk River aquifer, Alberta, Canada. *Applied Geochemistry*, 6:447–464.

Fabryka-Martin, J., Davis, S. N., Elmore, D., and Kubic, P. W. (1989) In situ production and migration of ^{129}I in the Stripa granite, Sweden. *Geochimica et Cosmochimica Acta*, 53:1817–1823.

Fabryka-Martin, J., Davis, S. N., Roman, D., Airey, P. L., Elmore, D., and Kubik, P. W. (1988) Iodine-129 and chlorine-36 in uranium ores: 2. Discussion of AMS measurements. *Chemical Geology (Isotope Geoscience Section)*, 72:7–16.

Fabryka-Martin, J., and Davis, S. N. (1987) Applications of ^{129}I and ^{36}Cl in hydrology. *Nuclear Instruments and Methods in Physics Research*, B29:361–371.

Fabryka-Martin, J. T., Bentley H. W., Elmore, D., and Airey, P. L. (1985) Iodine-129 as an environmental tracers. *Geochimica et Cosmochimica Acta*, 49:337–347.

Fairbank, W. M. Jr. (1987) Photon burst mass spectrometry. *Nuclear Instruments and Methods in Physics Research*, B249:407–414.

Faure, G. (2001) *Origin of Igneous Rocks: The Isotopic Evidence*. Springer, Berlin, 496 p.

Fehn, U., Snyder, G., Matsumoto, R., Muramatsu, Y., and Tomaru, H. (2003) Iodine dating of pore waters associated with gas hydrates in the Nanki area, Japan. *Geology*, 31:521–524.

Fehn, U., Snyder, G., and Egeberg, P. K. (2000) Dating of pore waters with ^{129}I: Relevance for the origin of marine gas hydrates. *Science*, 289:2332–2335.

Fehn, U., Peters, E. K., Tullai-Fitzpatrick, S., Kubic, P. W., Sharma, P., Teng, R. T. D., Gove, H. E., and Elmore, D. (1992) ^{129}I and ^{36}Cl concentrations in waters of the eastern Clear Lake area, California: Residence times and source ages of hydrothermal fluids. *Geochimica et Cosmochimica Acta*, 56:2069–2079.

Fehn, U., Tullai-Fitzpatrick, S., Teng, R. T. D., Gove, H. E., Kubic, P. W., Sharma, P., and Elmore, D. (1990) Dating oil field brines using ^{129}I. *Nuclear Instruments and Methods in Physics Research*, B52:446–450.

Ferronskij, V. I., Romanov, V. V., Vlasova, L. S., Kolesov, G. P., Zavileiskij, S. A., and Dubinchuk, V. T. (1994) Cyclic model for seasonal recharge and discharge of ground and spring water in the Valday experimental basin (humid zone) using environmental isotope data. In *Mathematical Models and Their Applications to Isotope Studies in Groundwater Hydrology*. IEAE-TECDOC-777, IAEA, Vienna, 255–282.

Fette, M., Kipfer, R., Schubert, C. J., Hoehn, E., and Wehrli, B. (2005) Assessing river–groundwater exchange in the regulated Rhone River (Switzerland) using stable isotopes and geochemical tracers. *Applied Geochemistry*, 20:701–712.

Fetter, C. W. (1999) *Contaminant Hydrogeology*. Prentice Hall, Upper Saddle River, NJ, 500 p.

Fitts, C. R. (2004) *Groundwater Science*. Academic Press, Amsterdam, 450 p.

Florkowski, T., and Rózanski, K. (1986) Radioactive noble gases in the terrestrial environment. In: Fritz, P., and Fontes, J. Ch. (eds.) *Handbook of Environmental Geochemistry*, Vol. 2, Elsevier, New York, 481–506.

Fogg, G. E. (2002) Groundwater vulnerability and the meaning of groundwater age dates. Geological Society of America, Birdsall-Dreiss Distinguished Lecture.

Fontes, J. Ch. (1985) Some considerations on groundwater dating using environmental isotops. In: *Memoirs of 18th IAH Congress*, Hydrogeology in the services of Mans, Part 1. Cambridge, 118–154.

Fontes, J. Ch. (1981) Palaeowaters. In: *Stable Isotope Hydrology: Deuterium and Oxygen-18 in the Water Cycle*. IAEA technical report series No. 210, 273–302.

Fontes, J. Ch., and Garnier, J. M. (1979) Determination of the initial 14C activity of the dissolved carbon: A review of the existing models and a new approach. *WRR*, 15:399–413.

Forster, M., Ramm, K., and Maier, P. (1992a) Argon-39 dating of groundwater and its limiting conditions. In: *Isotope Techniques in Water Resource Development and Management*, IAEA, Vienna, 203–214.

Forster, M., Loosli, H. H., and Weise, S. (1992b) ^{39}Ar, ^{85}Kr, ^{3}He and ^{3}H isotope dating of ground water in the Bocholt and Segeberger Forst aquifer systems. In: Mattess et al. (eds.), *Progress in Hydrogeochemistry*, Springer, Berlin, 467–475.

Forster, M., and Maier, O. (1987) The Munich facility for argon and krypton preparation and low level measurement of argon-39 and krypton-85 in the hydrosphere. In: Garcia-Leon, M., and Madurga, G. (eds.) *Low-Level Measurements and Their Application to Environmental Radioactivity*, World Scientific Publications, Singapore. 563 p.

Forster, M., Moser, H., and Loosli, H. H. (1984) Isotope hydrological study with ^{14}C and argon-39 in the bunter sandstone of the Saar region. In: *Isotope Hydrology 1983*, IAEA, Vienna, 515–533.

Frederickson, G. C., and Criss, R. E. (1999) Isotope hydrology and residence times of the unimpounded Meramec River Basin, Missouri. *Chemical Geology*, 157:303–317.

Freeze, R. A., and Cherry, J. A. (1979) *Groundwater*, Prentice-Hall, Englewood Cliffs, New Jersey, 604 p.

Fritz, P., and Fontes, J. Ch. (eds.) (1989) *Handbook of Environmental Isotope Geochemistry, Vol. 3, The Marine Environment, A*, Elsevier, New York, 428 p.

Fritz, P., and Fontes, J. Ch. (eds.) (1986) *Handbook of Environmental Isotope Geochemistry, Vol. 2, The Terrestrial Environment.* Elsevier, New York, 557 p.

Fritz, P. and Fontes, J. Ch. (eds.) (1980) *Handbook of Environmental Isotope Geochemistry, Vol. 1, The Terrestrial Environment.* Elsevier, New York, 545 p.

Fröhlich, K., Ivanovich, M., Hendry, M. J., Andrews, J. N., Davis, S. N., Drimmie, R. J., Fabryka-Martin, J., Florkowski, T., Fritz, P., Lehmann, B., Loosli, H. H., and Nolte, E. (1991) Application of isotopic methods to dating of very old groundwaters: Milk River aquifer, Alberta, Canada. *Applied Geochemistry*, 6:465–472.

Fröhlich, K., Franke, T., Gellermann, G., Hebert, D., and Jordan, H. (1987) Silicon-32 in different aquifer types and implications for groundwater dating. *Proc. Intl. Symp. Isotopic Techniques in Water Resource Development*, International Atomic Energy Agency, Vienna, 30 March–3 April, 1987, 149–163.

Fuge, R. (2005) Soils and iodine deficiency. In: Selinus et al. (eds.) *Essential of Medical Geology: Impact of the Natural Environment on Public Health.* Elsevier Academic Press, Amsterdam, 417–433.

Fulda, C., and Kinzelbach, W. (2000) Sulphur hexafluoride (SF_6) as a new age-dating tool for shallow groundwater: Methods and first results. In: Dassargues, A. (ed.). *Tracers and Modelling in Hydrogeology*, IAHS publication no. 262. IAHS Press, Oxfordshire, UK, 181–185.

Fulda, C., and Kinzelbach, W. (1997) Datierungen junger grundwässer im Gebiet Sindelfingen—Stuttgart mit Hilfe eines neuen tracers—Schwefelhexafluoride. Die Stuttgarter Mineralwasser-Herkunft und Genese. Amt Für Umweltschutz, Stuttgart.

Gascoyne, M., and Sheppard, M. I. (1993) Evidence of terrestrial of deep groundwater on the Canadian sheild from helium in soil gas. *Environmental Science and Technology*, 27(12):2420–2486.

Gaye, C., Aggarwal, P., Gourcy, L., Gröning, M., Kulkarni, K., Suckow, A., and Wallin, B. (2005) Isotope methodologies for the protection and management of groundwater resources within the International Atomic Energy Agency water program. *Geophysical Research Abstracts*, 7:10641 (EGU05-A-10641).

Gellermann, R., Börner, I., Franke, T., and Fröhlich, K. (1988) Preparation of water samples for ^{32}Si determinations. *Isotopenpraxis*, 24:114–117.

Geyer, G., and Shergold, J. H. (2000) The quest for internationally recognized divisions of Cambrian time. *Episodes*, 23:188–195.

Geyh, M. A. (2000) An overview of ^{14}C analysis in the study of groundwater. *Radiocarbon*, 42:99–114.

Gifford, S. K., Bentley, H., and Graham, D. L. (1985) Chlorine isotopes as environmental tracers in Columbia River basalt ground water, *Memoirs, Hydrogeology of Rocks of Low Permeability*, International Association of Hydrgeologists, Tucson, AZ, VII(1):417–429.

Ginn, T. R. (1999) On the distribution of multi-component mixtures over generalized exposure time in subsurface flow and reactive transport: Foundations and formulations for groundwater age, chemical heterogeneity, and biodegradation. *WRR*, 35:1395–1407.

Gleick, P. H. (ed.) (1993) *Water in Crisis.* Oxford University Press, New York, 473 p.

Glynn, P. D., and Plummer, L. N. (2005) Geochemistry and the understanding of ground-water systems. *Hydrogeology Journal*, 13:263–287.

Gonfiantini, R., Fröhlich, K., Araguas-Araguas, L., and Rozanski, K. (1998) Isotopes in Groundwater hydrology. In: Kendall, C., and McDonnell, J. J. (eds.) *Isotope Tracers in Catchment Hydrology*. Elsevier, Amsterdam, 203–246.

Goode, D. J. (1996) Direct simulation of groundwater age. *WRR*, 32:289–296.

Gove, H. E., Fehn, U., Teng, R. T. D., and Beasley, T. M. (1994) Measurements of ^{129}I at the Idaho Falls National Engineering Laboratory: Rochester, N.Y., Nuclear Structure Research Laboratory Annual Report, 1994, University of Rochester, 62–63.

Gradstein, F., Ogg, J., and Smith, A. (2004) *A Geologic Time Scale*. Cambridge University Press, Cambridge, 589.

Graham, D. W. (2002) Noble gas isotope geochemistry of Mid-Ocean Ridge and Ocean Island Basalts: Characterization of mantle source reservoirs. In: Porcelli, D., Ballentine, C. J., and Wieler, R. (eds.) *Reviews in Mineralogy and Geochemistry, Volume 47: Noble Gases in Geochemistry and Cosmochemistry*. The Mineralogical Society of America, Washington, 247–317.

Haitjema, H. M. (1995) On the residence time distributions in idealized groundwatersheds. *J. Hydrol.*, 172:127–146.

Hanshaw, B. B., and Back, W. (1974) Determination of regional hydraulic conductivity through ^{14}C dating of groundwater. In: *Memoirs of the Intl. Association of Hydrogeologists*, 10:195–196.

Hanshaw, B. B., Back, W., and Rubin, M. (1965) Radiocarbon determinations for estimating groundwater flow velocities in Central Florida. *Science*, 148:494–495.

Harlow, G. E., Jr., Nelms, D. L., and Puller, J. C. (1999) Ground-water dating to assess aquifer susceptibility, In: *Proc. Water Resource Management*: Source-of-supply challenges—1999 Water resources conference, Norfolk, Virginia, Sept. 26–29, 1999, Denver, American Water Works Association [on CD-ROM].

Harvey, C. F., and Gorelick, S. M. (1995) Temporal moment generating equations: Modeling transport and mass transfer in heterogeneous aquifers. *WRR*, 31:1895–1912.

Heaton, T. H. E. (1981) Dissolved gases: Some applications to groundwater research. *Transactions of the Geological Society of South Africa.*, 84:91–97.

Heaton, T. H. E., and Vogel, J. C. (1981) Excess air in groundwater. *J. Hydrol.*, 50:201–216.

Heidinger, M., Loosli, H. H., Bertleff, B., Eichinger, L., Göppel, M., Oster, H., and Traub, R. (1997) Kombination von isotopenmethoden zum verstädnis von ausgewählten grundwassersystemen. *Proceedings of isotopenkolloquium Freiberg*.

Herczeg, A. L., Leaney, F. W. J., Stadler, M. F., Allan, G. L., and Fifield, L. K. (1997) Chemical and isotopic indicators of point-source recharge to a karst aquifer, South Australia. *J. Hydrol.*, 192:271–299.

Hofer, M., Amaral, H., Brennwald, M. S., Hoehn, E., Holzner, C. P., Klump, S., Scholtis, A., and Kipfer, R. (2005) Estimating groundwater residence times by graphical analysis of conductivity time series in the Thur valley, Switzerland. *Geophysical Research Abstracts*, Vol. 7, 03409, EGU05-A-03409.

Hoehn, E., and von Gunten, H. R. (1989) Radon in groundwater: A tool to assess infiltration from surface waters to aquifers. *WRR*, 25:1795–1803.

Höhener, P., Werner, D., Balsiger, C., and Pateris, G. (2003) Worldwide occurrence and fate of chlorofluorocarbons in groundwater. *Critical Reviews in Environmental Science and Technology*, 33:1–29.

Holmes, A. (1913) *The Age of the Earth*. Harper Brothers, New York.

Hölting, B., Haertlé, T., Hohberger, K.-H., Nachtigall, K. H., Villinger, E. Weinzierl, W., and Wrobel, J.-P. (1995) Konzept zur ermittlung der schutzfunktion der grundwasserüberdeckung—Geol. Jb., C, 63, 5–24; Hannover [Concept for the determination of the protective effectiveness of the cover above the groundwater against pollution.—Ad-hoc working group on hydrogeology, 28 p.; Hannover].

Hunt, A. G., Naus, C. A., Nordstrom, D. K., and Landis, G. P. (2004) Use of 3-helium/tritium ages to determine chlorofluorocarbon degradation associated with naturally occurring acid drainage, Questa, New Mexico. Geological Society of America Abstracts with Programs, 36(5):574.

Huyakorn, P. S., and Pinder, G. F. (1983) *Computational Methods in Subsurface Flow*. Academic Press, New York, 473 p.

IAEA (2004a) Isotope hydrology and integrated water resources management. Unedited Proc. of Int. Symp., 19–23 May 2003, IAEA, Vienna, 531 p.

IAEA (2004b) *Management of Waste Containing Tritium and Carbon-14*. IAEA Technical reports series No. 421, Vienna, 109 p.

IAEA (1994) *Mathematical Models and Their Applications to Isotope Studies in Groundwater Hydrology*. IEAE-TECDOC-777, IAEA, Vienna, 283 p.

IAEA (1976) *Interpretation of Environmental Isotope and Hydrochemical Data in Groundwater Hydrology*. IAEA, Vienna, 228 p.

IAEA (1974) *Isotope Techniques in Groundwater Hydrology 1974, Vol. II*, IAEA, Vienna, 499 p.

IAEA (1970) *Isotope Hydrology* 1970. IAEA, Vienna, 918 p.

IAEA (1967) *Isotopes in Hydrology*. IAEA, Vienna, 740 p.

Imbrie, J., and 17 others (1992) On the structure and origin of major glaciation cycles. 1. Linear responses to Milankovitch forcing. *Paleoceanography*, 7:701–738.

Ivanovich, M., and Harmon, R. S. (eds.) (1992) *Uranium Series Disequilibrium: Applications to Earth, Marine, and Environmental Sciences*, Clarendon, Oxford, England, 910 p.

Ivanovich, M., Fröhlich, K., and Hendry, M. J. (1991) Uranium-series radionuclides in fluids and solids, Milk River aquifer, Alberta, Canada. *Applied Geochemistry*, 6:405–418.

Johnston, C. T., Cook, P. G., Frape, S. K., Plummer, L. N., Busenberg, E., and Blackport, R. J. (1998) Groundwater age and nitrate distribution within a glacial aquifer beneath a thick unsaturated zone. *Ground Water*, 36:171–180.

Jury, W. A., and Roth, K. (1990) *Transfer Functions and Solute Movement Through Soil: Theory and Applications*. Birkhauser, Boston, 289 p.

Kalin, R. M. (2000) Radiocarbon dating of groundwater systems. In: Cook, P. G., and Herczeg, A. L. (eds.), *Environmental Tracers in Subsurface Hydrology*. Kluwer Academic Publishers, Boston, 111–145.

Kantelo, M. V., Tiffany, B., and Anderson, T. J. (1982) Iodine-129 distribution in the terrestrial environment surrounding a nuclear-fuel reprocessing plant after 25 years of operation. In: *Environmental Migration of Long-Lived Radionuclides*: Knoxville, Tenn., National Atomic Energy Agency International Symposium, July 27–31, 1981, Proceedings. 495–500.

Katz, B. G. (2004) Sources of nitrate contamination and age of water in large karstic springs of Florida. *Environmental Geology*, 46:689–706.

Katz, B. G., Lee, T. M., Plummer, L. N., and Busenberg, E. (1995) Chemical evolution of groundwater near a sinkhole lake, northern Florida, 1, Flow patterns, age of groundwater, and influence of lake water leakage. *WRR*, 31:1549–1564.

Kaufman, A., Magaritz, M., Paul M., Hillaire-Marcel, C., Hollus, G., Boaretto, E., and Taieb, M. (1990) The ^{36}Cl ages of the brines in the Magadi-Natron basin, East Africa. *Geochimica et Cosmochimica Acta*, 54:2827–2834.

Kaufman, M. I., Rydell, H. S., and Osmond, J. K. (1969) ^{234}U/^{238}U disequilibrium as an aid to hydrologic study of the Floridan aquifer. *J. Hydrol.*, 9:374–386.

Kazemi, G. A. (2005a) Age dating old groundwater. In: Lehr, J. H., and Keeley, J. (eds.) *Water Encyclopedia: Oceanography; Meteorology; Physics and Chemistry; Water Law; and Water History, Art, and Culture*. John Wiley and Sons, Hoboken, New Jersey, 388–390.

Kazemi, G. A. (2005b) Chlorine-36 and very old groundwater. In: Lehr, J. H., and Keeley, J. (eds.) *Water Encyclopedia: Oceanography; Meteorology; Physics and Chemistry; Water Law; and Water History, Art, and Culture*. John Wiley and Sons, Hoboken, New Jersey, 416–420.

Kazemi, G. A. (2005c) Chlorofluorocarbons. In: Lehr, J. H., and Keeley, J. (eds.) *Water Encyclopedia: Oceanography; Meteorology; Physics and Chemistry; Water Law; and Water History, Art, and Culture*. John Wiley and Sons, Hoboken, New Jersey, 420–423.

Kazemi, G. A. (1999) Groundwater factors in the management of dryland and stream salinity in the upper Macquarie Valley, New South Wales, Australia. Ph.D. thesis, University of Technology, Sydney, Australia, 252 p.

Kazemi, G. A., and Milne-Home, W. A. (1999) The application of Strontium, Oxygen 18 and Deuterium isotopes in the estimation of the solute budget and hydrology of a small watershed in Eastern Australia. Eos Trans. *AGU, 80 (17)* Spring Meet. Suppl. S139.

Kazemi, G. A., Rathur, A. Q., and King, N. (1998a) Isochrone map of the unconfined aquifer, Superficial Formation, Perth, Western Australia. In: *Proc. of IAH 28th Congress "Gambling with Groundwater"*, Las Vegas, Nevada, USA, 633–638.

Kazemi, G. A., Milne-Home, W. A., and Keshwan, M. (1998b) Groundwater age dating with CFC in the Macquarie River Basin, NSW, Australia; Implications for dryland salinity management. In: Brahana et al. (eds.) *Proc. IAH 28th Congress "Gambling with Groundwater,"* Las Vegas, USA, Sep–Oct. 1998, 549–554.

Kellett, J. R., Evans, W. R., Allan, G. L., and Fifield, L. K. (1993) Reinterpretation of ^{36}Cl age data: physical processes, hydraulic interconnections, and age estimates in ground water systems-Discussion. *Applied Geochemistry*, 8:653–658.

Kharkar, D. P., Nijampurkar, V. N., and Lal, D. (1966) The global fallout of ^{32}Si produced by cosmic rays. *Geochimica et Cosmochimica Acta*, 30:621–631.

Kigoshi, K. (1971) Alpha-recoil thorium-234: Dissolution into water and the uranium-234/uranium-238 disequilibrium in nature. *Science*, 173:47–48.

Kilius, L. R., Zhao, X.-L., and Barrie, L. A. (1995) Anomalous I-129 concentrations in Arctic air samples, In: Strand, P., and Cook, A. (eds.) *Environmental Radioactivity in the Arctic*: Oslo, Norway, International Conference on Environmental Radioactivity in the Arctic.

Kimmel, G. E., and Braids, O. C. (1974) Leachate plumes in a highly permeable aquifer. *Ground Water*, 12:388–393.

Kinzelbach, W. (1992) *Numerische Methoden zur Modellierung des Transports von Schadstoffen im Grundwasser*. Oldenbourg, Munich, 343 p.

Kipfer, R., Aeschbach-Hertig, W., Peeters, F., and Stute, M. (2002) Noble gases in lakes and groundwaters. In: Porcelli, D., Ballentine, C., and Wieler, R. (eds.) Noble gases in geochemistry and cosmochemistry. Vol. 47 of *Rev. Mineral Geochemistry*, Mineralogical Society of America, Geochemical Society, 615–700.

Krauss, L. M., and Chaboyer, B. (2003) Age estimates of globular clusters in the Milky Way: Constraints on cosmology. *Science*, 299:65–69.

Kronfeld, J., and Adams, A. S. (1974) Hydrologic investigations of the groundwters of central Texas using U-234/U-238 disequilibrium. *J. Hydrol.*, 22:77–88.

Kronfeld, J., Grodsztajn, E., Muller, H. N., Redin, J., Yanin, A., and Zach, R. (1975) Excess ^{234}U: An Aging Effect in Confined Waters. *Earth Planet Sci Lett.*, 27:342–345.

Kroonenberg, S. B. (2005) Dating Caspian Sea level change—State of the art. In: *Abstracts of the International Conference on Rapid Sea Level Change: A Caspian perspective*, Rasht, Iran, May 2005, Kroonenberg, S. B. (ed.) The University of Guilan Press, Rasht, 71–72.

Kumar, U. S., Jacob, N., Rao, S. M., and Murthy, J. S. R. (2000) Development and verification of a tritium and water balance model for estimating the water retention times of lakes. In: Dassargues, A. (ed.), *Tracers and Modeling in Hydrogeology*. IAHS publication no. 262. IAHS Press, Oxfordshire, UK, 545–552.

Kummel, B. (1970) *History of the Earth*. Second edition, W. H. Freeman and Company, San Francisco, 707 p.

Kulongoski, J. T., Hilton, D. R., and Izbicki, J. A. (2005) Source and movement of helium in the eastern Morongo groundwater Basin: The influence of regional tectonics on crustal and mantle helium fluxes. *Geochimica et Cosmochimica Acta*, 69:3857–3872.

LaBolle, E. M., Quastel, J., and Fogg, G. E. (1998) Diffusion theory for transport in porous media: Transition-probability densities of diffusion processes corresponding to advection-dispersion equations. *WRR*, 34:1685–1693.

Laier, T., and Wiggers, L. (2004) Residence time and nitrate in groundwater in the Mariager Fjord Catchment—Results of CFC age dating. EGU 1st General Assembly, April 26–30, 2004, Nice, France. Abstract No, EGU04-A-06236.

Lal, D. (2001) Some comments on applications of cosmogenic radionuclides for determining groundwater flow, with special reference to ^{32}Si and ^{10}Be. Paper presented at the SAHRA—Recharge workshop held at New Mexico Tech (Socorro), March 22–23, 2001.

Lal, D., Jull, A. J. T., Pollard, D., and Vacher, L. (2005) Evidence for large century timescale changes in solar activity in the past 32 Kyr, based on in-situ cosmogenic ^{14}C in ice at Summit, Greenland. *Earth Planet Sci Lett.*, 234:335–349.

Lal, D. (1999) An overview of five decades of studies of cosmic ray produced nuclides in oceans. *The Science of Total Environment*, 237/238:3–13.

Lal, D., Nijampurkar, V. N., Rajagopalan G., and Somayajulu, B. L. K. (1979) Annual fallout of ^{32}Si, ^{210}Pb, ^{22}Na, ^{35}S and ^{7}Be in rains in India. *Proc. Ind. Acad. Sci*, 88A (II):29–40.

Lal, D., Nijampurkar, V. N., Somayajulu, B. L. K., Koide M., and Goldberg, E. D. (1976) Silicon-32 specific activities in coastal waters of the world oceans. *Limnol. Oceanogr*, 21:285–293.

Lal, D., Nijampurkar, V. N., and Rama, S. (1970) Silicon-32 hydrology. *Proc. Symp. Isotope Hydrology*, International Atomic Energy Agency, Vienna, Mar. 9–13, 1970, pp. 847–868.

Lal, D., and Peters, B. (1967) Cosmic ray produced radioactivity on the earth. In: *Handbuch der Physik*. Berlin, Springer-Verlag, 46:551–612.

Lal, D., Goldberg, E. D., and Koide, M. (1959) Cosmic ray produced ^{32}Si in nature. *Phys. Rev. Lett.*, 3:380.

Langmuir, D. (1997) *Aqueous Environmental Geochemistry*. Prentice Hall, Saddle River, New Jersey, 600 p.

Larson, K. R., Keller, C. K., Larson, P. B., and Allen-King, R. M. (2000) Water resources implications of ^{18}O and ^{2}H distributions in a basalt aquifer system. *Ground Water*, 38:947–953.

Lawrence, E., Poeter, E., and Wanty, R. (1991) Geohydrologic, geochemical, and geologic controls on the occurrence of radon in groundwater near Conifer, Colorado, USA. *J. Hydrol.*, 127:367–386.

Le Gal La Salle, C., Fontes, J. Ch., Andrews, J. N., Schroeter, P., Karbo, A., and Fifield, K. L. (1995) Old groundwater circulation in the Iullemeden basin (Niger): preliminary results of an isotopic study. USA. In: *Application of Tracers in Arid-Zone Hydrology* (Adar, E. M., and Leibundgut, C., eds.). IAHS publications no. 232, IAEA, Vienna, 129–139.

Leavy, B. D., Philips, F. M., Elmore, D., Kubic, P. W., and Gladney, E. (1987) Measurement of ^{36}Cl/Cl in young volcanic rocks: An application of accelerator mass spectrometry to geochronology. *Nuclear Instruments and Methods in Physics Research*, B29:246–250.

Lehmann, B. E., Love, A., Purtschert, R., Collon, P., Loosli, H. H., Kutschera, W., Beyerle, U., Aeschbach-Hertig, W., Kipfer, R., Frape, S. K., Herczeg, A., Moran, J., Tolstikhin, I. N., and Groning, M. (2003) A comparison of groundwater dating with ^{81}Kr, ^{36}Cl and ^{4}He in four wells of the Great Artesian Basin, Australia. *Earth Planet Sci Lett.*, 211:237–250.

Lehmann, B. E., and Purtschert, R. (1997) Radioisotope dynamics—the origin and fate of nuclides in groundwater. *Applied Geochemistry*, 12:727–738.

Lehmann, B. E., Davis, S. N., and Fabryka-Martin, J. T. (1993) Atmospheric and subsurface sources of stable and radioactive nuclides used for groundwater dating. *WRR*, 29:2027–2040.

Lehmann, B. E., Loosli, H. H., Rauber, D., Thonnard, N., and Willis, R. D. (1991) ^{81}Kr and ^{85}Kr in groundwater, Milk River Aquifer, Alberta, Canada. *Applied Geochemistry*, 6:419–423.

Lehr, J. H., and Keeley, J. (eds.) (2005) *Water Encyclopedia*. John Wiley and Sons, Hoboken, New Jersey, 4112 p.

Leibundgut, Ch., and Zupan, M. (1992) Determination of residence time in Lake Bled. In: Hotzel, H., and Warner, A. (eds.), *Tracer Hydrology*, Balkema, Rotterdam, 3–10.

Levy, S., Chipera, S., Woldegabriel, G., Fabryka-Martin, J., and Roach, J. (2005) Flow-path textures and mineralogy in tuffs of the unsaturated zone. In: Haneberg,

W. C., Mozley, P. S., Moore, J. C., and Goodwin, L. B. (eds.), Faults and subsurface fluid flow in the shallow crust. *Geophysical Monograph* 113, AGU, Washington, 159–184.

Libby, W. F. (1953) The potential usefulness of natural tritium. *Proceeding of National Academy of Science of the USA*, 39:245–247.

Libby, W. F. (1952) *Radiocarbon Dating*. University of Chicago Press, Chicago.

Lindsey, B, D., Phillips, S. W., Donnelly, C. A., Speiran, G. K., Plummer, L. N., Böhlke, J. K., Focazio, M. J., Burton, W. C., and Busenberg, E. (2003) Residence time and nitrate transport in groundwater discharging to streams in the Chesapeake Bay Watershed. USGS Water—Resources Investigation Report 03-4035, 201 p.

Lloyd, J. W. (1981) Environmental isotopes in groundwater. In: Lloyd, J. W. (ed.) *Case Studies in Groundwater Resources Evaluation*, 113–132.

Lodemann, M., Fritz, P., Wolf, M., Ivanovich, M., Hansen, B. T., and Nolte, E. (1997) On the origin of saline fluids in the KTB (continental deep drilling project of Germany). *Applied Geochemistry*, 12:831–849.

Loew, S. (2004) Nuclear waste disposal in Switzerland: Science, politics and uncertainty. *Hydrogeology Journal*, 12:121–122.

Loosli, H. H. (1992) Applications of ^{37}Ar, ^{39}Ar and ^{85}Kr in hydrology, oceanography, and atmospheric studies: Current state of the art. In: Isotopes of Noble Gases as Tracers in Environmental Studies; *Proceedings of a Consultants Meeting*, International Atomic Energy Agency, Vienna, 73–85.

Loosli, H. H. (1989) Argon-39: A tool to investigate ocean water circulation and mixing. In: Fritz, P., and Fontes, J. Ch. (eds.), *Handbook of Environmental Geochemistry*, Vol. 3, Elsevier Science, New York, 385–392.

Loosli, H. H., Lehmann, B. E., and Smethie, W. M. (2000) Noble gas radioisotopes: ^{37}Ar, ^{85}Kr, ^{39}Ar, ^{81}Kr. In: *Environmental Tracers in Subsurface Hydrology*, Cook, P. G., and Herczeg, A. L. (eds.) Kluwer Academic Publishers, Boston, 379–396.

Loosli, H. H., Lehmann, B. E., and Balderer, W. (1989) Argon-39, argon-37 and krypton-85 isotopes in Stripa groundwaters. *Geochimica et Cosmochimica Acta*, 53:1825–1829.

Loosli, H. H. (1983) A dating method with ^{39}Ar. *Earth Planet Sci Lett.*, 63:51–62.

Loosli, H. H., and Oeschger, H. (1979) Argon-39, carbon-14 and krypton-85 measurements in groundwater samples. In: *Isotope Hydrology*, IAEA, Vienna, 931–947.

Loosli, H. H., and Oescher, H. (1969) ^{37}Ar and ^{81}Kr in the atmosphere. *Earth Planet Sci Lett.*, 7:67–71.

Love, A. J., Herczeg, A. L., Sampson, L., Cresswell, R. G., and Fifield, K. L. (2000) Sources of chlorine and implications for 36Cl dating of old groundwater, southwestern Great Artesian Basin. *WRR*, 36:1561–1574.

Love, A. J., Herczeg, A. L., Leaney, F. W., Stadter, M. F., Dighton J. C., and Armstrong, D. (1994) Groundwater residence time and palaeohydrology in the Otway Basin, South Australia: ^{2}H, ^{18}O and ^{14}C data. *J. Hydrol.*, 153:157–187.

Lovelock, J. E. (1971) Atmospheric fluorine compounds as indicators of air movements. *Nature* 230:379.

Lu, Z.-T., and Wendt, K. D. A. (2003) Laser based methods for ultrasensitive trace-isotope analyses, *Review of Scientific Instruments*, 74:1169–1179.

Luther, K. H., and Haitjema, H. M. (1998) Numerical experiments on the residence time distributions of heterogeneous groundwatersh eds. *J. Hydrol.*, 207:1–17.

MacDonald, A. M., Darling, W. G., Ball, D. F., and Oster, H. (2003) Identifying trends in groundwater quality using residence time indicators: an example from the Permian aquifer of Dumfries, Scotland. *Hydrogeology Journal*, 11:504–517.

Maloszewski, P., and Zuber, A. (1996) Lumped parameter models for interpretation of environmental tracer data. In: *Manual on Mathematical Models in Isotope Hydrogeology*. IAEA-TECDOC-910, IAEA, Vienna, 9–58.

Maloszewski, P., and Zuber, A. (1993) Principles and practice of calibration and validation of mathematical models for the interpretation of environmental tracer data in aquifers. *Advances in Water Resources*, 16:173–190.

Maloszewski, P., and Zuber, A. (1991) Influence of matrix diffusion and exchange reactions on radiocarbon ages in fissured carbonate aquifers. *WRR*, 27:1937–1945.

Maloszewski, P., Rauert, W., Stichler, W., and Herrmann, A. (1983) Application of flow models in an alpine catchment area using tritium and deuterium data. *J. Hydrol.*, 66:319–330.

Maloszewski, P., and Zuber, A. (1982) Determining the turnover time of groundwater systems with the aid of environmental tracers, 1, Models and their applicability. *J. Hydrol.*, 57:207–231.

Mann, L. J., and Beasley, T. M. (1994) Background concentration of ^{129}I in ground and surface water, eastern Snake River Plain, Idaho-1992. *Journal of the Idaho Academy of Sciences*, 30:75–87.

Manning, A. H., Solomon, D. K., and Thirosl, S. A. (2005) ^{3}H/^{3}He age data in assessing the susceptibility of wells to contamination. *Ground Water*, 43:353–367.

Marine, I. W. (1979) The use of naturally occurring helium to estimate groundwater velocities for studies of geologic storage of radioactive waste. *WRR*, 15:1130–1136.

Marks, W. E. (2005) Water Clocks. In: Lehr, J. H., and Keeley, J. (eds.), *Water Encyclopedia: Oceanography; Meteorology; Physics and Chemistry; Water Law; and Water History, Art, and Culture*. John Wiley and Sons, Hoboken, New Jersey, 704–707.

Marques, J. M., Andrade, M., Aires-Barros, L., Graca, R. C., Eggenkamp, H. G. M., and Antunes da Silva, M. (2001) ^{87}Sr/^{86}Sr and ^{37}Cl/^{35}Cl signatures of CO_2-rich mineral waters (N-Portugal): Preliminary results. In: *New Approaches Characterizing Groundwater Flow*. (Seiler, K. P., and Wohnlich, S., eds.), Swets and Zeitlinger Lisse, 2:1025–1029.

Marty, B., Dewonck, S., and France-lanord, C. (2003) Geochemical evidence for efficient aquifer isolation over geological timeframes. *Nature*, 425:55–58.

Marty, B., Criaud, A., and Fouillac, C. (1988) Low enthalpy geothermal fluids from the Paris sedimentary basin. 1. Characteristics and origin of gases. *Geothermics*, 17:619–633.

Marty, B., Torgersen, T., Meynier, V., O'Nions, R. K., and de Marsily, Gh. (1993) Helium isotope fluxes and groundwater ages in the Dogger aquifer, Paris basin. *WRR*, 29:1025–1035.

Mattle, N. (1999) Interpretation von tracer-daten in grundwässern mittels boxmodellen und numerischen strömungs-transportmodellen. Ph.D. thesis, University of Bern, Switzerland.

Mayo, A. L., Morris, T. H., Peltier, S., Petersen, E. C., Payne, K., Holman, L. S., Tingey, D., Fogel, T., Black, B. J., and Gibbs, T. D. (2003) Active and inactive groundwater flow systems: Evidence from a stratified, mountainous terrain. *GSA Bulletin*, 115:1456–1472.

Mazor, E. (2004) *Applied Chemical and Isotopic Groundwater Hydrology*. 3rd edition, Marcel Dekker, Inc, New York, 453 p.

Mazor, E. (1991) *Applied Chemical and Isotopic Groundwater Hydrology*. Open University Press, 271 p.

Mazor, E., and Nativ, R. (1994) Stagnant groundwater stored in isolated aquifers: Implications related to hydraulic calculations and isotopic dating-Reply. *J. Hydrol.*, 154:409–418.

Mazor, E., and Bosch, A. (1992) Helium as a semi-quantitative tool for groundwater dating in the range of 10^4–10^8 years. In: *Isotopes of Noble Gases as Tracers in Environmental Studies*. IAEA, Vienna, 163–178.

Mazor, E., and Nativ, R. (1992) Hydraulic calculation of groundwater flow velocity and age: Examination of the basic premises. *J. Hydrol.*, 138:211–222.

McFarlane, D. J. (1984) The effect of urbanisation on groundwater quantity and quality in Perth, Western Australia. Ph.D. thesis, University of Western Australia, Australia.

McGuire, K. J., McDonnell, J. J., Weiler, M., Kendall, C., McGlynn, B. L., Welker, J. M., and Seibert, J. (2005) The role of topography on catchment-scale water residence time. *WRR*, 41:W05002, DOI:10.1029/2004WR003657.

McGuire, K. J., Weiler, M., and McDonnell, J. J. (2004) Understanding the variability of residence time distributions by simulating artificial hill slope tracers experiments. *Geophysical Research Abstracts*, Vol. 6, EGU04-A-04513.

McMahon, P. B., Böhlke, J. K., and Christensen, S. C. (2004) Geochemistry, radiocarbon ages, and palaeorecharge conditions along a transect in the central High Plains aquifer, southwestern Kansas, USA. *Applied Geochemistry*, 19:1655–1686.

Mendonca, L. A. R., Frischkon, H., Santiago, M. F., and Filho, J. M. (2005) Isotope measurements and groundwater flow modeling using MODFLOW for understanding environmental changes caused by a well field in semiarid Brazil. *Environmental Geology*, 47:1045–1053.

Merrihue, C., and Turner, G. (1966) Potassium-argon dating by activation with fast neutrons. *J. of Geophysical Research*, 71:2852–2857.

Metcalfe, R., Hooker, P. J., Darling, W. G., and Milodovski, A. E. (1998) Dating Quaternary groundwater flow events: A review of available methods and their application. In: Parnell, J. (ed.), *Dating and Duration of Fluid Flow and Fluid–Rock Interaction*. Geological Society, London, Special publications, No. 144, 233–260.

Michalak, A. M., and Kitanidis, P. K. (2004) Estimation of historical groundwater contaminant distribution using the adjoint state method applied to geostatistical inverse modeling. *WRR*, 40:W08302, DOI:10.1029/2004WR003214.

Michel, R. L. (2000) Sulphur-35. In: Cook, P. G., and Herczeg, A. L. (eds.), *Environmental Tracers in Subsurface Hydrology*. Kluwer Academic Publishers, Boston, 502–504.

Michel, R. L., and Turk, J. T. (2000) Use of ^{35}S to study rates of suplphur migration in the Flat Tops Wilderness Area, Colorado. In: *Isotopes in Water Resources Management*. IAEA, Vienna, 293–301.

Milton, G. M., Kramer, S. J., Kotzer, T. G., Milton, J. C. D., Andrews, H. R., Chant, L. A., Cornett, R. J., Davies, W. G., Greiner, B. F., Imahori, Y., Koslowsky, V. T., and McKay, J. W. (1997) ^{36}Cl—A potential paleodating tool. *Nuclear Instruments and Methods in Physics Research*, B123:371–377.

Mitchell, J. G. (1968) The argon-40/argon-39 method for potassium-argon age determination. *Geochimica et Cosmochimica Acta*, 32:781–790.

Modica, E., Buxton, H. T., and Plummer, L. N. (1998) Evaluating the source and residence times of groundwater seepage to streams, New Jersey Coastal Plain. *WRR*, 34:2797–2810.

Moeller, S., Berger, L., Salvador, J. G., and Helitzer, D. (2002) How old is that child? Validating the accuracy of age assignments in observational surveys of vehicle restraint use. *Injury Prevention*, 8:248–251.

Montzka, S. A., Butler, J. H., Elkins, J. W., Thompson, T. M., Clark, A. D., and Lock, L. T. (1999) Present and future trends in the atmospheric burden of ozone-depleting halogens. *Nature*, 398:690–694.

Mook, W. G. (1972) On the reconstruction of the initial ^{14}C content of groundwater from the chemical and isotopic composition. In: *Proc 8th Intl. Conf. on Radiocarbon Dating*, Vol. 1, Royal Society of New Zealand, Wellington, 342–352.

Moran, J. E., Hudson, G. B., Eaton, G. F., and Leif, R. (2004a) A contamination vulnerability assessment for the Sacramento area groundwater basin. Lawrence Livermore National Laboratory internal report, UCRL-TR-203258, 44 p.

Moran, J. E., Hudson, G. B., Eaton, G. F., and Leif, R. (2004b) A contamination vulnerability assessment for the Santa Clara and San Mateo county groundwater basins. Lawrence Livermore National Laboratory internal report, UCRL-TR-201929, 49 p.

Moran, J. E., Hudson, G. B., Eaton, G. F., and Leif, R. (2002a) A contamination vulnerability assessment for the Livermore-Amador and Niles Cone groundwater basins. Lawrence Livermore National Laboratory internal report, UCRL-AR-148831, 25 p.

Moran, J. E., Oktay, S. D., and Santschi, P. H. (2002b) Sources of iodine and iodine 129 in rivers. *WRR*, 38:10.1029, 24-1–24-10.

Moran, J. E., Fehn, U., and Teng, R. T. D. (1998) Variations in the $^{129}I/I$ ratios in recent marine sediments: Evidence for a fossil organic component. *Chemical Geology*, 152:193–203.

Moran, J. E., Fehn, U., and Hanor, J. S. (1995) Determination of source ages and migration patterns of brines from the US Golf Coast basin using ^{129}I. *Gochimica et Cosmochimica Acta*, 59:2055–5069.

Morgenstern, U. (2000) Silicon-32. In: Cook, P.G., and Herczeg, A. L. (eds.), *Environmental Tracers in Subsurface Hydrology*. Kluwer Academic Publishers, Boston. 498–502.

Morgenstern, U., and Ditchburn, R. G. (1997) ^{32}Si as natural tracer: Measurement and global distribution. Presented to the International conference on isotopes in the Solar System, Ahmedabad, India, November 1997.

Morgenstern, U., Taylor, C. B., Parrat, Y., Gaggeler, H. W., and Eichler, B. (1996) ^{32}Si in precipitation: Evolution of temporal and spatial variation and as a dating tool for glacial ice. *Earth Planet Sci Lett.*, 144:289–296.

Morgenstern, U., Gellermann, R., Hebert, D., Börner, I., Stolz, W., Vaikmäe, R., Rajamäe, R., and Putnik, H. (1995) ^{32}Si in limestone aquifers. *Chemical Geology*, 120:127–134.

Morikawa N., Kazahaya K., Yasuhara M., Inamura A., Nagao K., Sumino H., and Ohwada M. (2005) Estimation of groundwater residence time in a geologically active region by coupling ^{4}He concentration with helium isotopic ratios. *Geophysical Research Letters*, **32**, L02406.

Moysey, S., Davis, S. N., Zreda, M., and Cecil, L. D. (2003) The distribution of meteoric ^{36}Cl/Cl in the United States: A comparison of models. *Hydrogeology Journal*, 11:615–627.

Murad, A. A., and Krishnamurthy, R. V. (2004) Factors controlling groundwater quality in Eastern United Arab Emirates: A chemical and isotopic approach. *J. Hydrol.*, 286:227–235.

Murphy, E. M., Ginn, T. R., and Phillips, J. L. (1996) Geochemical estimates of paleorecharge in the Pasco Basin: Evaluation of the chloride mass balance technique. *WRR*, 32:2853–2868.

Nativ, R., Hötzel, H., Reichert, B., Günay, G., Tezcan, L., and Solomon, K. (1997) Separation of groundwater-flow components in a karstified aquifer. In: Kranjc, A. (ed.), *Tracer Hydrology*, Balkema, Roterdam, 269–272.

NEA (1995) The environmental and ethical basis of geological disposal of long-lived radioactive wastes / a collective opinion of the Radioactive Waste Management Committee of the OECD Nuclear Energy Agency—Paris: Nuclear Energy Agency Organisation for Economic Co-operation and Development, cop. 30 p.

Nelms, D. L., Harlow, G. E., Jr., Plummer, L. N., and Busenberg, E. (2003) Aquifer susceptibility in Virginia, 1998–2000. USGS WRIR 03-4278, 20 p.

Nelms, D. L., Harlow, G. E., Jr., Plummer, L. N., and Busenberg, E. (2002) Susceptibility of fractured-rock aquifers in Virginia based on ground-water dating techniques [abs.], National Ground Water Association fractured-rock aquifers 2002, Denver, Colo., p. 51.

Nelms, D. L., and Harlow, G. E., Jr. (2001) Preliminary results from chlorofluorocarbon-based dating for determination of aquifer susceptibility in Virginia [abs.], In: Poff, J. (ed.), *Proc. of the Virginia Water Research Symposium 2001*, Protecting our water resources for the next generation: Where do we go from here: Virginia Water Resources Research Center P7-2001, p. 17.

Nelms, D. L., Plummer, L. N., and Busenberg, E. (1999) Ground-water dating to predict aquifer susceptibility in crystalline terranes of Virginia [abs.]: *Geological Society of America Abstracts with Programs*, 31(3):78.

Neupauer, R., and Wilson, J. L. (2001) Adjoint-derived location and travel time probabilities for a multidimensional groundwater system. *WRR*, 37:1657–1668.

Neupauer, R., and Wilson, J. L. (1999) Adjoint method for obtaining backward-in-time location and travel time probabilities of a conservative groundwater contaminant. *WRR*, 35:3389–3398.

Nijamparkar, V. N., Amin, B. S., Kharkar, D. P., and Lal, D. (1966) Dating groundwater of ages younger than 1000–1500 years using natural ^{32}Si. *Nature*, 210:478–480.

Notle, E., Krauthan, P., Korschinek, G., Maloszewski, P., Fritz, P., and Wolf, M. (1991) Measurements and interpretations of ^{36}Cl in groundwater, Milk River aquifer, Alberta, Canada. *Applied Geochemistry*, 6:435–445.

Oeschger, H., Guglemann, A., Loosli, H. H., Schotterer, U., Siegenthaler, U., and Wiest, W. (1974) ^{39}Ar dating of groundwater. In: *Isotope Techniques in Groundwater Hydrology 1974*, IAEA, Vienna, 179–190.

Olley, J. M., Roberts, R. G., and Murray A. S. (1997) A novel method for determining residence times of rivers and lake sediments based on disequilibrium in the thorium decay series. *WRR*, 33:1319–1326.

Oremland, R. S., Lonergan, D. J., Culbertson, C. W., and Lovely, D. R. (1996) Microbial degradation of hydrochlorofluorocarbons ($CHCl_2F$ and $CHCl_2CF_3$) in soils and sediments. *Applied and Environmental Microbiology*, 62:1818–1821.

Osenbrück, K., Lippmann, J., and Snntag, C. (1998) Dating very old pore waters in impermeable rocks by noble gas isotopes. *Geochimica et Cosmochimica Acta*, 62:3041–3045.

Osmond, J. K., and Cowar, J. B. (2000) U-Series nuclides as tracers in groundwater hydrology. In: Cook, P. G., and Herczeg, A. (eds.) *Environmental Tracers in Subsurface Hydrology*. Kluwer Academic Publishers, Boston, 145–173.

Osmond, J. K., and Cowar, J. B. (1974) Mixing volume calculations, sources and aging trends of Floridan aquifer water by uranium isotopic methods. *Geochimica et Cosmochimica Acta*, 38:1083–1100.

Osmond, J. K., Rydell, H. S., and Kaufman, M. I. (1968) Uranium disequilibrium in Groundwater: An isotope dilution approach in hydrologic investigations. *Science*, 162:997–999.

Oster, H., Sonntag, C., and Münnich, K. O. (1996) Groundwater age dating with chlorofluorocarbons. *WRR*, 32:2989–3001.

Ottonello, G. (1997) *Principles of Geochemistry*. Columbia University Press, New York, 894 p.

Oudijk, G., and Schmitt, L. M. (2000) Age dating of a chlorinated solvent plume in groundwater. In: Dassargues, A. (ed.) *Tracers and Modelling in Hydrogeology*, IAHS publication no. 262. IAHS Press, Oxfordshire, UK, 255–261.

Ozima, M., and Podosek, F. (2001) *Noble Gas Geochemistry*. Second edition. Cambridge University Press, Cambridge, 286 p.

Park, J., Bethke, C. M., Torgerson, T., and Johnson, T. M. (2002) Transport modeling applied to the interpretation of groundwater ^{36}Cl age. *WRR*, 38: DOI:10.1029/2001WR000399.

Patterson, C. C. (1956) Age of meteorites and the Earth. *Geochimica et Cosmochimica Acta*, 10:230–237.

Payne, B. R. (1988) The status of isotope hydrology today. *J. Hydrol.*, 100:207–237.

Pearson, F. J., Balderer, W., Loosli, H. H., Lehmann, B. E., Matter, A., Peters, Tj, Schmassmann, H., and Gautschi, A. (eds.) (1991) Applied isotope hydrogeology. A case study in northern Switzerland. *Studies in Environmental Science*, 43, Elsevier, Amsterdam, 439 p.

Pearson, F. J., Noronha, C., and Andrews, R. (1983) Mathematical modeling of the distribution of natural ^{14}C, ^{234}U, and ^{238}U in a regional ground-water system. *Radiocarbon*, 25:291–300.

Pearson, F. J., and White, D. E. (1967) Carbon-14 ages and flow rates of water in Carrizo Sand, Atascosa County, Texas. *WRR*, 3:251–261.

Peeters, F., Beyerle, U., Aeschbach-Hertig, W., Holocher, J., Brennwald, M. S., and Kipfer, R. (2003) Improving noble gas based paleoclimate reconstruction and groundwater dating using $^{20}Ne/^{22}Ne$ ratios. *Geochimica et Cosmochimica Acta*, 67:587–600.

Petit, J. R., and 18 others (1999) Climate and atmospheric history of the past 420,000 years from the Vostok ice core, Antarctica. *Nature*, 399:429–436.

Phillips, F. M. (2000) Chlorine-36. In: Cook, P., and Herczeg, A. (eds.) *Environmental Tracers in Subsurface Hydrology*. Kluwer Academic Publishers, Boston, 299–348.

Phillips, F. M. (1995) The use of isotopes and environmental tracers in subsurface hydrology. *Rev. Geophys.* Vol. 33 Suppl.

Phillips, F. M. (1994) Chlorine-36 in fossil rat urine: A key to the chronology of groundwater. Birdsall Dreiss Lecture Series.

Phillips, F. M., Tansey, M. K., and Peeters, L. A. (1989) An isotope investigation of groundwater in the central San Juan Basin, New Mexico: Carbon 14 dating as a basis for numerical flow modelling. *WRR*, 25:2259–2273.

Phillips, F. M., Mattick, J. L., Duval, T. A., Elmore, D., and Kubik, P. W. (1988) Chlorine-36 and tritium from nuclear-weapons fallout as tracers for long term liquid and vapor movement in desert soils. *WRR*, 24:1877–1891.

Phillips, F. M., Bentley, H. W., Davis, S. N., Elmore, D., and Swanick, G. B. (1986) Chlorine-36 dating of very old groundwater, 2, Milk River Aquifer, Alberta, Canada. *WRR*, 22:2003–2016.

Phillips, S. W., and Lindsey, B. D. (2003) The influence of groundwater on nitrogen delivery to the Chesapeake Bay. USGS FS-091-03, 6 p.

Phillips, S. W., Focazio, M. J., and Bachman, L. J. (1999) Discharge, nitrate load, and residence time of groundwater in the Chesapeake Bay. USGS FS-150-99, 6 p.

Pint, C. D., Hunt, R. J., and Anderson, M. P. (2003) Flowpath delineation and ground water age, Allequash Basin, Wisconsin. *Ground Water*, 41:895–902.

Pinti, D. L., and Marty, B. (1998) The origin of helium in deep sedimentary aquifers and the problem of dating very old groundwaters. In: Parnell, J. (ed.) *Dating and Duration of Fluid Flow and Fluid–Rock Interaction*. Geological Society, London, Special publications, 144:53–68.

Pinti, D. L., Marty, B., and Andrews, J. N. (1997) Atmospheric noble gas evidence for the preservation of ancient waters in sedimentary basins. *Geology*, 25:111–114.

Plummer, L. N., and Sprinkle, C. L. (2001) Radiocarbon dating of dissolved inorganic carbon in groundwater from confined parts of the Upper Floridan aquifer, Florida, USA. *Hydrogeology Journal*, 9:127–150.

Plummer, L. N., Busenberg, E., Böhlke, J. K., Nelms, D. L., Michel, R. L., and Schlosser, P. (2001) Groundwater residence times in Shenandoah National Park, Blue Ridge Mountains, Virginia, USA: A multi-tracer approach. *Chemical Geology*, 179:93–111.

Plummer, L. N., and Busenberg, E. (2000) Chlorofluorocarbons. In: Cook, P. G., and Herczeg, A. L. (eds.) *Environmental Tracers in Subsurface Hydrology*. Kluwer Academic Publishers, Boston, 441–478.

Plummer, L. N., Rupert, M. G., Busenberg, E., and Schlosser, P. (2000) Age of irrigation water in groundwater from the Snake River Plain aquifer, South-Central Idaho. *Ground Water*, 38:264–283.

Plummer, L. N., Michel, R. L., Thurman, E. M., and Glynn, P. D. (1993) Environmental tracers for age dating young groundwater. In: Alley, W. M. (ed.) *Regional Ground-Water Quality*, Van Nostrand Reinhold, New York, 227–254.

Plummer, L. N., Prestemon, E. C., and Parkhurst, D. L. (1994) An interactive code (NETPATH) for modeling NET geochemical reactions along a flow PATH—Version 2.0. Water-Resources Investigations Rep., 94-4169. USGS, Washington, DC.

Plummer, L. N., Busby, J. F., Lee, R. W., and Hanshaw, B. B. (1990) Geochemical modelling of the Madison aquifer in parts of Montana, Wyoming and South Dakota. *WRR*, 26:1981–2014.

Porcelli, D., and Ballentine, C. J. (2002) Models for the distribution of terrestrial noble gases and evolution of the atmosphere. In: Porcelli, D., Ballentine, C. J., and Wieler, R. (eds.) *Reviews in Mineralogy and Geochemistry, Volume 47: Noble Gases in Geochemistry and Cosmochemistry*. The Mineralogical Society of America, Washington, 412–480.

Porcelli, D., Ballentine, C. J., and Wieler, R. (eds.) (2002) *Reviews in Mineralogy and Geochemistry, Volume 47: Noble Gases in Geochemistry and Cosmochemistry*. The Mineralogical Society of America, Washington, 844 p.

Price, R. M., Top, Z., Happell, J. D., and Swart, P. K. (2003) Use of tritium and helium to define groundwater flow conditions in Everglades National Park. *WRR*, 39:1267, 13-1–13-12.

Priyadarshi, N. (2005) Groundwater dating with radiocarbon. In: Lehr, J. H., and Keeley, J. (eds.) *Water Encyclopedia: Groundwater*. John Wiley and Sons, Hoboken, New Jersey, 64–65.

Purdy, C. B., Helz, G. R., Mignerey, A. C., Kubik, P. W., Elmore, D., Sharma, P., and Hemmick, T. (1996) Aquia Aquifer dissolved Cl^- and $^{36}Cl/Cl$: Implications for flow velocities. *WRR*, 32:1163–1171.

Purdy, C. B. (1991) Isotopic and chemical tracer studies of ground water in the Aquia Formation, southern Maryland, including ^{36}Cl, ^{14}C, ^{18}O, and ^{3}H. Ph.D. thesis, University of Maryland, USA.

Rademacher, L. K., Clark, J. F., Clow, D. W., and Hudson, G. B. (2005) Old groundwater influence on stream hydrochemistry and catchment response times in a small Sierra Nevada catchment: Sagehen creek, California. *WRR*, 41: W02004, doi:10.1029/2003WR002805.

Rademacher, L. K., Clark, J. F., and Hudson, G. B. (2002) Temporal changes in stable isotope composition of spring waters: Implications for recent changes in climate and atmospheric circulation. *Geology*, 30:139–142.

Rademacher, L. K., Clark, J. F., Hudson, G. B., Erman, D. C., and Erman, N. A. (2001) Chemical evolution of shallow groundwater as recorded by springs, Sagehen basin; Nevada County, California. *Chemical Geology*, 179:37–51.

Rao, U. (1997) Sources, reservoirs and pathways of anthropogenic ^{129}I in western New York. Rochester, Ph.D. thesis, University of Rochester, USA, 147 p.

Reardon, E. J., and Fritz, P. (1978) Computer modeling of ground water ^{13}C and ^{14}C isotope compositions. *J. Hydrol.*, 36:201–224.

Reilly, T. E., Plummer L. N., Phillips, P. J. and Busenberg E. (1994) The use of simulation and multiple environmental tracers to quantify ground-water flow in a shallow aquifer. *WRR*, 30:421–433.

Remenda, V. H., Cherry, J. A., and Edwards, T. W. D. (1994) Isotopic composition of old groundwater from Lake Agassiz: Implications for late Pleistocene climate. *Science*, 266:1975–1978.

Richter, J., Szymezak, P., Abraham, T., and Jordan, H. (1993) Use of combination of lumped parameter models to interpret groundwater isotopic data. *Journal of Contaminant Hydrology*, 14:1–13.

Ringwood, A. E. (1979) *Origin of the Earth and Moon*. Springer-Verlag, New York, 295 p.

Robertson, W. D., and Cherry, J. A. (1989) Tritium as an indicator of recharge and dispersion in a groundwater system in central Ontario. *WRR*, 25:1097–1109.

Robkin, M. A., and Shleien, B. (1995) Estimated maximum thyroid doses from ^{129}I releases from the Hanford Site for the years 1994–1995. *Health Physics*, 69:917–922.

Rodgers, P., Solulsby, C., and Waldron, S. (2004) Stable isotope tracers as diagnostic tools in upscaling flow path understanding and residence time estimates in a mountainous mesoscale catchment. *Geophysical Research Abstracts*, Vol. 6, EGU04-A-00243.

Rogers, S. G. (1921) Helium-bearing natural gas. US Geological Survey Professiona Paper No. 121, 113 p.

Rose, T. P., Denneally, J. M., Smith, D. K., Davisson, M. L., Hudson, G. B. and Rego, J. A. H. (1997) Chemical and isotopic data for ground water in southern Nevada. Report UCRLID-12800, Lawrence Livermore National Laboratory, Livermore, California, 36 p.

Rovery, C. W., and Niemann, W. L. (2005) Do conservative solutes migrate at average pore-water velocity? *Ground Water*, 43:52–62.

Rozanski, K. (1985) Deuterium and oxygen-18 in European groundwaters: Links to atmospheric circulation in the past. *Chemical Geology (Isotope Geoscience Section)*, 52:349–363.

Rozanski, K., and Florkowski, T. (1979) Krypton-85 dating of groundwater. *Isotope Hydrology 1978*. Vol. II, IAEA, Vienna, 949–961.

Rübel, A. P., Sonntag, C., Lippmann, J., Pearson, F. J., and Gautschi, A. (2002) Solute transport in formations of very low permeability: profiles of stable isotope and dissolved noble gas contents of pore water in the Opalinus Clay, Mont Terri, Switzerland. *Geochimica et Cosmochimica Acta*, 66:1311–1321.

Rupert, M. G. (2001) Calibration of the DRASTIC groundwater vulnerability mapping method. *Ground Water*, 39:625–630.

Russell, A. D., and Thompson, G. M. (1983) Mechanisms leading to enrichment of the atmospheric fluorocarbons CCl_3F and CCl_2F_2 in groundwater. *WRR*, 19:57–60.

Sakaguchi, A., Ohtsuka, Y., Yokota, K., Sasaki, K., Komura, K., and Yamamoto, M. (2005) Cosmogenic radionuclide ^{22}Na in the Lake Biwa system (Japan): Residence time, transport and application to the hydrology. *Earth Planet Sci Lett.*, 231:307–316.

Santella, N., Schlosser, P., Ho, D. T., and Stute, M. (2002) Elevated SF_6 concentration in soil air near New York city and its effect on the utility of SF_6 for dating groundwater. Goldschmidt conference, Davos, Switzerland. *Geochimica et Cosmochimica Acta*, 66:A667.

Sapienza, G., Hilton, D. R., and Scribano, V. (2005) Helium isotopes in peridotite mineral phases from Hyblean Plateau xenoliths (south-eastern Sicily, Italy). *Chemical Geology*, 219:114–129.

Schink, D. R. (1968) Radiochemistry of Silicon. US Atomic Energy Commission, 75 p.

Schlosser P., Shapiro, S. D., Stute, M., Aeschbach-Hertig, W., Plummer, L. N., and Busenberg, E. (1998) Tritium/^{3}He measurements in young groundwater: Chronologies for environmental records. In *Isotope Techniques in the Study of Environmental Change*, IAEA, Vienna, 165–189.

Schlosser, P., Stute, M., Dorr, H., Sonntag, C., and Münnich, K. O. (1988) Tritium/^{3}He dating of shallow groundwater. *Earth Planet Sci Lett.*, 89:353–362.

Schoeberl, M. R. (1999) Ozone and stratospheric chemistry. In: King, M. D. (ed.) *EOS Science Plan*. NASA, 309–337.

Schwehr, K. A., Santschi, P. H., Moran J. E., and Elmore D. (2005) Near-conservative behavior of ^{129}I in the orange county aquifer system, California. *Applied Geochemistry*, 20:1461–1472.

Selaolo, E. T., Hilton, D. R., and Beekman, H. E. (2000) Groundwater recharge deduced from He isotopes in the Botswana Kalahari. In: Sililo et al. (eds.) *Groundwater: Past Achievements and Future Challenges*. Balkema, Rotterdam, 313–318.

Selinus, O., Alloway, B., Centeno, J. A., Finkelman, R. B., Fuge, R., Lindh, U., and Smedley, P. (2005) *Essentials of Medical Geology: Impact of the Natural Environment on Public Health*. Elsevier Academic press, Amsterdam, 812 p.

Shapiro, A. M., and Cvetkovic, V. D. (1988) Stochastic analysis of solute arrival time in heterogeneous porous media. *WRR*, 24:1711–1718.

Sheets, R. A., Bair, E. S., and Rowe, G. L. (1998) Use of ^{3}H/^{3}He ages to evaluate and improve groundwater flow models in a complet buried-valley aquifer. *WRR*, 34:1077–1089.

Sivan, O., Yechieli, Y., Herut, B., and Lazar, B. (2005) Geochemical evolution of timescale of seawater intrusion into the coastal aquifer of Israel. *Geochimica et Cosmochimica Acta*, 69:579–592.

Sivan, O., Herut, B., Yechieli, Y., Boaretto, E., Heinemeier, J., and Heut, B. (2001) Radiocarbon dating of porewater-Correction for diffusion and diagenetic processes. *Radiocarbon*, 43:765–771.

Smethie, Jr. W. M., Solomon, D. K., Schiff, S. L., and Mathieu, G. G. (1992) Tracing groundwater flow in the Borden aquifer using krypton-85. *J. Hydrol.*, 130: 279–297.

Smethie, Jr. W. M., Ostlund, H. G., and Loosli, H. H. (1986) Ventilation of the deep Greenland and Norwegian seas: Evidence from krypton-85, tritium, carbon-14 and argon-39. *Deep Sea Research*, 33:675–703.

Smith, D. B., Downing, R. A., Monkhouse, R. A, Otlet, R. L., and Pearson, F. J. (1976) The age of groundwater in the Chalk of the London Basin. *WRR*, 12:392–404.

Solomon, D. K. (2000) ^{4}He in groundwater. In: Cook, P. G., and Herczeg, A. L. (eds.) *Environmental Tracers in Subsurface Hydrology*. Kluwer Academic Publishers, Boston, 425–439.

Solomon, D. K., and Cook, P. G., (2000) ^{3}H and ^{3}He. In: Cook, P. G., and Herczeg, A. L. (eds.) *Environmental Tracers in Subsurface Hydrology*. Kluwer Academic Publishers, Boston, 397–424.

Solomon, D. K., Hunt, A., and Poreda, R. J. (1996) Sources of radiogenic helium-4 in shallow aquifers: Implications for dating young groundwater. *WRR*, 32:1805–1813.

Solomon, D. K., Poreda, R. J., Schiff, S. L., and Cherry, J. A. (1992) Tritium and helium 3 as groundwater age tracers in the Borden aquifer. *WRR*, 28:741–755.

Solomon, D. K., Poreda, R. J., Cook, P. G., and Hunt, A. (1995) Site characterization using ^{3}H/^{3}He ground-water ages, Cape Cod, MA. *Ground Water*, 33:988–996.

Solomon, D. K., Schiff, S. L., Poreda, R. J., and Clarke, W. B. (1993) A validation of the ^{3}H/^{3}He method for determining groundwater recharge. *WRR*, 29:2951–2962.

Spalding, D. B. (1958) A note on mean residence-times in steady flows of arbitrary complexity. *Chemical Engineering Science*, 9:74–77.

Strassmann, K. M., Brennwald, M. S., Peeters, F., and Kipfer, R. (2005) Noble gases in the porewater of lacustrine sediments as palaeolimnological proxies. *Geochimica et Cosmochimica Acta*, 69:1665–1674.

Stuiver, M., and Polach, H. A. (1977) Discussion on reporting of ^{14}C data. *Radiocarbon*, 19:335–363.

Sturchio, N. C., Du, X., Purtschert, R., Lehmann, B. E., Sultan, M., Patterson, L. J., Lu, Z.-T., Müller, P., Bigler, T., Bailey, K., O'Connor, T. P., Young, L., Lorenzo, R., Becker, R., El Alfy, Z., El Kaliouby, B., Dawood, Y., and Abdallah, A. M. A. (2004) One million year old groundwater in the Sahara revealed by krypton-81 and chlorine-36. *Geophysical Research Letters*, 31:L05503, doi:10.1029/2003GL019234.

Stute, M., and Schlosser, P. (2000) Atmospheric noble gases. In: Cook, P. G., and Herczeg, A. L. (eds.) *Environmental Tracers in Subsurface Hydrology*. Kluwer Academic Publishers, Boston, 349–377.

Sudicky, E. A., and Frind, E. (1981) Carbon-14 dating of groundwater in confined aquifers: Implication of aquitard diffusion. *WRR*, 17:1060–11064.

Sueker, J. J., Turk, J. T., and Michel, R. L. (1999) Use of cosmogenic sulfur-35 for comparing ages of water from three alpine-subalpine basins in the Colorado Front Range. *Geomorphology*, 27:61–74.

Sultan, M., Sturchio, N. C., Gheith, H., Abdel Hady, Y., and El Anbeawy, M. (2000) Chemical and isotopic constraints on the origin of Wadi El-Tarfa groundwater, Eastern Desert, Egypt. *Ground Water*, 38:743–751.

Swanson, S. K., Bahr, J. M., Schwar, M. T., and Potter, K. W. (2001) Two-way cluster analysis of geochemical data to constrain spring source waters. *Chemical Geology*, 179:73–91.

Szabo, Z., Rice, D. E., Plummer, L. N., Busenbeg, E., Drenkard, S., and Schlosser, P. (1996) Age dating of shallow groundwater with chlorofluorocarbons, tritium, helium and flow path analysis, southern New Jersey Coastal Plain. *WRR*, 32:1023–1038.

Talma, A. S., Weaver, J. M. C., Plummer, L. N., and Busenberg, E. (2000) CFC tracing of groundwater in fractured rock aided with ^{14}C and ^{3}H to identify water mixing. In: Sililo et al. (eds.) *Groundwater: Past Achievements and Future Challenges*. Balkema, Rotterdam, 635–639.

Tamers, M. A. (1975) Validity of radiocarbon dates on groundwater. *Geophysical Survey*, 2:217–239.

Tamers, M. A. (1966) Ground water recharge of aquifers as revealed by naturally occurring radiocarbon in Venezuela. *Nature*, 212:489–492.

Taugs, R., and Moosmann, L. (2001) Calibration of a groundwater flow model for water management in Hamburg, Germany, using groundwater age measurements and hydrochemical data. In: Sililo et al. (eds.) *Groundwater: Past Achievements and Future Challenges*. Balkema, Rotterdam, 647–650.

Tesoriero, A. J., Spruill, T. B., H. E., Mew Jr., Farrell, K. M., and Harden, S. L. (2005) Nitrogen transport and transformations in a coastal plain watershed: Influence of geomorphology on flow paths and residence times. *WRR*, 41: W02008, doi:10.1029/2003WR002953.

Testa, S. M. (2005) Dating groundwaters with tritium. In: Lehr, J. H., and Keeley, J. (eds.) *Water Encyclopedia: Groundwater.* John Wiley and Sons, Hoboken, New Jersey, 69–72.

Thompson, G. M., Hayes, J. M., and Davis, S. N. (1974) Fluorocarbon tracers in hydrology. *Geophysical Research Letters*, 1:177–180.

Thonnard, N., McKay, L. D., Cumbie, D. H., and Joyner, C. P. (1997) Status of laser-based krypton-85 analysis development for dating of young groundwater. Geological Society of America, 1997 annual meeting, Abstracts and programs, 29(6):78.

Thonnard, N., Willis, R. D., Wright, M. C., Davis, W. A., and Lehmann, B. E. (1987) Resonance ionization spectrometery and the detection of ^{81}Kr. *Nuclear Instruments and Methods in Physics Research*, B29:398–406.

Tolman, C. F. (1937) *Groundwater*, McGraw Hill, New York, 593 p.

Tolstikhin, I. N., Lehmann, B. E., and Loosli, H. H., and Gautschi, A. (1996) Helium and argon isotopes in rocks, minerals, and related groundwaters: A case study in northern Switzerland. *Geochimica et Cosmochimica Acta*, 60:1497–1514.

Tolstikhin, I. N., and Kamensky, I. L. (1969) Determination of groundwater age by the T-^{3}He method. *Geochemistry International*, 6:810–811.

Torgersen, T. (1994) Hydraulic calculation of groundwater flow velocity and age: examination of the basic premises-Comment. *J. Hydrol.*, 154:403–408.

Torgersen, T., Habermehl, M. A., Phillips, F. M., Elmore, D., Kubik, P., Jones, B. G., Hemmick, T., and Gove, H. E. (1991) Chlorine-36 dating of very old ground water: III. Further studies in the Great Artesian Basin, Australia. *WRR*, 27:3201–3214.

Torgersen, T., Kennedy, B. M., Hiyagon, H., Chiou, K. Y., Reynolds J. H., and Clarke, W. B. (1989) Argon accumulation and the crustal degassing flux of ^{40}Ar in the Great Artesian Basin, Australia. *Earth Planet Sci. Lett.*, 92:43–56.

Torgersen, T., and Clarke, W. B. (1985) Helium accumulatioin in groundwater, I, An evaluation of sources and the continental flux of crustal ^{4}He in the Great Artesian Basin, Australia. *Geochimica et Cosmochimica Acta*, 49:1211–1218.

Turner, J. V., Rosen, M. R., Fifield, L. K., and Allan, G. L. (1995) Chlorine-36 in hypersaline palaeochannel groundwaters of Western Australia. In: *Application of Tracers in Arid-Zone Hydrology* (Adar, E. M., and Leibundgut, C., eds.). IAHS publications no. 232, Vienna, 15–33.

UNESCO (1992) *Proc. of Int. Workshop on Hydrological Considerations in Relation to Nuclear Power Plant.* IHP IV (UNESCO Chernobyl program), UNESCO, 409 p.

van der Leeden, F. (1971) *Ground Water: A Selected Bibliography. Water Information Center*, Port Washington, New York, 116 p.

van Grosse, A. V., Johnston, W. M., Wolfgang, R. L., and Libby, W. F. (1951) Tritium in nature. *Science*, 113:1–2.

van Herwaarden, O. A. (1994) Spread of pollution by dispersive groundwater flow. *SIAM Journal on Applied Mathematics*, 54:26–41.

van Kooten, J. J. A. (1995) An asymptotic method for predicting the contamination of a pumping well. *Advances in Water Resources*, 18:295–313.

Varni, M., and Carrera, J. (1998) Simulation of groundwater age distributions. *WRR*, 34:3271–3281.

Vehagen, B., Mazor, E., and Sellschop, J. P. F. (1974) Radiocarbon and tritium evidence for direct rain recharge to groundwaters in the northern Kalahari. *Nature*, 249:643–644.

Vengosh, A., Gill, J., Lee Davisson, M., and Bryant H. G. (2002) A multi-isotope (B, Sr, O, H, and C) and age dating (^{3}H/ ^{3}He and ^{14}C) study of groundwater from Salinas valley, California: Hydrochemistry, dynamics and contamination processes. *WRR*, 38:10.1029, 9-1–9-17.

Verniani, F. (1966) Total mass of the atmosphere. *J. Geophysical Research*, 71:835–891.

Vigier, N., Bourdon, B., Lewin, E., Dupré, B., Turner, S., Chakrapani, G. J., van Calsteren, P., and Allègre, C. J. (2005) Mobility of U-series nuclides during basalt weathering: An example from the Deccan Traps (India). *Chemical Geology*, 21:69–91.

Vitvar, T. (1998a) Water residence time and runoff generation in a small prealpine catchment. Ph.D. thesis, ETH, Zurich, 111 p.

Vitvar, T. (1998b) Water residence times in a small prealpine catchment (Rietholzbach, Northeastern Switzerland). In: *Proc. of IAH 28th Congress "Gambling with Groundwater"* Las Vegas, Nevada, USA, 617–622.

Vogel, J. C. (1967) Investigation of groundwater flow with radiocarbon. In: *Isotopes in Hydrology*, IAEA, Vienna, 355–369.

Vogel, J. C., Thilo, L., and van Dijken, M. (1974) Determination of groundwater recharge with tritium. *J. Hydrol.*, 23:131–140.

Voigt, H. J., Heinkele, T., Jahnke, C., and Wolter, R. (2004) Characterization of groundwater vulnerability to fulfill requirements of the water framework directive of the European Union. *Geofísica Internacional*, 43:567–574.

von Buttlar, H. (1959) Groundwater studies in New Mexico using tritium as a tracer, II. *J. of Geophysical Research*, 64:1031–1038.

von Buttlar, H., and Wendt, I. (1958) Groundwater studies in New Mexico using tritium as a tracer. *Trans. Am. Geophys. Union*, 39:660–668.

Voss, C. (2005) The future of hydrogeology. *Hydrogeology Journal*, 13:1–6.

Vrba, J., and Zaporozec, A. (eds.) (1994) *Guidebook on Mapping Groundwater Vulnerability. International Contributions to Hydrogeology*, Vol. 16, 131, Heise Publisher, Hannover.

Wagner, G. A. (1998) *Age Determination of Young Rocks and Artifacts*. Springer, Berlin, 466 p.

Wagner, M. J. M., Dittrich-Hannen, B., Synal, H. A., Sutter, M., and Schotterer, U. (1996) Increase of 129I in the environment. *Nuclear Instruments and Methods in Physics Research*, B113:490–494.

Wang, H. F., and Anderson, M. P. (1982) *Introduction to Groundwater Modeling*. Freeman, New York, 237 p.

Wanninkhof, R., Mullholland, P. J., and Elwood, J. W. (1990) Gas exchange rates for a first-order stream determined with deliberate and natural tracers. *WRR*, 26:1621–1630.

Warner, M. J., and Weiss, R. F. (1985) Solubilities of chlorofluorocarbons 11 and 12 in water and seawater. *Deep Sea Research*, 32:1485–1497.

Wassenaar, L. I., and Hendry, M. J. (2000) Mechanisms controlling the distribution and transport of ^{14}C in a clay-rich till aquitard. *Ground Water*, 38:343–349.

Wayland, K. G., Hyndman, D. W., Boutt, D., Pijanowski, B. C., and Long, D. T. (2002) Modelling the impact of historical land uses on surface-water quality using groundwater flow and solute-transport models. *Lakes and Reservoirs: Research and Management*, 7:189–199.

Weise, S. M., Stichler, W., and Bertleff, B. (2001) Groundwater inflows into an excavated artificial lake (gravel pit) indicated by apparent $^{3}H/^{3}He$ ages. In: Seiler, K. P., and Wohnlich, S. (eds.) *Proc. XXXI IAH Congress "New Approaches Characterizing Groundwater Flow,"* Swets and Zeitlinger Lisse, Munich, 225–228.

Weiss, W., Sartorius, H., and Stockburge, H. (1992) Global distribution of atmospheric ^{85}Kr. In: *Isotopes of Noble Gases as Tracers in Environmental Studies*. IAEA, Vienna, 29–62.

Weissmann, G. S., Zhang, Y., Labolle, E. M., and Fogg, G. E. (2002) Dispersion of groundwater age in an alluvial aquifer system. *WRR*, 38:1198, 6-1–16-13, doi:10.1029/2001WR000907.

Weyhenmeyer, C. E., Burns, S. J., Waber, H. N., Aeschbach-Hertig, W., Kipfer, R., Loosli, H. H., and Matter, A. (2000) Cool glacial temperatures and changes in moisture source recorded in Oman groundwaters. *Science*, 287:842–845.

Wigley T. M. L., Plummer, L. N., and Pearson, Jr, F. J. (1978) Mass transfer and carbon isotope evolution in natural water systems. *Geochimica et Cosmochimica Acta*, 42:1117–1139.

Wigley, T. M. L. (1976) Effect of mineral precipitation on isotopic composition and ^{14}C dating of groundwater. *Science*, 263:219–221.

Wiles, D. R. (2002) *The Chemistry of Nuclear Fuel Waste Disposal*. Polytechnic International Press, Montreal, Canada.

Wilson, R. D., and Mackay, D. M. (1993) The use of sulfur hexafluoride as a tracer in saturated sandy media. *Groundwater*, 31(5):719–724.

Winston, W. E., and Criss, R. E. (2004) Dynamic hydrologic and geochemical response in a perennial karst spring. *WRR*, 40: W05106, doi:10.1029/2004WR003054.

WMO (2001) World meteorological organization global atmosphere watch, No. 143. Global atmosphere watch measurement guide, WMO TD No. 1073. 83 p.

Wolterink, T. J. (1979) Identifying sources of sub-surface nitrate pollution with stable nitrogen isotopes. US Environmental Protection Agency, EPA-600/4-79-050, 150 p.

Woodbury, A. D., and Rubin, Y. (2000) A full-Bayesian approach to parameter inference from tracer travel time moments and investigation of scale effects at the Cape Cod experimental site. *WRR*, 36:159–171.

Xu, Y., and Beekman, H. F. (eds.) (2003) Groundwater recharge estimation in southern Africa. UNESCO IHP Series No. 64, UNESCO Paris. ISBN 92-9220-000-3.

Yeager, K. M., Santschi, P. H., Phillips, J. D., and Herbert, B. E. (2002) Sources of alluvium in a coastal plain stream based on radionuclide signature from the ^{238}U and ^{232}Th decay series. *WRR*, 38:1242, 24-1–24-11.

Yechieli, Y., Sivan, O., Lazar, B., Vengosh, A., Ronen, D., and Herut, B. (2001) Radiocarbon in seawater intruding into the Israeli Mediterranean coastal aquifer. *Radiocarbon*, 43:773–781.

Yechieli, Y., Ronen, D., and Kaufman, A. (1996) The source and age of groundwater brines in the Dead Sea area, as deduced from ^{36}Cl and ^{14}C. *Geochimica et Cosmochimica Acta*, 60:1909–1916.

York, D., and Farquhar, R. M. (1972) *The Earth's Age and Geochronology*, Pergamon Press, Oxford, 178 p.

Young, D. A. (1982) *Christianity and the Age of the Earth*, Artisan, California, 188 p.

Zahn, A., Neubert, R., Maiss, M., and Platt, U. (1999) Fate of long lived trace species near the Northern Hemisphere tropopause: Carbon dioxide, methane, ozone, and sulfur hexafluoride. *J. Geophysical Research*, 104(D11):13, 923–13.

Zaikowski, A., Kosanke, B. J., and Hubbard, N. (1987) Noble gas composition of deep brines from the Palo Duro Basin, Texas. *Geochimica et Cosmochimica Acta*, 51:73–84.

Zhao, X., Fritzel, T. L. B., Quinodoz, H. A. M., Bethke, C. M., and Torgerson, T. (1998) Controls on the distribution and isotopic composition of helium in deep groundwater flows. *Geology*, 26:291–294.

Zimmerman, U., Münnich, K. O., Roether, W., Kreutz, W., Schubach, K., and Siegel, O. (1966) Tracers determine movement of soil moisture and evapotranspiration. *Science*, 152:346–347.

Zito, R. R., and Davis, S. N. (1981) Subsurface production of argon and dating of ground water, Appendix 10.12, In: Davis, S. N. (ed.) *Workshop on Isotope Hydrology Applied to the Evaluation of Deeply Buried Repositories for Radioactive Wastes, Office of Nuclear Waste Isolation*. Battelle Memorial Institute, 169–193.

Zito, R. R., Donahue, D. J., Davis, S. N., Bentley, H. W., and Fritz, P. (1980) Possible sun surface production of carbon-14. *Geophysical Research Letters*, 77:235–238.

Zoellmann, K., Kinzelbach, W., and Fulda, C. (2001) Environmental tracer transport (^{3}H and SF_6) in the saturated and unsaturated zones and its use in nitrate pollution management. *J. Hydrol.*, 240:187–205.

Zongyu, C., Zhenlong, N., Zhaoji, Z., Jixiang, Q., and Yunju, N. (2005) Isotopes and sustainability of ground water resources, North China plain. *Ground Water*, 43:485–493.

Zreda, M. G., Phillips, F. M., Elmore, D., Kubic, P. W., Sharma, P., and Dorn, R. I. (1991) Cosmogenic chlorine-36 production in terrestrial rocks. *Earth Planet Sci. Lett.*, 105:94–109.

Zubari, W. K., Madany, I. M., Al-Junaid, S. S., Al-Noaimi, S. (1994) Trends in the quality of groundwater in Bahrain with respect to salinity, 1941–1992. *Environment International*, 20:739–746.

Zwahlen, F. (ed.) (2004) Vulnerability and risk mapping for the protection of carbonate (karst) aquifers, final report COST Action 620, European Commission, Directorate-General for Research, EUR 20912: 21–24; Luxemburg.

Zwietering, T. N. (1959) The degree of mixing in continuous flow systems. *Chem. Eng. Sci.*, 11:1–15.

Zwingmann, H., Clauer, N., and Gaupp, R. (1998) Timing of fluid flow in a sandstone reservoir of the north German Rotliegend (Permian) by K-Ar dating of related hydrothermal illite. In: Parnell, J. (ed.) *Dating and Duration of Fluid Flow and Fluid–Rock Interaction*. Geological society, London, Special Publications, 144:91–106.

APPENDIX 1

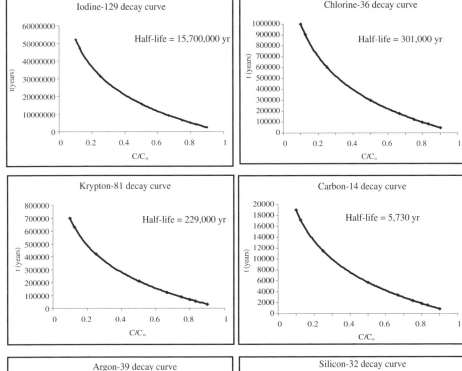

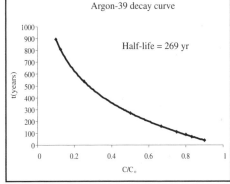

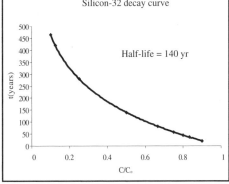

Groundwater Age, by Gholam A. Kazemi, Jay H. Lehr, and Pierre Perrochet
Copyright © 2006 John Wiley & Sons, Inc.

APPENDIX 2

SOME USEFUL INFORMATION FOR GROUNDWATER DATING STUDIES AND TABLE OF CONVERSION OF UNITS

- Atomic weight = Number of protons + Number of neutrons (A = Z + N). For carbon-14, we write $^{14}_{6}C$, which means that the atomic weight of carbon-14 is 14 and it has 6 protons.
- An α particle is heavy and consists of two proton and two neutrons and has an energy typically between 4–7 MeV (Million electron volt). A β particle is much lighter than α and is like an electron with an energy typically between 0.5 and 1.5 MeV. Gamma (γ) rays are photons like light or x-rays and have energy of 0.1–1.5 MeV. Most gamma rays can be stopped by 5–10 cm of lead.[1]
- Decay constant (λ) = ln2/half-life = −ln0.5/half-life. λ unit is time^{-1}.
- [e = natural exponent = 2.718, ln(e) = 1, ln1 = 0, ln($a \times b$) = ln$a$ + ln$b$, ln(a/b) = ln$a$ − ln$b$].
- Mean life = 1/λ = half-life/ln2.
- Activity = Number of atoms × λ (second^{-1}). Standard unit for activity is Becquerel.
- Avogadro's Number = 6.023 × 10^{23} = Number of atoms per mole = 1 mole.
- STP = Standard temperature and pressure = 0°C (273.15 K, 32°F) and 1 atmosphere (101.325 kPa, 760 mm Hg).
- Fmole = femtomole = 10^{-15} mole.
- pptv = part per trillion by volume.

Groundwater Age, by Gholam A. Kazemi, Jay H. Lehr, and Pierre Perrochet
Copyright © 2006 John Wiley & Sons, Inc.

- R_a = Atmospheric ratio, i.e., ratio of isotopes of one element in the atmosphere.
- $^{40}Ar^*$ = Radiogenic argon-40, i.e.,* sign indicates radiogenic origin.

Multiply	by	To obtain
Curie (Ci)	10^{12}	Picocurie (pCi)
Curie (Ci)	3.7×10^{10}	Becquerel (Bq)
Curie (Ci)	2.2×10^{12}	Disintegration per minute (dpm)
Becquerel (Bq)	10^3	Millibecquerel
Becquerel (Bq)	1	Disintegration per second (dps)
Becquerel (Bq)	0.0166	Disintegration per minute (dpm)
Tritium unit (TU)	0.1181	Bq/Kg
Tritium unit (TU)	6.686×10^{10}	Tritium atoms/kg of water
Tritium unit-^{3}He	2.487	Picocm^3 ^{3}He/kg of water
MeV	10^6	eV
Femtomole/L	1	pptv
Gram/L	2.6868×10^{23}/AW*	Atom/L
1 cm^3 STP	2.6868×10^{23}	Atoms

[1] Wiles, D. R. (2002) The chemistry of nuclear fuel waste disposal. Polytechnic International Press, Montreal, Canada.
* Atomic weight.

APPENDIX 3

CONCENTRATION OF NOBLE GASES (USED IN GROUNDWATER DATING) AND SOME IMPORTANT CONSTITUENTS OF THE ATMOSPHERE

Name	Formula	Percentage (by volume) in air
Nitrogen	N_2	78.08
Oxygen	O_2	20.95
Water vapour	H_2O	0.1–4
Argon	Ar	0.934 ± 0.01
Neon	Ne	0.002
Carbon dioxide	CO_2	3.7×10^{-4}
Helium	He	$(5.24 \pm 0.05) \times 10^{-6}$
Krypton	Kr	$(1.14 \pm 0.01) \times 10^{-6}$
Radon	Rn	6×10^{-20}
CFC-11	CCl_3F	253×10^{-12}*
CFC-12	CCl_2F_2	534×10^{-12}*
CFC-113	$C_2F_3Cl_3$	81×10^{-12}*
Sulfur hexafluoride	SF_6	5.78×10^{-12}*

Source: Mazor, E. (2004) *Applied Chemical and Isotopic Groundwater Hydrology*. Third edition, Marcel Dekker, Inc, New York, 453 p.; Porcelli, D., Ballentine, C. J., and Wieler, R. (eds.) (2002) *Reviews in Mineralogy and Geochemistry, Volume 47: Noble Gases in Geochemistry and Cosmochemistry*. The Mineralogical Society of America, Washington, 844 p.

* Global average CFCs values and SF_6 average value at Nivot Ridge, CO, USA, in 2004 (data from E. Busenberg, USGS).

Groundwater Age, by Gholam A. Kazemi, Jay H. Lehr, and Pierre Perrochet
Copyright © 2006 John Wiley & Sons, Inc.

INDEX

Absolute age, 3–4
Absorption, 8
Abstraction, in wells, 8
Accelerator mass spectrometer (AMS):
  ^{14}C Dating Centre, 32
  characterized, 32, 174, 194
  with cyclotron, 167
  laboratories, 31
  technique, 148, 150
Accumulating tracers, 15
Accumulation rate, 184–185
Acidification, 194
Active elements, 30
Active zone groundwater, 10–11
Adelaide Laboratory, 32
Adsorption, 201
Advection, 12, 204
Advective age, 22–23, 207–208
Advective backward particle-tracking solution, 209
Advective capture zone, 208
Advective flow paths, 249
Advective systems, 227–235
Advective-dispersive equations, 14, 209–210
Advective-dispersive systems, 240–244
Aerosols, 160
Africa, age-dating research:
  Dakhla Basin, Sahara Desert (Egypt), 170
  Elandsfontein (Witwatersrand, South Africa), 58
  Kalahari aquifer (Botswana), 53, 57–58
  Nile River (Egypt), 53
  Nubian Sandstone Aquifer (Egypt), 59, 63, 170
  Stampriet aquifer (South Africa), 189
  Table Mountain Group sandstone (South Africa), 65–66
  Uitenhage aquifer (South Africa), 189
Age:
  absolute, 3–4
  contour lines, 18
  defined, 1–2
  determination in geology, *see* Age determination
  distributions, 221
  gradient, 11
  groundwater, *see* Groundwater age

Groundwater Age, by Gholam A. Kazemi, Jay H. Lehr, and Pierre Perrochet
Copyright © 2006 John Wiley & Sons, Inc.

human, 2
relative, 3–4
Age-dating techniques:
  decay curves of isotopes, 266–298
  event markers, 15–16
  future directions for, 264–265
  historical perspective, 25–50
  methodologies, current proposals for, 254–258
  noble gas concentration, 302
  old groundwater, 134–164
  overview of, 30
  problematic, 264
  publications, 38–40
  research studies, 36–37
  rock samples, 4
  simulations, translation into practice, 259–260
  unit conversion table, 300–301
  very old groundwater, 165–204
  worldwide groundwater age maps, 260–262
  worldwide practices, 260
  young groundwater, 81–133
Age determination:
  absolute age, 3–4
  geological time table, 4–6
  relative age, 3–4
  residence-time distributions:
    advective-dispersive systems, 240–244
    advective systems, 227–235
    backward flow fields, 235–239, 242
    characterized, 227
    forward flow fields, 235–239, 242
Age gap, 9
Age isochrones, 224
Age map, world groundwater, 263
Age mass:
  balance, 213
  characterized, 11–12
  pulse-injection Cauchy condition, 241
  transport equation, 210–211
Age monitoring, for prevention of overexploitation and contamination of aquifers, 54
Age of Rain, 26–27
Age range, 121

Age transport modeling principles, 227
Aggarwal, P. K., 52
Agricultural watersheds, nitrate contamination, 70–71
Agrochemicals, 259
Air bubbles, 108
Air temperature, 67–68
Alamogordo area aquifer (NM), 91–92
Allequash Basin (WI), 61
Alluvial aquifers, 58, 75
Alluvium, 199
Alpha:
  decay, 199–200
  emission, 140, 255–256
Alunite, 156
Alvarado, Corcho, 127
Ambient groundwater, 161, 256
Ammonia, 106
Ammonium hydroxide, 194
Analytical age, residence-time distributions, 229
Analytical facilities, worldwide, 31
Andrews, A. N., 35
Andrews, John Napier, 37, 43
Anhydrite, 153, 190, 259
Antalya Plateau (Turkey), 97
ANTARES (Australian Nuclear Science and Technology), 32
Apparent age, 23
Apparent atmospheric concentration, CFCs, 110
Applied Chemical and Isotopic Groundwater Hydrology, 3E (Mazor), 38–39
Applied Geochemistry, 43–44
Applied Isotope Hydrogeology (Pearson et al.), 38
Aquifer, *see specific aquifers*
  classification of, 6, 262–263
  contamination vulnerability, 68
  deep, confined, 84
  deep old, 78
  flow paths, 14
  geometrics, 64
  hydraulic conductivity, 59
  hydraulic properties, 10, 75
  hydrodispersion components of, 205
  hydrodispersive multilayer, 246–248
  hydrodynamic models of, 59

INDEX 305

isochrone and life expectancy maps, 18–19
lithology, 143
mass transport in, 12
matrix, 258
pore-velocity estimation, 226
porosity of, 59, 223
overexploitation and contamination of, 54
production from, 18
protection classification, 73–74
recharge rate, 11
renewable reservoirs, 52
river water infiltration, 72
self-purification of, 72
shallow unconfined, 78
storage, 77
transmissivity, 77
unconfined shallow, 83
with uniform and localized recharge, 245–246
Aquifer-aquitard systems, 222–227
Aquitaine basin (France), 59
Aquitards, 61–63, 222–227
$^{40}Ar/^{36}Ar$, 189–191
Archean period, 5
Archaeology/archaeological objects, 3, 28
Argon, 100, 111, 125. *See also* Argon-36; Argon-39; Argon-40
Argon-36 (^{36}Ar), 171
Argon-39 (^{39}Ar):
  advantages of age-dating, 144, 163
  age-dating by, 142–143
  atmospheric, 142–144
  characterized, 30, 139–140, 263
  dating case studies, 145–146
  disadvantages of age-dating, 144, 163
  groundwater analysis, 142–143, 163
  production of, 141–142
  sampling, 141–142, 163
  sources of, 140
Argon-40 (^{40}Ar):
  advantages in age-dating, 202
  age-dating by, 189–190
  atmospheric, 188–189
  characterized, 30, 35, 188–189
  dating case studies, 190–191

disadvantages of age-dating, 202
groundwater analysis, 189
production of, 189–190
results reports, 189
sampling, 189
splitting of, 71
^{40}Ar-^{40}K dating technique, 190
Argonne National Laboratory, 34
Argonne Tandem Linac Acclerator System (ATLAS), 141
Arid zones, 52
Arizona AMS and Geosciences Laboratory, 34
ARTEMIS, 32
Artesia area aquifer (NM), 92
Asia, age-dating research:
  Kobe Area (Japan), 186–187
  Nankai Hydrate Field (Japan), 197–199
Asymmetric distributions, 205
Atmosphere/atmospheric:
  acid deposition, 76–77
  bomb tests, 193
  krypton content, 100, 166–167
  mean residence time, 15
  precipitation, 85, 129
  residence range time in hydrologic cycle, 15
  tracers, 65
  water in, 1
Atom Trap Trace Analysis (ATTA), 45, 167, 170, 255
Australia, age-dating research:
  ANTARES (Australian Nuclear Science and Technology), 32
  Australian National University, Department of Nuclear Physics, 32
  Buckinbah Creek Watershead (BCW) 113, 115, 117, 155–156
  Clare Valley, 58, 117
  Eyre Peninsula, 117
  Gambier Embayment, 77
  Gambier Limestone aquifer, 127
  Macquarie Valley, 79
  Murray Basin, 180
  Otway Basin, 77
  Paleochannel groundwater, 180
  Superficial Aquifer, 19–20

Australian National University, Department of Nuclear Physics, 32
Autonomous University of Barcelona, Environmental Radioactivity Laboratory, 33
Average age, 26, 215–217
Avogadro's number, 173, 193
Awards and recognitions, 26, 31, 33
Azoic era, 5

Back, William, 17
Back-calculation, flow rate, 59
Back-extracting sampling, 194
Back-tracking method, 208
Backward flow fields, 209, 235–239, 242
Backward travel time probability, 210
Bacterial degradation, 151
Bacteriophages, 259
BALANCE, 152
Basalts, 144, 159, 173
Bayesian hydrogeological parameters, 210
Bay of Bengal, 97–98
Becquerel, Henry, 198
Bedrock, radioactive waste research, 75–76
Begemann, Friedrich, 26
Below ground (BG) surface, contaminant transport rate, 62
Bentley, H. W., 35–36
Beta Analytic, Inc., 54
Beta:
 decay, 103, 199–200
 emission, 174, 195
BETA Laboratory, 34
Bethke, C. M., 36
Bhabha Atomic Research Center, Isotope Hydrology division, 32
Bicarbonate, 148, 152
Big Bang, 84
Bimodal-singular age distribution, 14
Biosphere, 15, 126
Bioturbation, 195
Birdsall Dreiss Lecturer, 31, 35
Blackwater Draw Formations, 158
Blake Ridge Atlantic Ocean, 199
Bluegrass Spring, Meramec River Basin (MI), 131, 133

Blue Ridge aquifer, 74
Bocholt aquifer (Germany), 145
Böhlke, J. K., 37
Bomb(s):
 ^{36}Cl, 125–127
 peak tritium, 90–91
 pulse tritium, 88, 90
 tritium, 85–86, 99, 196
Books, as information resource, 38–40
Borden Aquifer (Canada), 59, 62, 105, 127
Boundary conditions, residence-time distributions, 233, 235, 243–244
Br/Cl ratio, 160
Breakthrough curve, 209
Bremen University, Institute of Environmental Physics, 32
Bromide-81, 167
Buckinbah Creek Watershed (BCW) (Australia), 113, 115, 117, 155–156
Bunter sandstone aquifer (Germany), 145, 203
Busenberg, E., 35–36

Cadmium contamination, 72
Calcites, 151–153, 161, 259
Calcium-42 (^{42}Ca), 140
Cambrian period, 5
Campana, M. E., 37
Canada, age-dating research:
 Borden Aquifer, 59, 62, 105, 127
 Cigar Lake, 176–177
 Deep marine brines (Canadian Shield), 199
 Cone Mine, Yellowknife, Northwest Territories, 159
 East Bull Lake aquifer, 189
 Milk River aquifer, 44, 145, 105, 169, 190, 199, 203
 Precambrian Shield aquifer, 186
 Strugeon Falls aquifer, 55–56, 186
Capture reaction, 140, 173, 176
Capture zones, 208, 219, 249–251
Carbonate/carbonate materials, 151–152, 155, 173, 259
Carbonate rock aquifers, 144
Carbon cycle, 146
Carbon dioxide (CO_2), 105–106, 146, 149, 151–152, 183

INDEX 307

Carbon-14 (^{14}C):
  advantages of age-dating, 154, 163
  age-dating by, 26, 28, 35, 74, 82, 89, 149–154
  characterized, 4, 30, 53, 146
  contaminant transport rate, 62–63
  dating case study, 155–156
  disadvantages of age-dating, 154, 163
  geochemical reactions, 151
  groundwater analysis, 148–149, 154
  groundwater flow velocity research, 59
  isochrone map, 18
  mixing trends, 66
  production of, 16, 146–148
  reactive ions, 161–162
  results report, 148–149
  sampling, 148–149, 154
  seawater-level fluctuations, 78
  sources of, 146–147
Carboniferous period, 4–5
Carbon monoxide (CO), 146
Carbon-13 (^{13}C), 146
Carbon-12 (^{12}C), 146
Carrera, J., 35, 37
Carrizo Sand Aquifer (TX), 59–60, 203
Caspian Sea fluctuations, 77
Castro, M. C., 37
$^{13}C/^{12}C$, 259
CEDEX, laboatory of isotope hydrology, 33
Cenozoic-Quaternary era, 5
Cenozoic-Tertiary era, 5
CFC-11, 112
CFC-12/CFC-113, 111
CFC-12/CFC-11, 111
CF-113, 123
CFC-113, 112
CH_4, 105
Chemical age formula, 163
Chemical engineering, 210
Chemical erosion, 199
Chemical Geology, 44
Chemical timescale, 257
Chemical waste repositories, 211
Chernobyl accident, 16, 191–193
Chesterville Branch watershed (MD), 70–71
Chloramines, 170
Chlorates, 170

Chloride/chlorine ($^{36}Cl/Cl$):
  age-dating case studies, 125–129
  atmospheric, 124
  characterized, 30, 81, 83, 124–125, 132, 170, 174, 180–181
  recharge rate, 125, 128
Chloride (Cl), movement rate, 177
Chloride-36 (^{36}Cl):
  age-dating very old groundwaters:
    advantages of, 177, 202
    atmospheric, 175
    case studies, 176–181
    characterized, 170–171
    disadvantages of, 177, 202
    groundwater analysis, 173–174, 177
    groundwater dating by, 169, 174–176
    ingrowth, 176
    production of, 171–173
    results report, 173–174
    sampling, 173–174, 177
    sources of, 172, 177
  characterized, 30, 35
  production of, 16
Chlorofluorocarbons (CFCs):
  advantages of age-dating, 113
  age-dating techniques, 21, 35, 74, 105–106, 108–109
  air temperature, 110–111
  analyzing, 108, 111, 113
  atmospheric, 107, 110–111, 113
  case studies, 113
  characterized, 30, 132, 156, 255, 106–108
  concentration from watersheds, 116
  contamination, 114
  dating case studies, 113
  dating error sources, 111–113
  degradation, 112–113
  desorption, 112
  disadvantages of age-dating, 113
  environmental issues, 107
  excess air, 100, 110, 113–114
  flow rate research, 64
  groundwater contamination, 35
  hydrodynamic dispersion, 115
  microbial degradation, 112, 114
  mixing trends, 66–67, 115
  modification processes, 114–115
  nitrate contamination research, 70

production of, 16
recharge elevation, 114
recharge temperature, 110–111, 113–114
sampling, 108–109, 111, 113
solubility of, 110
sorption, 112, 114
unsaturated zone, 114
urban air, 114
Chlorofluorocarbon Laboratory, 31
Cigar Lake (Canada), 176–177
Clare Valley (Australia), 58, 117
Classical age modeling, 206
Clay, 55, 75
Claystones, 190
Cl/Br ratio, 163
Climate, significance of, 67–68, 129
Clock tracer, 15
Coastal Plain shallow regional aquifer, 74
Coastal Ranges (CA), 160
Colorado Front Range (CO), 258
Communications skills, significance of, 53–54
Computer software applications, Tracermodel1, 65
Conductivity time series, graphical analysis of, 258
Conduit flow, 65
Cone Mine, Yellowknife, Northwest Territories (Canada), 159
Conferences/forums, 263
Confined aquifers, 145, 157–159, 169, 180–181
Connate water, 10
Conservative ions:
  case study of, 161–164
  characterized, 160–161
Contaminants:
  concentration of, 69–70
  human-induced, 69
  transport, 210
Contamination:
  event, 217
  implications of, 54
  plume, 71–72, 125
  prevention programs, 73
Cook, P. G., 35, 37
Co-precipitation, 194

Cornaton, F., 36
Cosmic radiation, 144, 146
Cosmic rays, 171
Cosmic-ray spallation, 192, 257
Cosmogenic, generally:
  production rates, 90, 196
  reactions, 191
Cretaceous period, 5, 169
Cross-formational flow, 23, 175
CRPG, 32
Crustal helium, 182–183, 185
Crystal lattices, 99, 183, 200
Crystalline bedrock aquifer (Sweden), 116–117
Crystalline metamorphic bedrock, 75
CSIRO (Commonwealth Scientific and Industrial Research Organization), 108

Dakhla Basin, Sahara Desert (Egypt), 170
Dam construction, 59
Darcy Lecturer, 31, 35
Darcy velocity, 223, 228
Darcy's law, 9, 19, 21, 78, 233, 235
Date, defined, 2
Dating and Duration of Fluid Flow and Fluid-Rock Interaction, 39
Dating tracer, *Tracers*
Davis, S. N., 35, 37
Dead chlorine, 175
Dead groundwaters, 10–11
Decay:
  age, 19, 21
  first-order kinetic rate law, 150
  principles, 195, 198
De-dolomitization, 153
Deep aquifers, 74, 156–157, 201
Deep marine brines (Canadian Shield), 199
Deep natural gas reservoir, 97
Deep-sea sediments, dating techniques, 4
Deforestation, impact of, 70
Degassing process, 101, 105, 141, 152, 183
Degradation, 8, 69, 112–114, 151, 153
Deionization, 194
Delmarva Peninsula, 97, 105–106
Deltaic sediments, 97
Deming area aquifer (NM), 92

INDEX

Denitrification, 71
Deposition, 183
Desorption, 112
Deuterium, 30, 84, 163
Devonian period, 5, 139
Diatomic gas, 84
Diffuse flow, 65
Diffusion, 15, 63, 84, 99, 127, 151–152, 175, 183, 204, 209
Diffusion coefficient, 23, 96
Diffusion-dispersion tensor, hydrodynamic, 240
Dilution, 175, 207, 252–253
Dilwyn aquifer, 77
Dinitrification, 69, 71
Dirac function, 241, 243
Direct discharge, 8
Direct measurement, ^{39}Ar, 141
Discharge area, 13–14, 22. *See also specific aquifers*
Discrete numerical modeling, 233
Disequilibrium series, 30. *See also Uranium disequilibrium series (UDS)*
Dispersion, 12, 15, 23, 84, 90, 96, 105, 127, 204, 207, 209, 248
Dispersion-diffusion effects, 243
Dispersive mixing, 206
Dispersive model (DM), 207
Dispersivity, 171
Dissolution, 152–153
Dissolved inorganic content (DIC), 62–63
Dissolved organic matter, 194
Distillation, 141, 194
Divergence theorem, 240
DNAPL, 70
Dockum Group sandstone (TX/NM), 158–159
Dogger aquifer (France), 187–188
Dolomite, 153, 190
Downgradient, 22, 201
Downhole Bennett pumps, 108
Downstream, water quality of, 69
DRASTIC, 73–74
Drinking water:
    contamination plume, 71
    pollution evaluation, 68
Drug industry, 84

Dry land salinity, 51, 80, 113, 115, 155
Dumfries sandstone and breccia aquifer (Scotland), 67, 69, 123
Dupuit assumptions, 229, 231

Earth:
    age of, 3–4, 28
    crust, 181
    internal heat, 199
    origin of, 3
    sciences, 16
    temperature of, 11
East Bull Lake aquifer (Canada), 189
Eastern Clear Lake (CA), 199
Eastern Europe, age-dating research:
    Antalya Plateau (Turkey), 97
    Lake Bled (Slovenia), 16
Eastern River Plain (ESRP) aquifer (ID), 67
East Kalahari (Botswana), 97
East Midland Triassic sandstone aquifer (United Kingdom), 128, 161–164
Ecosystem health, 77
Edwards and Trinity Group, 158
Elandsfontein (Witwatersrand, South Africa), 58
Electrical conductivity (EC), 8, 258
Electrolysis, 170
Electron capture detector (ECD), 108
Elevation contours, 19
Endemic goiter, 191
End-member equation, 65–67
Entropy, 209
Environmental geochemistry, 191
Environmental geological maps, 260
Environmental isotopes, 32, 68
Environmental Isotopes in Hydrogeology (Clarke/Fritz), 38–39
Environmental tracer data, 84, 206
Environmental Tracers in Subsurface Hydrology (Cook/Herczeg), 38
Eocene period, 5
Equilibrium, 2, 76, 85, 200
Equivalent atmospheric concentration (EAC), 110
Eriksson, E., 36
Erosion baseline, 77
Estonia, age-dating using ^{32}Si, 139
Ethyl chloride, 106

Europe, age-dating research:
  Aquitaine basin (France), 59
  Bocholt aquifer (Germany), 145
  Bunter sandstone aquifer (Germany), 145, 203
  Dogger aquifer (France), 187–188
  Fontainebleau sands aquifer, Paris Basin (France), 145
  Infra-Molassic Sands aquifer (France), 18, 59
  Innviertel, Molasse Basin aquifer (Austria), 186
  Oberrheingraben aquifer (Germany), 105
  Paris Basin (France), 145, 185–188
  Tertiary Basalt aquifer (Germany), 123
  Upper Limestone aquifer (Germany), 123
Evaporation process, 176
Evaporative discharge, 8
Evapotranspiration, 124–125
Everglades National Park aquifers (FL), 61, 63, 97
Excess air, 96–97, 100, 105, 110–114, 119–120, 182, 257
Exponential distributions, 207, 231
Exponential mixing, 206
Exponential model (EM), 207
Eyre Peninsula (South Australia), 117

Fabryka-Martin, J. T., 37
Fault systems, 205
F/Br ratio, 160
F/Cl ratio, 163
Fehn, U., 37
Fertilizers, 70–72, 253
Fick's law, 240
Filtering, 23
Final discharge point, 71
Finite difference, 233–234
Finite element techniques, 233
First moment, 14, 209, 220, 225, 238, 240
First-order kinetics, 201
Fission, 4, 84, 93, 100, 192
Flat Top Wilderness region (CO), 258
Flow models, parameter constraints, 51, 64–65

Flow net analysis, 83
Flow paths, 51, 59–61, 156, 204–205, 211, 235, 237
Flow system, significance of, 129, 175
Flow velocities, 11, 22, 51, 55, 57–59, 161, 206, 222–224
Fluorocarbon, 107
Flushing time, 23–24
Fluvio-glacial deposits, 99
Flux rate, 24
Fokker-Planck equation, 210
Fontainebleau sands aquifer, Paris Basin (France), 145
Fontes, Jean-Charles, 37, 68
Food chains, 147, 151
Forward flow field, 209, 235–239, 242
Forward travel time probability, 210
Fossil fuels, 146, 149, 151
Fossil water, 53
Fractional capture zone, 219, 249–250
Fractured rock aquifers, 23, 51, 58, 65, 74, 97, 134, 143
Fractured water, 112
Fracture porosity, 205
Freon, 11, 107
Frequency distribution, 205, 213, 217
Freshwater aquifers, 168
Fuel reprocessing plants, 193
Fuel stations, 72
Fusion devices, 171

Gabbro-anorthosite, 186
Gainsborough aquifer (United Kingdom), 189
Gambier Embayment (Australia), 77
Gambier Limestone aquifer (Australia), 127
Ganges-Brahmaputra Flood Plain (BENGAL basin), age-dating case study, 97–99
Gas chromatography, 108, 141
Gas hydrates, 77, 199
Geochemical conditions, 202
Geochemical processes, 77
Geochemical reactions, 160–161
Geochemical reservoirs, 192
Geochemistry, 16, 161, 263
Geochron Laboratories, 34

INDEX   311

Geochronology, *see*
   Hydrogeochronology
   absolute age, 3–4
   geological time table, 4–6
   relative age, 3–4
Geological maps, 260
Geological Society of America (GSA), 31, 35
Geological time table, 4–6
Geologic history dating methodology, 258–259
Geo-microbial processes, 77
Geothermal system research, 171
Geothermal waters, 151
Germany, age-dating using ^{32}Si, 138–139
Glacial age, 175
Glacial-interglacial cycles, 157–158
Glaciers:
   mean residence time, 15
   residence range time in hydrologic cycle, 15
Glatt Valley aquifer (Switzerland), 145
Global components:
   aging process, 223
   climate system, 77
   distributions, 244
   equilibrium tritium inventory, 85
   internal age distribution, 240
   residence-time distribution, 219–221
Goode, D. J., 36
Granatic Leuggern groundwater (Switzerland), 145
Granite, 186
Granitic Stripa groundwater (Sweden), 145
Great Artesian Basin (GAB) (Australia), 168–170, 178–180, 186, 190, 198–199, 258
Great Hungarian Plain, 157
Great Miami buried-valley aquifer (OH), 64
Groundwater, *see* Groundwater age
   chronology, 16
   flow rate, 18, 73
   hydrodynamics, 23
   hydrologic cycle, 1, 10
   hydrology, 16
   migration rate, 73
   mining, 54
   pollution, 51
   residence time, *see* Groundwater residence time; Residence-time distributions
   seepage, 118
   volume of world resources, 10
Groundwater age:
   age mass, 11–12
   data applications:
      age monitoring, prevention of overexploitation and contamination of aquifers, 54
      communications tools, 53–54
      flow models, parameter constraints, 51, 64–65
      flow path identification, 51, 59–61
      flow velocity calculation, 51, 58–59
      mixing, identification between end members, 51–52, 65–67
      pollution evaluation, 68–71
      radioactive waste disposal facilities, performance assessments, 75–76
      rate assessment of groundwater and contaminant transport, 61–63
      recharge rate estimation, 55–58
      research study, pre-Holocene climate, 67–68
      reservoir renewability, 51–53
      shallow aquifers, vulnerability maps, 73–75
      site-specific, 51, 76–80
      transport model, parameter constraints, 51, 64–65
      travel time of plume, calculation of, 51, 71–73
   data interpretation, 35
   dating techniques, 4, 15
   defined, 6
   dispersion of, 12
   distribution of, 12–14
   mean, 12–14
   modeling studies, 36
   old groundwater, 9–10, 134–164
   residence time, 7–9
   science, *see* Groundwater dating science
   transport of, 12
   very old groundwater, 9–10, 165–204

worldwide maps, 260–263
young groundwater, 9–10, 81–133
Groundwater/contaminant flow rate, 73
Groundwater dating science:
　characterized, 16–17, 19–20
　enhancement strategies, 263–264
　future directions for, 264–265
　major problems in, 264
　methodologies:
　　need for, 254–255
　　proposed, 254–259
　proposed groundwater age map, 260, 262–263
　simulation applications, 259–260
　worldwide practices, 260–262
Groundwater Hydrology (Todd), 28
Groundwater residence time:
　average, in hydrologic cycle, 14–15
　characterized, 6–7
　defined, 23
　life expectancy and, 17–18
　time, reservoir theory and, 24
Groundwater Tracers (Davis/Murphy), 35
Guidebook on the Use of Chlorofluorocarbons in Hydrology (Plummer et al.), 40
Gulf Coast brines (LA), 199
Gypsum, 156, 190

^{1}H, 84
^{2}H, 84
Half-life:
　^{39}Ar, 144
　^{40}Ar, 189
　^{14}C, 149, 152
　^{36}Cl, 170, 173, 176–177
　^{129}I, 191, 196–197
　^{81}Kr, 169
　^{85}Kr, 100, 103
　^{226}Ra, 255
　^{222}Rn, 255–256
　^{35}S, 257
　^{32}Si, 139
　of tritium, 84, 103
　^{234}U, 200
　uranium, 199, 201
Handbook of Environmental Isotope Geochemistry, 38, 68

^{4}He/^{40}Ar, 183, 190–191
Heavy isotope waters, 159
Helium-4 (^{4}He):
　advantages in age-dating, 185, 202
　age-dating by, 75–76, 81, 92, 95, 97, 99, 132, 169 184–185, 255
　atmospheric, 182
　characterized, 30, 35, 181
　dating case studies, 185–188
　disadvantages in age-dating, 185, 202
　excess air, 182
　groundwater analysis, 183–184
　production of, 181–183, 187
　results report, 183–184
　sampling, 183–185
　sources of, 181–183
Helium-ingrowth analysis, 87
Helium-3 (^{3}H), 30, 78, 87, 89
Henry Darcy Distinguished Lecture Series, 31, 35
^{3}He/^{4}He, sources of, 182
^{3}H/^{3}He:
　advantages of age-dating, 96
　analysis, 94
　atmospheric, 93, 95
　characterized, 92, 96
　dating case studies, 97–99
　dating groundwater by, 94–96, 132
　dating research, 56–58, 61, 74, 91
　disadvantages of age-dating, 96–97
　excess air impact, 96–97
　mantle, 93, 95
　nucleogenic, 93, 95
　pollution evaluation, 68
　recharge temperature, 96
　results report, 94
　sampling, 94, 96
　sources of, 93
　travel time research, 64, 73, 97
　tritium decay, 93–94
　tritiogenic, 96
High Plains aquifer, 158–159
Historical perspectives:
　aquifers, in extensive dating studies, 50
　dating publications, 36–37
　earliest publications, 25–31
　historical tracers, 15
　important publications, 37–50

major contributors, 31–36
worldwide laboratories, 31
Holmes, Arthur, 4, 6
Holocene era, 5, 157–158, 160, 257
Horizontal age gradient, 11
Horizontal aquifer, 208, 217
Horizontal flow path, 61
Horizontal flow velocity, 51, 55, 58–59, 224, 229–230
Horizontal-radial flow, 231
Housing development, 54
Human population, residence-time distribution analogy, 221
Hunt, A. G., 36
Hutton, James, 3
Hydraulic age, 21, 29–30, 67, 72–73, 77–78, 83
Hydraulic conditions, implications of, 64
Hydraulic conductivity, 19, 64, 76–77, 161
Hydraulic gradients, 9, 19, 21, 77, 223
Hydrocarbons, 183, 197
Hydrochlorofluorocarbons (HCFCs), 16, 107, 259
Hydrochronology, 15
Hydro-climatological conditions, 75
Hydrodispersive flow fields, 242–243
Hydrodispersive processes/systems, 211, 219, 227
Hydrodynamic dispersion, 12
Hydrodynamics, 23, 59, 62, 115, 240
Hydrogen, characterized, 84
Hydrogen peroxide, 194
Hydrogeochemical studies, 156, 201
Hydrogeochronology, 15–17
Hydrogeological configurations, 206
Hydrogeological maps, 260
Hydrogeological system, 233
Hydrogeological time table, 6
Hydrogeologists, functions of, 65
Hydrogeology, 3, 8, 16–17, 53, 77, 83, 254
Hydrogeology Journal, 265
Hydrographs, 61, 80
Hydroisotope GMBH, 32
Hydrological methods, classical, 60
Hydrologic cycle, 1, 10, 15, 75, 88
Hydrologic time, 15–17
Hydrologists, functions of, 54

Hydrology, historical perspectives, 107
Hydrothermal fluids, 199
Hypogenic ^{14}C, 148

I/Cl ratio, 160
IAEA, 52, 65, 68, 265
Ice:
 caps, 15
 cores, Argon-39 age-dating, 144
 dating of, 171
Idaho National Engineering and Environmental Laboratory (INEEL), 198
Igneous rock, 163, 191, 193
$^{129}I/I$, 191, 196
Illites, 258
Imnavit Creek (AK), 257
Inactive groundwater zone, 11
India, age-dating studies, 32, 139
Indirect methodology in age-dating:
 old groundwaters:
  advantages of, 132
  case study, 131, 133
  disadvantages of, 132
  stable water isotopes, 129–132
 very old groundwaters:
  advantages of, 163
  case study, 161–164
  deuterium, 157–160
  disadvantages of, 163
  ions, conservative and reactive, 157, 160–164
  overview of, 157
  oxygen-18, 157–160
Industrial development, 54
Industrial sites, wastewater discharge, 73
Inert elements, 30
Infiltration rate, 205–206, 249, 252–253
Infra-Molassic Sands aquifer (France), 18, 59
INGS, 138
Initial value, 265
Inlets, residence-time distributions, 219
Innviertel, Molasse Basin aquifer (Austria), 186
In situ reactions, 191
Institute of Geological and Nuclear Sciences, 33

Interdisciplinary groundwater age science, 15–17
Intermediate aquifer, recharge rate calculation, 58
Internal age, 238
International Association of Tracer Hydrology, 16
International Union of Pure and Applied Chemistry, 146
Iodine, see Iodine-129
  deficiency disorder, 191
  geochemistry of, 196
Iodine-129 (^{129}I):
  advantages of dating technique, 196, 202
  age-dating by, 195–196
  anthropogenic, 257
  atmospheric, 192
  characterized, 30, 191–192
  dating case studies, 197–199
  disadvantages of dating technique, 196, 202
  groundwater analysis, 194–195
  production of, 192–193, 195–196
  results reports, 194–195
  sampling, 194–195
  source of, 193
Ion chromatography, 174
Ionium, dating techniques, 4
Ion selective electrode, 174
Irradiation, 86, 140, 144, 146
Island of Funen (Denmark), 97, 105, 124
Isobutane, 106
Isochrone maps, 18–19, 71
Isolines, 209
Isotech Labs, 33
Isotope(s), see Isotopic; *specific isotopes*
  geology, 144
  half-life of, 83
  hydrology, 54, 263
  laboratories, 31–32
Isotope Tracers in Catchment Hydrology (Kendall/McDonnell), 39
Isotopic:
  age, 19, 21, 77, 216
  dating, 16
  decay principles, 21
  exchange, 78, 151, 153
  measurements, 14

IsoTrace, 32
Iwatsuki, T., 37

Johnson, T. M., 37
Journals/journal papers, as information resource, 29, 43–45
Jurassic era, 5, 186, 258

^{39}K, 143
Kalahari aquifer (Botswana), 53, 57–58
Kalin, R., 37
Kamen, Martin, 146
Kamensky, I. L., 36
Kaolinite, 156
Karstic aquifers, 145
Karstic limestone, 138
Karstic Springs (FL), 97, 124
Katz, B. G., 36
Keuper claystones, 190
K-feldspar, 161
Kirkwood-Cohansey aquifer system (NJ), 91, 97
Klaproth, Martin, 198
Kobe Area (Japan), 186–187
Kolmogorov equation, 210
Krueger Enterprises, 34
Krypton-81 (^{81}Kr):
  advantages of age-dating, 168, 202
  age-dating by, 167–168
  characterized, 30, 100
  dating case studies, 169–170
  disadvantages of age-dating, 168, 202
  groundwater analysis, 166–167
  production of, 166
  results report, 166–167
  sampling, 166–167
Krypton-85 (^{85}Kr):
  advantages of age-dating, 105
  age-dating by, 30, 103–104, 132, 255, 263
  analysis, 101–103
  atmospheric, 99–101, 103
  case studies, 105–106
  characterized, 99–100, 105
  dating case studies, 105–106
  disadvantages of age-dating, 105
  excess air, 105
  flow path research, 59–60
  low-level counting, 102

INDEX

manmade, 100
production of, 16, 100–101, 103
radioactive decay, 103–104
recharge temperature, 105
sampling, 101–105
transport of, 105
Kurtosis, 209

Laboratories, *see specific laboratories*
establishment of, 263, 265
financial issues, 264
Lagoon systems, wastewater discharge in, 73
Lake Agassiz, 67
Lake Bled (Slovenia), 16
Lakes:
artificial, 77
^{36}Cl levels in, 126
heavy isotopes in, 160
residence time distribution, 24
Lamont-Doherty Earth Observatory, 87
Landfills, 84, 87
Land-use practices, implications of, 72
Laplace transform, 242–243
Large-scale flow, estimation of, 64–65
Laser resonance ionization mass spectrometry (RIMS), 102–103, 167
Laurentide Ice Sheet, 159
Lawrence Radiation Laboratory, University of California-Berkeley, 146
LDEO, Columbia University, 34
Leachate/leasing, 75, 84, 87, 195
Lead contamination, 72
Lehmann, B. E., 35–36
Leibniz-Labor Laboratory, 32
Leucogranite, 186
Libby, Willard Frank, 25–26, 37, 149
Life expectancy:
backward, 236
distributions, 221, 238
of humans, 10
implications of, 17–18, 208–209, 244
maps, 18–19
residence-time distributions, 230–231, 239
Lifetime, defined, 1–2
Limestone, 138, 139, 161, 186, 190

Linear model (LM), 207
Linsental aquifer (Switzerland), 97
Liquid scintillation counting (LSC), 87–88
Lithium, 86, 161
Lithology, 186
Lithosphere, 192–193
Local average ages, 212
Lockyear, J. Norman, 181
Locust Grove groundwater system (MD), 64, 122–123
Long, A., 36
Longitudinal dispersivity coefficient, 77
Loosli, H. H., 35–36
Love, A. J., 36
Low-level counting (LLC), 166
Low protection aquifers, 73
Low-water conditions, 249, 251
Lumped-parameter models, 206
Lutry spring (Switzerland), 207

Macquarie Valley (Australia), 79
Magmatic water, 10
Magneto-optical trap, 167
Maloszewski, P., 35, 37
Mantle helium, 93, 95, 182–183, 185
Marine carbon/carbonate aquifers, 139
Marine water, 10
Marl, 161, 186
Mass balance, 152, 213, 233, 237
Mass spectrometers/mass spectrometry, 87, 138, 183, 189, 200
Mathematical modeling, 14, 206, 211
Matrix water, 112
Mazor, E., 36
Mean:
age data, 209, 222
life expectancy, 209
residence time, 1, 11, 13, 208–209, 215, 222
transit time, 23
Medical diagnostics, 84
Mediterranean seawater, 78
Meinzer, Oscar Edward, 35
Meinzer Award, 31, 35
Mesozoic era, 5
Metamorphic water, 10
Meteoric water, 186
Meteorites, 144, 192

Meteorology, 84
Methane, 77, 146, 152, 197
Methanogenesis, 151–152
Methyl chloride, 106
Methylene chloride, 106
Michigan State University, national superconducting cyclotron, 33
Microbes, 54
Microbial degradation, 112, 153
Microbiological processes, 77
Milk River aquifer (Canada), 44, 145, 105, 169, 190, 199, 203
Milky Way Galaxy, 146
Mineralization, 205
Mineralogical techniques, dating methodologies, 259
Mineral precipitation, 152–153
Mining exploration, 180
Miocene Ogallala Formations, 158
Miocene period, 5, 186, 188
Mixing:
    age transport, 207
    characterized, 15, 23, 59, 96, 206, 211
    defined, 12–13
    identification between end members, 51–52, 65–67
    problems with, 112–113, 115, 265
    residence-time distributions, 219, 234–235, 248, 252
Model age, 23
Modeling, benefits of, 83. *See also* Mathematical modeling
Modica, E., 36
Morgan Creek watershed (MD), 70–71
Mudstone, 161
Multilayer aquifer system, 205
Multimodal age distributions, 205, 209
Multiple environmental tracers, 68
Multitracer dating, 83
Münnich, K. O., 26, 28, 35, 37
Muon capture reaction, 173
Murray Basin (Australia), 180
Murray Group Limestone, ^{36}Cl production, 173

NAGRA hydrogeologic study, 76, 145
Nankai Hydrate Field (Japan), 197–199
NAPL contamination, 255

National Institute for Environmental Studies, Environmental Chemistry Division, 32
National Ocean Sciences AMS Facility, 34
Natural gas fields, 181
Natural purification, 72
Nature, 44
Nazar Begiry, 3
^{21}Ne, 165, 259
Near-surface contaminating sources, 74
Neon, 111
NERC-SUERK, cosmogenic isotope laboratory, 33
NETPATH, 82, 153
Neumann condition, 241
Neutron(s):
    activation analysis, 100, 194
    capture reaction, 140, 173, 176
    production rate, 143
    radiation, 86
New Mexico aquifers, 91–92
New Zealand aquifers, 139
NIH (India), 32
Nile River (Egypt), 53
Nitrate(s), 69–70, 72, 253
Nitrogen, 84, 108–109, 146, 148, 183
Noble gas, 35, 111, 183, 256–257, 263
Noble Gas laboratory, University of Utah, 189
Nonrenewable groundwater resources, 53
North China Plain, 157
Northern Hemisphere, 85, 89–90, 104, 120
Nubian Sandstone Aquifer (Egypt), 59, 63, 170
Nuclear fuel(s):
    re-processing facilities, 192, 198
    waste disposal repositories, 145, 192, 211
Nuclear industry, 84
Nuclear physics techniques, 148
Nuclear reactors, 86, 100
Nuclear weapons testing, 100, 140, 146–148, 179, 192
Nucleogenic helium, 93, 95

Oberrheingraben aquifer (Germany), 105
Ocean(s):
　argon content, 144
　hydrologic cycles, 1, 15, 88
　iodine content, 191, 196
　krypton content, 166
　mean residence time, 15
　residence range time, 15, 24
　tritium content, 88
　uranium content, 200
Oceanography, 84
Off-gases, 86
^{18}O-^2H, 30
Oil field brines(TX/LA/Gulf of Mexico), 199
Old groundwaters:
　age-dating:
　　argon-39, 134, 139–146, 163
　　carbon-14, 134, 146–156, 163
　　characterized, 134–135, 163
　　indirect methods, 157–164
　　silicon-32, 134–139, 163
　pollution evaluation, 69
　vulnerability maps, 74
Old regional background water, 67
Oligocene period, 5
Oman, 68
Opalinus Clay, Mont Terri (Switzerland), 188, 190
Orange County aquifer (NY), 257
Ordovician period, 5, 186
Osmosis, 78
Oster, H., 36
Otway Basin (Australia), 77
Outcrop, flow velocity research, 59–60
Outlets, residence-time distributions, 219–221
Overdevelopment, 54
Overexploitation, 54
Oxidation, 87, 129, 151–152, 157, 200
Oxide, 259
Oxidization, 194
Oxygen-18, 30, 67, 129, 131

Paleocene period, 5
Paleochannel groundwater (Australia), 180
Paleotemperature, 257
Paleowaters, 1239
Paleozoic era, 5
Palo Duro Basin, 190
Paris Basin (France), 185–188
Particle-tracking techniques, 207–208
^{206}Pb, 199
Pearson, F. J., 36
Percent modern Carbon (pmC), 149
Perchlorates, 170
Percolating water, 73–74, 108
Permeability, 157, 223, 225
Permian period, 4–5, 161
Permian sandstone aquifer (Scotland), 123
Phillips, F. M., 35–36
Photon burst mass spectrometry (PBMS), 167
Photosynthesis, 147, 151
Physics, dating methodologies, 16
Physics Institute, University of Bern, 141
Piezometer/piezometer nests, 55–58, 90, 115, 125
PINSTECH, Institute of Nuclear Science and Technology, 33
Piston-exponential mixing, 206
Piston-flow:
　age, 22–23
　model (PM), 206–207, 224
　residence-time distributions, 216, 223–226
　theory, 175
Pleistocene era, 5, 52, 67, 77, 157–158, 160, 186, 257
Pleistocene-Holocene era, 78
Plume, see Contamination plume
　groundwater flow path research, 59
　travel time calculation, 51, 71–73
Plummer, L. N., 35, 37
Plutonium, 100
Pollutants/pollution:
　aquifer contamination, 54
　evaluation, 68–71
　protection strategies, see Contamination prevention programs
Pollution-free water, groundwater age-data, 69
Polonium-218, 255

318 INDEX

Pore velocity, residence-time distributions, 226–227, 232–234
Pore volume, residence-time distributions, 212–213
Pore water, dating techniques, 62, 190–191, 195, 197
Pore Waters (Switzerland), 188
Poreda, R. J., 37
Porosity, 11, 19, 59, 202, 205, 207, 223, 229–230, 233, 237. *See also* Pore velocity; Pore volume; Pore water
Porous media, 210
Potassium-Argon dating, 4
Precambrian crystalline basement, 99
Precambrian era, 5, 186
Precambrian Shield aquifer (Canada), 186
Precipitation:
 atmospheric, 175, 181–182
 isotopic concentration in, 157–158
 noble gas solubility, 256
 as reservoir recharge, 52
 water isotopes and, 129–130
Pre-Holocene era, 67–68
Prevention strategies, *see* Contamination prevention programs
PRIME Lab, Purdue University, 33, 174, 194
Probabilistic capture zone, 219
Protection zones, 211, 249
Proterozoic period, 5
Protium, 84
Proton-emission reaction, 140
Publications:
 books, 38–40, 263–264
 journals, 43–45
 reports, 45–50
 theses, Ph.D. and M.Sc., 40–43
Pullman-Moscow Basin (WA), 159
Pumping rate, 54
Pumping tests, 75, 83
Pumping well, 208, 249, 251, 253
Purge-and-trap gas chromatography, 108
Purification, types of, 72
Purtschert, R., 37

Quartz, 161, 259
Quartzite, 186
Quaternary multilayer aquifer, 105

Quaternary period, 75, 77, 157, 248, 259
Quaternary sand aquifer, 105, 124
Queens University, Belfast (Ireland), 33

Radioactive and Stable Isotope Geology (Attendorn/Bowen), 40
Radioactive decay:
 characterized, 15–16, 63, 201
 ^{14}C levels, 151
 ^{36}Cl production, 174
 krypton-85, 103–104
 principle, 4
 of uranium, 199
Radioactive equilibrium, 200
Radioactive isotopes, 140
Radioactive tracers, 15
Radioactive waste disposal, 192, 198
Radioactive waste disposal facilities, 75–76, 257
Radioactivity counting, 144
Radiocarbon, 32, 44, 58, 146–147, 149, 159, 161, 163
Radiocarbon After Four Decades. An Interdisciplinary Perspective (Taylor/Kra), 40
Radiocarbon Dating (Libby), 25–31
Radiogenic helium, 182, 184–185
Radiogenic Isotope Geology (Dickin), 40
Radiometric age, 19, 21, 209
Radionucleotides, 28–29
Radionuclides, 76, 143, 191
Radiotracers, 89
Radium-226, 255
Radon-222 (^{222}Rn), 83, 87, 255–256
Rainfall:
 $^{36}Cl/Cl$ levels, 125
 iodine-129 content, 192
 isotopic content, 129–131
 measurement of, 51–52
 quality of, 160
 recharge, 71
 tritium concentration, 85–86, 89, 92
Rain water, *see* Rainfall
 ^{39}Ar content, 142
 characterized, 81
 ^{14}C levels in, 150, 154
 ^{36}Cl levels in, 172, 175
 CFC dating methods, 111
 $^3H/^3He$ age-dating, 99

INDEX

infiltration of, 112 i
^{129}I content, 192
isotope content, 130
percolation of, 108
Ramsay, William, Sir, 99, 139–140
Random walk, 210
Rangeland water, 67
Rayleigh, Lord, 139
^{85}Rb, 103
Reactive ions:
  case study of, 161–164
  characterized, 160–161
Rebirth, groundwater molecules, 8
Recharge:
  rate, 11, 21, 24, 55–58, 74, 206–207, 212, 252
  temperature, 96, 105, 110–111, 113–114
  zone, 14, 54, 201
Redox front, 201
Refrigerants, 106–107
Regeneration, groundwater molecules, 8
Regional Ground-Water Quality (Alley), 39
Regolith, 180
Regression analysis, 163–164
Relative age, 3–4, 82
Renewability, 51–53
Renewable reservoirs, 52–53
Reports, as information resource, 45–50
Reservoir:
  mean residence time, 15
  renewability, 51–53
  theory, *see* Reservoir theory
  volume, isotopic content, 131
Reservoir theory:
  aquifer-aquitard systems, 222–227
  characteristic times, 214–217
  characteristic volumes, 217–219
  components of, 14, 24, 45, 210–212
  human population analogies, 221
  multiple inlets and outlets, 219–221
  relation between internal age and residence times, 212–214
  transport processes, 238
Residence time:
  of ^{36}Cl, 171
  ^{36}Cl/Cl case study, 128

distribution, *see* Residence-time distributions
  of iodine, 191, 196
  nitrate contamination research, 70
  reactive ions, 161
  relation between internal ages and, 212–214
  in reservoirs, 24
  significance of, 129
  water isotopes, 129–131
Residence-time distributions:
  analytical modeling of, 227–228
  groundwater age transport basics, 211–244
  implications of, 204–206
  modeling studies, 36
  overview, 206–211
  simulation studies, 36
  state-of-the-art, 206–211
  typical examples, 244–253
Residence-time probability density, 213
Resource management, 54
Retardation factor, 201, 224, 226
Retford aquifer (United Kingdom), 189
Reverse flow-field transport equations, 219
Rhone river basin (Switzerland), 58
Rhyolite, 173
Rietholzbach catchment (Switzerland), 105
River(s):
  ^{36}Cl levels in, 126
  hydrologic cycle, 1, 15
  ^{129}I content, 199
  mean residence time, 15
  residence range time, 15
River channels, 180
River Glatt Valley (Switzerland), 256
River Töss (Switzerland), 97
Rocks, absolute age determination, 4.
  See also specific types of rocks
Rocky Mountains, 181
Rotliegend sandstone reservoirs, 258
Ruben, Sam, 146
Rubidium, 161
Rubidium-Strontium dating, 4

Sagehen Creek basin (CA), 80, 97
Salinas Valley (CA), 72

Saline waters/salinity, 156–157, 171, 177
Salinization, 70, 80
Salt Lake City (UT), contamination research, 75
Salt water methane, 97
San Joaquin Valley (CA), 159–160
Sandstone, 173, 186, 190. *See also specific sandstone aquifers*
Sanford, W. E., 37
Saprolite, 75
Saturated zones, 15, 94, 256
Savanah River Plant (SRP), 75
Savannah River Nuclear fuel reprocessing plant, 193
Scandinavia, age-dating research:
  Crystalline bedrock aquifer (Sweden), 116–117
  Glatt Valley aquifer (Switzerland), 145
  Granatic Leuggern groundwater (Switzerland), 145
  Granitic Stripa groundwater (Sweden), 145
  Island of Funen (Denmark), 97, 105, 124
  Linsental aquifer (Switzerland), 97
  Lutry spring (Switzerland), 207
  Opalinus Clay, Mont Terri (Switzerland), 188, 190
  Pore Waters (Switzerland), 188
  Rietholzbach catchment (Switzerland), 105
  River Glatt Valley (Switzerland), 256
  River Töss (Switzerland), 97
  Stripa aquifer (Sweden), 171, 189
  Stripa granite (Sweden), 105, 190, 199, 203
  Thur Valley (Switzerland), 258
  Zurzach thermal spring (Switzerland), 145
Scheele, Carl Wilhelm, 170
Schlosser, P., 37
Science, 44
Sea level, 77
Seasonal amplitude, water isotopes, 130
Seawater:
  characterized, 10
  hydrogen isotopes in, 84
  intrusion, timescale calculation, 51, 78

iodine content, 191
level fluctuations, 51, 77–78
Second temporal moment, 209, 215
Sedimentary rock aquifers, 144, 190–191, 193
Sediments, dating of, 171
Seeland Phreatic aquifer, 248–253
Seepage pits, wastewater discharge in, 73
Selected Bibliography of Groundwater (van der Leeden), 28
Self-purification, 72
Semi-arid region, 156
Shale, 166, 186
Shallow aquifer(s):
  characterized, 113, 217
  ^{14}C levels, 156
  nitrate contamination, 70
  renewable reservoirs, 53
  travel times, 56
  vulnerability maps, 73–75
Shallow groundwater, 64, 68, 78, 157, 180, 183, 257
Shallow wells, 138
Shallowness, significance of, 129
Shenandoah National Park, Blue Ridge Mountains (VA), 74, 96, 258
Sherwood sandstone, 161–163
Shoreline fluctuation, 77
Shroud of Turin, 149
Sierra Nevada, 67, 159
Sign illumination, 84
Silurian marine carbonate aquifers, 139
Silicon-32 (^{32}Si):
  advantages of age-dating, 138, 163
  analysis of, 136–137
  characterized, 30, 135, 255
  dating case studies, 138–139
  dating groundwater by, 137–138
  disadvantages of age-dating, 138, 163
  production of, 135–136
  sampling, 136–137
Silt, 55
Silurian period, 5, 139
Silver nitrate titration, 174
Simulation applications, 64, 259–260
Site-specific groundwater age data:
  dryland salinity, 79–80
  hydrograph separation, 80

overview of, 51
seawater fluctuation identification, 77–78
seawater intrusion, timescale calculation, 78
waste disposal, 78
Situation map, 248–249
Skewed distributions, 205
Skewness, 14, 209
Slope stability maps, 260
Slow groundwater velocity, 11
Slow infiltration processes, 52
Slow-moving storage, 54
Slug tracers, 15
Smethie, W. M., 37
Smith, William, 3
Snake Creek Watershed (SCW), 113
Snake River Plain aquifer (ID), 198
Snow water, 108
Sodium chloride, 170
Soil:
  in hydrologic cycle, 15
  isotopic content, 131
  salinization, 78
  water, 88, 111, 131
  zone, 151
Solid-state diffusion, 99
Solomon, D. J., 36
Solomon, D. K., 35
Sonntag, C., 36
Sorbing tracer, 224
Sorption processes, 112, 139, 201
Southeast Asia, age-dating research:
  Ganges-Brahmaputra Flood Plain (BENGAL basin), 97–99
  NIH (India), 32
Southern Hemisphere, 85, 90, 120
Spallation, 100
Spatial distribution, 234, 236, 239
Spontaneous fission, 4, 173, 192–193
Spring systems, 128, 130–131
$^{34}S/^{32}S$, 259
Stagnant groundwater, 10, 77
Stampriet aquifer (South Africa), 189
Standard deviation, 215–216
Standard Mean Ocean Water (SMOW), 167
Statistical distributions, 214–215, 231–232, 239

Steady-state:
  divergence-free flow field, 240
  equations, 211
  flow systems, 211, 219
  groundwater flow, 206
  reservoir, 216
  volumetric flow rate, 23
Storage time, 23, 26
Storativity, 23
Stratigraphy principles, 3
Stream hydrochemistry, 80
Streamtube age, 22–23
Stripa aquifer (Sweden), 171, 189
Stripa granite (Sweden), 105, 190, 199, 203
Strontium/strontium isotopes, 161, 258
Strugeon Falls aquifer (Canada), 55–56, 186
Subreservoirs, 24
Subsurface environment, 6, 8–9, 15, 67, 88–89, 140, 142
Subsurface groundwater system, 175
Subsurface hydrology, 209
Sulfate reduction, 153
Sulfide, 259
Sulfur dioxide, 106
Sulfur-35 (^{35}S), 257–258
Sulfur-36 (^{36}S), 174
Sulfur/sulfur hexafluoride (SF_6):
  advantages of age-dating, 121
  analysis of, 119, 122
  atmospheric, 101, 120–121
  characterized, 81, 83, 89
  dating case studies, 122–124
  dating groundwater by, 30, 81, 83, 89, 118–121, 132, 255
  disadvantages of age-dating, 121–122
  excess air, 119–120
  production of, 16
  ratio to CFCs, 113
  recharge, 120
  sampling, 119, 122
Summer rain, 129
Suntop Watershed (STW), 113
Superficial Aquifer (Australia), 19–20
Surface and Ground Water, Weathering, and Soils (Drever), 39
Surface contaminant sources, 72
Surface water, 59, 71, 88, 199

Surface watershed studies, 24
Surficial aquifers, 97, 105
Swamps, hydrologic cycle, 15
Sweetgrass (MT), 169
Swiss Federal Institute of Science and Technology (EAWAG), 33
Szabo, Z., 37

Table Mountain Group sandstone (South Africa), 65–66
Tamers, M. A., 37
Temperature:
 paleotemperature, 257
 recharge rate, 96, 105, 110–111, 113–114
 significance of, 8, 67–68, 103
Temporal moment(s):
 analysis, 209
 equations, 14
Teriogenic helium, 94
Terrestrial environment, 191
Terrigenic helium, 183
Tertiary age, 195, 198
Tertiary Basalt aquifer (Germany), 123
Testmark Laboratories, 32
^{234}Th, 201
Thermal springs, 145
Thermoluminescence dating method, 4
Thermonuclear:
 bomb testing, 147–148
 reactions, 191
 tests/testing, 85, 87, 90, 149–150
 weapons, 84
Theses, as information resource, 40–43
Thorium, 86, 173, 182, 184
Thur Valley (Switzerland), 258
Tillits deposits, 99
Time series analysis, 131
Time table, geological, 4–6
Todd, D. K., 28
Tolstikhin, I. N., 36
Topography map, 18–19
Torgersen, T., 35, 37
Tracer(s):
 breakthroughs, 216
 characterized, 15, 205, 224
 ^{36}Cl, 127
 conservative, 161
 ^{129}I as, 191

^{81}Kr, 166
^{86}Kr, 100
reactive, 161
^{32}S, 138
^{35}S, 257–258
tests, 75
tritium as, 90
Tracermodell, 65
Transit time, 236
Transmissivity maps, 19, 260
Transport:
 large-scale, 64–65
 model, parameter constraints, 51, 64–65
 processes, 12, 22
 rate assessment of, 61–63
 velocity, 224
Travel time:
 density function, 210
 flow models, 64
 probabilities, 209–210
 residence-time distributions, 224
 significance of, 6, 23–24, 56–57, 78, 205
 water isotopes, 130
Travers, Morris William, 99
Triassic era, 5, 158–159, 186–188, 190, 203
Trichloroethylene (TCE) contamination, 72
Tritiated water (THO), 84
Tritium:
 advantages of age-dating, 90
 age-dating by, 28–29, 33, 83, 87–90, 108, 139, 255
 analysis, 52–53, 86–87
 atmospheric, 88–89
 bomb peak, 55
 bomb testing, 85–86
 characterized, 26, 45, 84, 103, 163
 dating case studies, 90–92
 decay of, 89, 93–94
 disadvantages of age-dating, 90
 flow rate research, 64
 isochrone map, 18
 mixing trends, 66
 nitrate contamination research, 70
 production of, 16, 84–86
 recharge rate calculation, 61
 results report, 87
 safety guidelines, 86

INDEX

sampling, 86–87, 90, 92
  water sample (TU), 84–85
Tritium-helium groundwater, 75
Troposphere, 140
Troy Creek Watershed (TCW), 113
Turbulence, 22
Turnover time, defined, 23
Two-component groundwater mixture, 11

$^{234}U/^{238}U$, 200–201, 203
^{238}U, 173
UFZ Center for Environmental Research, Department of Isotope Hydrology, 32
Uitenhage aquifer (South Africa), 189
Ultra-high purity nitrogen, 108–109
Unconfined aquifers, 21–22, 248–253
Underground Laboratory of Physics Institute, Univerity of Bern, 102
Underground radioactive waste, 75
Underground storage tanks, 72
United Arab Emirates, $^{36}Cl/Cl$ case study, 128
United Kingdom, age-dating research:
  Dumfries sandstone and breccia aquifer (Scotland), 67, 69, 123
  Gainsborough aquifer (United Kingdom), 189
  Permian sandstone aquifer (Scotland), 123
  Retford aquifer (United Kingdom), 189
United States, age-dating research:
  Alamogordo area aquifer (NM), 91–92
  Allequash Basin (WI), 61
  Artesia area aquifer (NM), 92
  Bluegrass Spring, Meramec River Basin (MI), 131, 133
  Carrizo Sand Aquifer (TX), 59–60, 203
  Coastal Ranges (CA), 160
  Colorado Front Range (CO), 258
  Deming area aquifer (NM), 92
  Dockum Group sandstone (TX/NM), 158–159
  Eastern Clear Lake (CA), 199
  Everglades National Park aquifers (FL), 61, 63, 97

Flat Top Wilderness region (CO), 258
Great Miami buried-valley aquifer (OH), 64
Gulf Coast brines (LA), 199
Imnavit Creek (AK), 257
Karstic Springs (FL), 97, 124
Kirkwood-Cohansey aquifer system (NJ), 91, 97
Locust Grove groundwater system (MD), 64, 122–123
Morgan Creek watershed (MD), 70–71
Oil field brines(TX/LA/Gulf of Mexico), 199
Orange County aquifer (NY), 257
Pullman-Moscow Basin (WA), 159
Sagehen Creek basin (CA), 80, 97
Salinas Valley (CA), 72
San Joaquin Valley (CA), 159–160
Shenandoah National Park, Blue Ridge Mountains (VA), 74, 96, 258
Snake River Plain aquifer (ID), 198
Sweetgrass (MT), 169
United States National Ground Water Research and Educational Foundation (NGWREF), 35
University College Dublin, Radiocarbon laboratory, 32
University College London, groundwater tracing unit, 33
University of Arizona, SAHRA, 34
University of Bern, Institute of Physics, 33
University of California, Lawrence Livermore National, 34
University of Chicago, Argonne National Laboratory, 34
University of Groningen, Center for Isotope Research, 33
University of Miami, RSMAS, 34
University of Nebraska, Water Sciences, 34
University of Perdue:
  Earth and Atmospheric Sciences, 33
  PRIME, 33
University of Tennessee, IRIM, 34
University of Tokyo, Earthquake chemistry, 32

University of Utah, Noble Gas laboratory, 34–35
University of Witwatersrand, Environmental isotope, 33
Unsaturated soil, 210
Unsaturated zones, 6, 22, 74, 77, 96–97, 105, 108, 111–112, 120–121, 129, 138, 256
Upper Floridan aquifer, carbon-14 dating methods, 153–154
Upper Limestone aquifer (Germany), 123
Uranium:
  characterized, 30, 86, 100, 148, 173, 182, 184, 192–193, 199
  disequilibrium series, *see* Uranium disequilibrium series (UDS)
  ore, 176
  oxide, 199
  salts, 199
Uranium disequilibrium series (UDS):
  advantages of, 202
  characterized, 198–199
  dating case studies, 203
  dating groundwater by, 200–203
  disadvantages of, 202
  groundwater analysis, 200
  phenomenological model, 201, 203
  results reports, 200
  sampling, 200
Uranium-234 (^{234}U), 199–200
Uranium-235 (^{235}U), 192
Uranium-238 (^{238}U), 4, 192, 199, 255
Uranium-Krypton dating, 4
Uranium-rich rocks, 105
Uranium-Xenon dating, 4
Urban air, 114
USGS:
  functions of, 108, 171
  groundwater age report, 69
  laboratories, 34
  in Reston, VA, 34

Vadose zone/vadose zone tracing, 74, 171
Variance, 14, 209, 215
Velocity field, 22
VERA, 32
Vertical depth profile, 175
Vertical flow velocity, 51, 55, 96
Vertically well-mixed aquifer, 217
Vertical piston-flow age, 230
Vertical profile, ^{36}Cl concentration, 127–128
Very high protection aquifers, 73
Very old groundwater:
  age-dating methods:
    argon-39 (^{39}Ar), 134, 139–146
    argon-40 (^{40}Ar), 177, 188–191, 202, 255
    carbon-14 (^{14}C), 134
    characterized, 134–135, 165
    chloride-36 (^{36}Cl), 170–181, 202
    helium-4 (^{4}He), 177, 181–188, 202
    iodine-129 (^{129}I), 191–198, 202
    krypton-81 (^{81}Kr), 166–170, 177, 202, 255
    krypton-86 age-dating, 100
    silicon-32 (^{32}Si), 134–139
    uranium disequilibrium series, 198–203
  nonrenewable, 53
  radioactive waste research, 75
  renewable reservoir, 52–53
Very young groundwater, 54, 70, 258
Viruses, 54
Vogel, J. C., 37
Volatile organic compounds (VOCs), 74
Volcanic activities, 191
Volcanic rock, 155, 171
Volcanoclastic rock, 155
Volume, residence-time distributions and, 212–213, 217–219
Volumetric age-mass production rate, 233
Vulnerability maps/mapping, 73–75, 249, 260

Wadi El-Tarfa, 53
Waste disposal, in deep, old saline groundwater systems, 51, 78
Waste repositories, 236
Wastewater discharge, contaminant travel time research, 73
Watch industry waste, 84, 86
Water, generally:
  chemistry, 205
  clocks, 17
  engineers, functions of, 53

isotopes, in age-dating, 129–131, 133
protection, 206
quality, 69, 72, 206, 253
stable isotopes of, 129–131
vapor, 183
Water-level:
  fluctuations, measurement of, 52
  measurement studies, 75
Waterloo Moraine region, 97
Water-rock equilibrium concentration, 76
Water-rock interactions, 161
Watershed:
  characterized, 24
  dryland salinity, 79–80
  ^{3}He/^{4}He content, 181
  ^{3}H/^{3}He content, 97
  nitrate contamination research, 70–71
  pollution evaluation, 69
  residence time, 14–15
Weapon(s):
  plutonium production, 100
  testing, 171. *See also* Nuclear weapons testing
Well-head protection zones, 211, 236
Wells, contamination research, 75
Werner, Abraham, 3
Wigley, T. M. L., 35, 37
Winter rain, 129

Xenon (XE), 99, 259
Xenon-129 (^{129}Xe), 192, 195

Young groundwaters:
  age-dating:
    characterized, 81–84, 90
    chloride/chlorine (^{36}Cl/Cl), 81, 83, 124–129, 132
    chlorofluorocarbons (CFCs), 81, 83, 105–118, 132
    ^{3}H/^{3}He, 81, 91–99, 132
    helium-4 (^{4}He), 81, 92, 95, 97, 99, 132
    indirect methods, 129–133
    krypton-85 (^{85}Kr), 81, 99–106, 132
    sulphur/sulfur hexafluoride (SF_6), 81, 83, 89, 118–124, 132
    tritium, 83–92
  mixing process, 65
  pollution evaluation, 68–69
  renewable reservoir, 52
  vulnerability maps, 74
Yucca Mountains, 129

Zones of imperceptibly slow moving groundwater, 77
Zuber, A., 36
Zurzach thermal spring (Switzerland), 145